Essential Hydraulics: An Engineering Perspective

Essential Hydraulics: An Engineering Perspective

Edited by Roman Morris

SYRAWOOD
PUBLISHING HOUSE
New York

Published by Syrawood Publishing House,
750 Third Avenue, 9th Floor,
New York, NY 10017, USA
www.syrawoodpublishinghouse.com

Essential Hydraulics: An Engineering Perspective
Edited by Roman Morris

International Standard Book Number: 978-1-64740-427-7 (Hardback)

Cataloging-in-publication Data

Essential hydraulics : an engineering perspective / edited by Roman Morris.
p. cm.
Includes bibliographical references and index.
ISBN 978-1-64740-427-7
1. Hydraulic engineering. 2. Hydraulics. 3. Engineering. 4. Fluid mechanics. I. Morris, Roman.
TC153 .H93 2023
627--dc23

TABLE OF CONTENTS

PREFACE

Hydraulics refers to the study of fluids in motion or at rest. It is associated with fluid mechanics, which serves as a major theoretical groundwork. Hydraulics is concerned with the mechanical properties of liquids and focuses on the engineering applications of fluid properties. Hydraulic systems use an incompressible fluid such as water or oil to transfer forces from one location to another within the fluid. Hydraulic engineering applies the principles of fluid mechanics to solve issues involving the collection, storage, control, regulation, transportation, measurement and use of water. Dam design, hydropower, erosion, pipe flow, river channel behavior, pumps, computational fluid dynamics, turbines and flow measurement, and river hydrology are some of the industrial applications of hydraulics. This book is a compilation of chapters that discuss the most vital concepts and emerging trends related to the study of hydraulics from an engineering perspective. It aims to serve as a resource guide for students and experts alike and contribute to the growth of the discipline.

This book is a result of research of several months to collate the most relevant data in the field.

When I was approached with the idea of this book and the proposal to edit it, I was overwhelmed. It gave me an opportunity to reach out to all those who share a common interest with me in this field. I had 3 main parameters for editing this text:

1. Accuracy - The data and information provided in this book should be up-to-date and valuable to the readers.

2. Structure - The data must be presented in a structured format for easy understanding and better grasping of the readers.

3. Universal Approach - This book not only targets students but also experts and innovators in the field, thus my aim was to present topics which are of use to all.

Thus, it took me a couple of months to finish the editing of this book.

I would like to make a special mention of my publisher who considered me worthy of this opportunity and also supported me throughout the editing process. I would also like to thank the editing team at the back-end who extended their help whenever required.

Editor

1

Coastal Engineers Embrace Nature: Characterizing the Metamorphosis in Hydraulic Engineering in Terms of Four Continua

Jill H. Slinger [1,2,*] **and Heleen S. I. Vreugdenhil** [1,3]

1 Faculty of Technology Policy and Management, Delft University of Technology, Jaffalaan 5, 2628 BX Delft, The Netherlands; heleen.vreugdenhil@deltares.nl
2 Institute for Water Research, Rhodes University, Artillery Road, Grahamstown 6140, South Africa
3 Deltares, Boussinesqweg 1, 2629 HV Delft, The Netherlands
* Correspondence: j.h.slinger@tudelft.nl

Abstract: Hydraulic engineering infrastructures, such as reservoirs, dikes, breakwaters, and inlet closures, have significantly impacted ecosystem functioning over the last two centuries. Currently, nature-based solutions are receiving increasing attention in hydraulic engineering projects and research programs. However, there is a lack of reflection on the concomitant, fundamental changes occurring in the field of hydraulic engineering, and coastal engineering in particular, and what this could mean for sustainability. In this article, we signal the shift from conventional to ecosystem-based hydraulic engineering design and characterize this in terms of four continua: (i) the degree of inclusion of ecological knowledge, (ii) the extent to which the full infrastructural lifecycle is addressed, (iii) the complexity of the actor arena taken into account, and (iv) the resulting form of the infrastructural artefact. We support our arguments with two carefully selected, iconic examples from the Netherlands and indicate how the stretching ideals of ecosystem-based engineering could engender further shifts towards sustainability.

Keywords: ecosystem-based design; Building with Nature; nature-based solutions; infrastructure lifecycle; coastal and river engineering; impacts on nature and society; multi-actor systems; multifunctional flood defense; critical reflection

---

1. Introduction

Since its inception, hydraulic engineering has focused on the design of infrastructures to serve expressed societal needs within the natural aquatic environment. Traditionally, hydraulic designs focus on controlling or withstanding natural fluctuations to facilitate economic and social development and to create safe living spaces for humans. As such, hydraulic engineers are key enablers of the Anthropocene [1]. For example, dam construction to store water and ensure freshwater supply even under low river flow conditions, building breakwaters to allow safe access to harbors, or designing dikes for flood defense. After two centuries of hydrological modification of many systems across the globe, it is now widely understood that these infrastructures impact the natural regulatory services of ecosystems [2]. In recent years, the wetlands in Louisiana [3], the dunes near New York [4], and the mangrove forests in the Philippines [5] have demonstrated that they too reduce the effects of coastal flooding. The growth in understanding of the natural environment, its benefits for human society [1,2], and the inequity of the expressed societal needs [6] is changing the field of hydraulic engineering. Hydraulic engineers increasingly need to use knowledge about the natural and the social environments in designing hydraulic infrastructures.

It is within this context that concepts like "Building with Nature" [7,8], "Engineering with Nature" [9], "Working with Nature" [10], "ecological engineering" [11], "nature-based flood defenses" [12], "integrating green and gray" [13], and "cyclic floodplain rejuvenation" [14] are emerging. These concepts are rooted in a deep knowledge of the dynamic, abiotic processes acting in the aquatic environment and a respect for the biotic environment. They specifically seek to effectively use natural materials, interactions, and dynamic processes in the design, realization, operation, and maintenance of hydraulic infrastructures. They strive for more ecosystem-based hydraulic engineering while acknowledging social complexity.

However, this begs the following question: "How are these concepts changing hydraulic engineering?" In this article, we apply critical reflection [15] in describing and conceptualizing how the hydraulic engineering design process has changed in recent years as a consequence of the increased understanding of ecosystems and their potential contribution to infrastructure development. Whereas Temmerman et al. [16] primarily ascribe the shift towards ecosystem-based coastal engineering to an increase in global environmental risks along coasts, we focus on the design process. We characterize the metamorphosis in the field of hydraulic engineering in terms of four continua spanning (i) the degree of inclusion of ecological knowledge, (ii) the extent to which the full infrastructural lifecycle is addressed in the hydraulic engineering design, (iii) the complexity of the actor arena taken into account, and (iv) the resulting form of the infrastructural artefact. We then apply our characterization to two Dutch examples from a range of international cases [8], distinguishing the original conventional hydraulic engineering designs from the new nature-based engineering designs in these examples and so demonstrating how we consider hydraulic engineering is changing. Finally, we call upon hydraulic engineers to reflect critically on the phenomenon of ecosystem-based design in their field and on our characterization of its metamorphosing effects. Is the metamorphosis we have conceptualized and the insights on design practice going to lead to lasting change? What actions are required to embed the metamorphosis in design practice?

2. Methods

As policy scientists working closely with coastal and river engineers from 2013 onwards, we employed the method of critical reflection (Figure 1), which, through reflection about action, seeks to initiate reflection for action [15,17]. According to Fook [17], critical reflection can free us from restrictive ways of knowing and indicate potential avenues for change. Indeed, Morley [18] claims that critical reflection allows us to research ways to promote congruence between practice-based aims and the ways we actually engage in practice.

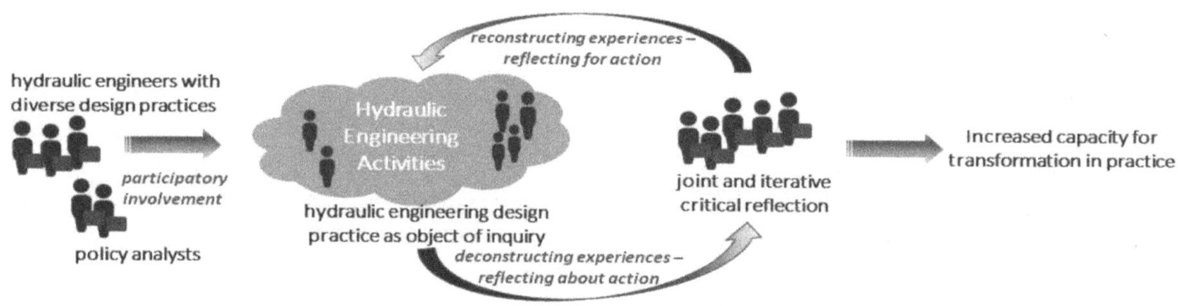

Figure 1. The method of critical reflection employed in this study.

Therefore, to initiate broader awareness of the changes in design practice amongst engineers, we first engaged in a process of deconstructive and subsequent reconstructive discussions (Appendix A), gradually formalizing our observations of the changes in engineering design practice (the action) as a metamorphosis in hydraulic engineering. This conceptualization has, as its basis, the experiences of the broad Dutch and allied international network of civil engineers who contributed substantive content to the series of activities listed in Table A1 of Appendix A. The range of participants and the agenda of the activities were often not under the control of the authors. Instead, the deconstructive and reconstructive processes listed under the individual activities in Table A1 were mostly undertaken as a single agenda item within a broader workshop or conference activity with opportunities for more personal engagement and feedback provided in each instance. Deconstructive processes involved questions on why choices were made or steps taken in an engineering design process, whereas reconstructive processes involved presenting partial explanations to diverse audiences and receiving correction and feedback. The conceptualization of a metamorphosis in hydraulic engineering in terms of four continua forms the first result of this critical reflection process.

Next, we selected two recent Building with Nature examples from a range of international projects in the EcoShape database [8] and in the Massive Open Online Course Engineering: Building with Nature 1x, in which over 20,000 professionals (primarily engineers) from 168 lands participated from 2016 to 2020 (www.edx.org). The iconic status of the examples from the coasts of South and North Holland [19,20] was important in their selection, as was the availability of publicly available information on the preceding, more conventional engineering infrastructures and the new infrastructures (e.g., https://www.rijkswaterstaat.nl/kaarten/kustlijnkaart.aspx and https://www.dezandmotor.nl/en/). The information allowed us to analyze the changes from conventional towards nature-based infrastructural design in terms of our four continuum conceptualization. This analysis and its visualization form the second result of our critical reflection process.

Finally, through this publication, we seek to advance the critical reflection process further and engender an active response on the part of engineers (reflection for action) [17,18]. We call upon them to apply this conceptualization to characterize new hydraulic engineering designs in terms of the four continua so that shifts in design practice become apparent.

3. Results: The Four Continuum Characterization

The output of an engineering design process is a design artefact, and it is the changing form of hydraulic engineering artefacts that has received the most attention from proponents of nature-based solutions [7–12]. However, our focus on the design process itself in addition to its product, the designed artefact, allowed us to distinguish four continua spanning (i) the degree of inclusion of ecological knowledge, (ii) the extent to which the full infrastructural lifecycle is addressed in the hydraulic engineering design, (iii) the complexity of the actor arena taken into account, and (iv) the resulting form of the infrastructural artefact. Although the continua are not fully independent of each other, for instance, the inclusion of more ecological knowledge in the design process is likely to result in a more nature-based artefact, they do reflect the design process and its product more fully.

3.1. Fields of Knowledge

First, we identified a change in the type of knowledge used in the design of the hydraulic engineering infrastructure. In contrast to traditional hydraulic engineering, nature-based hydraulic engineering requires the application of both ecological and engineering knowledge. Assessing the type of knowledge applied and the degree to which it is used is complex. Here, we adopted nine engineering design principles [21] and 11 ecological design principles [22] to assess the interdisciplinarity of the design. The trade-offs made in the character and functioning of the ambient natural environment to achieve the functional requirements of the hydraulic infrastructure are explicated in terms of these design principles, revealing the degree to which knowledge of the intrinsic character and functioning of the ecosystem is utilized in the engineering design process. This continuum therefore extends from only using hydraulic engineering knowledge, which includes geotechnical knowledge, to the full use of both the ecological and hydraulic engineering fields of knowledge, i.e., ecosystem-based engineering.

This continuum is depicted as extending from a single strand or field of knowledge on the left towards an interwoven multiplicity of knowledge fields on the right in the two uppermost tessellations of Figure 2.

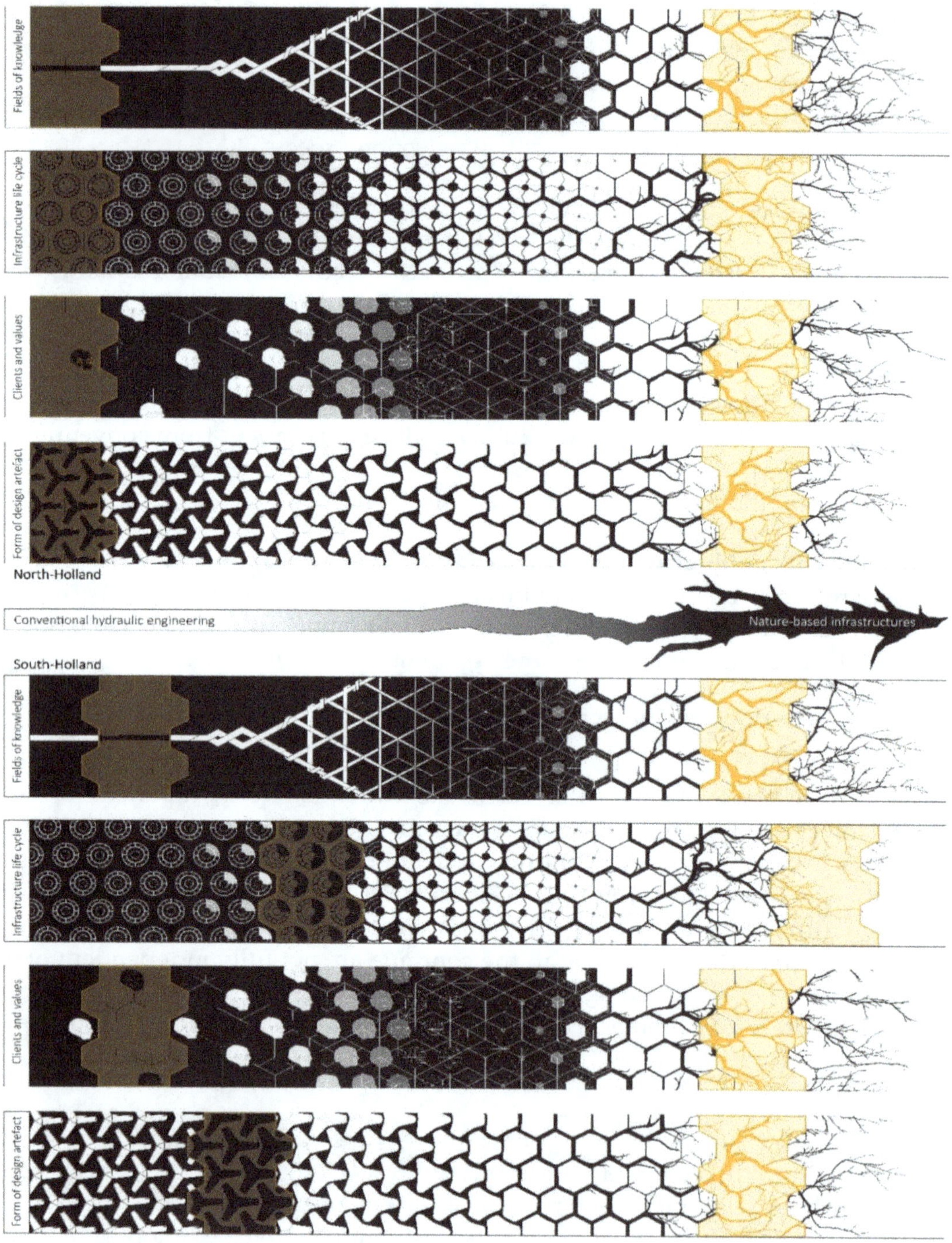

Figure 2. The metamorphosis in hydraulic engineering represented by the two examples of old infrastructures (brown) and new infrastructures (yellow) on the coasts of North Holland (**top**) and South Holland (**bottom**). The degree of adoption of a Building with Nature approach is characterized in terms of four continua: (i) fields of knowledge, (ii) infrastructure life cycle, (iii) clients and values, and (iv) form of the design artefact. A shift can be observed from more conventional hydraulic engineering designs marked in brown and located towards the left-hand side to nature-based infrastructures marked in yellow and located towards the right-hand side.

3.2. *Phases of the Infrastructure Life Cycle*

Second, we observed that in contemporary designs of hydraulic engineering infrastructures, later phases of the lifecycle of an infrastructure are specifically addressed. Therefore, the realization, operation, and maintenance phases are now taken into account in the design phase [23]. This represents an acknowledgement of the long lifetime of most hydraulic infrastructures, their effects in structuring society through their existence, operation, and use [24], and the ecologically based concepts of lifecycle analysis [25]. The degree to which the strategic goals associated with particular design choices are translated into more operational goals for the later phases in the construction and maintenance of an infrastructure have been assessed in a number of instances [26]. Accordingly, this continuum extends from the usual practice of designing for a single phase to designing for the lifetime of an infrastructure and so represents a departure from standard hydraulic design practice.

This continuum is depicted as extending from single-phased design on the left towards increasing inclusion of multiple phases (gradually filled circles) to lifecycle design (completely white circles) on the right of the upper middle tessellations of Figure 2.

3.3. *Clients and Values*

Third, we distinguished design for multiple actors across multiple values from design for a single client, often a representative of a public organization articulating societal needs. Geopolitical research has revealed that hydraulic infrastructures can serve the elite and not necessarily society as a whole [27,28]. Moreover, the complexity of settings in which many and conflicting stakeholder interests and values co-occur mean that networked solutions and a broader coalition of actors are required for the resolution of these wicked problems [29,30]. In many "Engineering with Nature" and "Building with Nature" projects executed to date, there have been coalitions of private and public actors in new configurations of cost and benefit sharing to ensure that the hydraulic infrastructures deliver increased living quality over their lifetime. This is exemplified in the Sand Motor [31] and the new Hondsbossche Sea Defense [32], in which the provision of high-quality, if temporary, recreational areas in addition to enhanced coastal safety and flexibility in adapting to climate change played a major role in the approval process. This continuum therefore extends from a single client orientation to designing for multiple actors across multiple values.

This continuum is depicted as extending from a single client at the left across an interconnected network of actors (increasing whiteness) to a completely connected multiplicity of actors and their values (complete whiteness) at the right of the two lower middle tessellations of Figure 2.

3.4. *The Form of the Design Artefact*

Fourth, we distinguished a change in the form of the resulting design artefact: the hydraulic infrastructure. Does the artefact, the outcome of the design process, exhibit attributes of a conventional hydraulic engineering infrastructure? Or does it also exhibit features of ecosystem-based design? Natural processes, interactions, and materials may be incorporated into the design, resulting in a hybrid form, such as a dike-in-dune construction [4], or the infrastructural artefact may comprise entirely of natural material and the design may make use of dynamic ecosystem processes. This continuum therefore extends from a conventional, static hydraulic infrastructure through hybrid infrastructural forms to a fully nature-based, dynamic artefact in the aquatic environment.

This continuum is depicted as extending from a single armor unit that is used in conventional breakwaters on the left through an increasingly abstract form to a completely white and natural image on the right of the lowest two tessellations of Figure 2.

4. Results: Visualizing the Metamorphosis in Hydraulic Engineering for Two Cases

Next, we analyzed two coastal examples from the Netherlands: on the North Holland coast and the South Holland coast. The key characteristics of the old Hondsbossche and Pettermer dike, the oldest sea dike in the Netherlands, are compared with the new sea defense infrastructure on the North Holland coast in Table 1. Likewise, in Table 2, the old sea defense strategy of beach and shoreface nourishment is compared with the innovation of the Sand Motor on the South Holland coast. In the North Holland example, the choice was made to move from a stone dike to an extensive, shallow sandy foreshore and dunes as primary flood protection infrastructure [32]. In the South Holland example, the strategy of regular sand nourishment of the coast was scaled up to that of a once-off mega-nourishment designed to meet flood protection requirements over decades [31].

Table 1. The sea defense on the North Holland coast demonstrating the metamorphosis in hydraulic engineering.

Old Hondsbossche and Pettermer Dike	New Hondsbossche Sea Defense [32]
Following the St. Elizabeth flood of 1421, the first earthen dikes were constructed near Petten.	Between 30 and 35 × 106 m3 sand was deposited on the seaward side of the existing dike in 2015.
In 1823, the dikes at Hondsbossche were rebuilt from stone, forming the first of the modern Dutch dikes.	Design consists of a shallow sandy foreshore and beach connected to a diverse dune habitat, shaped by wind and water over time.
In 1977, the stone dike was reinforced with armor units and raised to meet the safety standards of the time: an inundation risk of 1 in 10,000.	The dynamic dune and beach environment seaward of and covering the old dike is envisaged as a habitat for rare plants, birds, and other animals, supporting biodiversity.
Between 2001 and 2006, the existing dike was deemed a "Weak Link" in the flood defense chain as it would not continue to satisfy flooding safety standards over the next 50 years and so required improvement.	Aims of the new sea defense explicitly include flood protection, recreation, nature development, and knowledge acquisition extending over a period of 10 to 20 years.

Table 2. Beach nourishment on the South Holland coast demonstrating the metamorphosis in hydraulic engineering.

Beach and Shoreface Nourishment (1990–2010)	Sand Motor [31] (2011 Onwards)
Regular nourishments of the beach or shoreface with sand were undertaken at intervals of 3 to 5 years. The location of the nourishments was planned annually at a national level to prevent coastal retreat (maintain a base coastline position).	To combat coastal erosion, 21 × 106 m3 sand was deposited in a hook shape on the shoreface of the coast of South Holland during a single intervention in 2011.
The volume of sand required for a shoreface nourishment was roughly equivalent to that present in nearby underwater sand banks and the effects extended up to 3 km along the coast over the next 3 years. The total volume of sand required for the entire Dutch coast was 12 × 106 m3 in 2011.	Sand is redistributed from the mega-nourishment by tides, waves, and winds towards the beach and dunes along a 17 km stretch of coastline.
The ecosystem had not recovered before the next nourishment occurred, usually within a period of 3 to 5 years.	Project lifetime covers several decades during which the shoreface, beach, and dune habitats are not again disturbed by sand nourishment.
Needs of recreation and ecosystem requirements were taken into account in timing the nourishment (not in summer), but the primary aim was coastal defense.	Aims explicitly include coastal defense, nature, recreation, and education.

To demonstrate the metamorphosis in hydraulic engineering, the old and the new sea defense infrastructures on the North and South Holland coasts were positioned on the four continua. Their positioning on each of the four continua reflects the degree to which elements of a Building with Nature approach have been incorporated in the designs. Conventional hydraulic engineering infrastructures, which have few nature-based elements, are therefore located towards the left-hand side of each of the four continua. For instance, the old Hondsbossche and Pettermer dike, a conventional hydraulic engineering infrastructure, is located on the far left of each of the four continua in the upper part of Figure 2. This is not the case for the beach and shoreface nourishment strategy from 1990 to 2010 on the South Holland coast (lower part of Figure 2). Instead, the use of hydrodynamic and morphological knowledge in designing the shoreface nourishments is indicated by a positioning further from the left boundary in continuum 1, while the cognizance taken of cyclical bar dynamics in choosing shoreface nourishment justifies a position nearer to the middle of continuum 2. Taking account of the needs of beach pavilion owners and recreants regarding the timing of beach nourishments means that the position on continuum 3 is also a little way away from the left boundary. Finally, whereas the form of shoreface nourishments is highly dynamic, that of beach nourishments is less so, explaining why the position on continuum 4 lies between those of continuum 1 and 3 and that of continuum 2.

In contrast, the Sand Motor and the new Hondsbossche Sea Defense are located towards the right-hand side of each continuum, indicating that considerable ecological knowledge in addition to hydraulic engineering knowledge was employed in their design (continuum 1). It is noteworthy that the full lifecycle of the new infrastructure is incorporated in both of the new designs, but with the Sand Motor, it is even designed to disappear over time (continuum 2). The Sand Motor is therefore positioned further to the right than the new Hondsbossche Sea Defense on the second continuum (Figure 2). Aspects of spatial quality and education in addition to nature and engineering values are explicitly addressed, extending the functional requirements for the new infrastructural designs beyond standard hydraulic engineering criteria towards the right-hand side of continuum 3. In terms of the form of the artefact, the old North Holland dike (as static infrastructure) and the local South Holland beach and shoreface nourishments (slightly more dynamic) are located at the left of continuum 4. However, the highly dynamic, nature-based artefacts that now replace them are located towards the right of this continuum.

Inspecting the representation of the old and the new infrastructures on the four continua in Figure 2 allows us to detect at a glance the shifts, or metamorphoses, from conventional hydraulic engineering (on the left) towards nature-based infrastructures (on the right) that have occurred along the North and South Holland coasts.

5. Concluding Discussion

The exact position of an infrastructure on each of the continua cannot be determined unequivocally and will always be subject to debate. However, the value of distinguishing the four continua lies in (i) stimulating discussion amongst hydraulic engineers regarding the degree to which a hydraulic engineering solution can be termed ecosystem-based design and (ii) engendering reflection on this metamorphosis in the field of hydraulic engineering. Such discussions may act to curtail "greenwashing" and can stimulate the field of hydraulic engineering to move from the left-hand side of the continua towards the challenging ideals on the right-hand side of the continua.

Over time, we anticipate that the understanding of what it means to be an ecosystem-based hydraulic infrastructure will alter, and the value attached to ecosystem-based design ideals will increase. It is our conviction that infrastructures that would now score as marginally ecosystem-based design will later be considered as primal and early examples of a new type of sustainability engineering. However, whether hydraulic engineers continue to embrace nature will also depend on their conceptualization of themselves as engineers in the Anthropocene. Will they reflect critically on the phenomenon of ecosystem-based engineering design and their changing role in global sustainability?

We call upon hydraulic engineers to characterize each new Building with Nature/Working with Nature/Engineering with Nature design in terms of the four continua. Such action on the part of hydraulic engineers is a necessary first step in determining whether design practice is indeed changing and will provide a wider information base for reflection and continued learning.

Author Contributions: J.H.S. and H.S.I.V. developed the idea together and undertook the critical reflection process collaboratively. J.H.S. wrote the first draft, while H.S.I.V. analyzed available data and completed Tables 1 and 2. All authors discussed the findings and contributed their perspectives to the final manuscript. All authors have read and agreed to the published version of the manuscript.

Acknowledgments: We acknowledge Anne van Loenen (illustrations), Graciela del Carmen Nava Guerrero (contributor to ecological design principles), and thank colleagues from TU Delft, Deltares, and EcoShape for discussions. Sincere thanks to Ronald Waterman for inspiring this work.

Table A1. Authors' critical reflective experiences with coastal and river engineering over the period 2013 to 2020.

Year	Activity with Critical Reflection Process (in Brackets)	Substantive Content	Participants	Author
2013–2014	Series of discussions (deconstructive)	Perspectives on Building with Nature, whether and how the concept should be incorporated in engineering education.	Hydraulic engineer and ecologists both within and outside the Building with Nature community.	J.S.
2013–2014	Series of discussions (deconstructive)	Potential for multidisciplinary collaboration within the Faculty of Civil Engineering and Geosciences (CiTG), TU Delft.	Thought leaders on the staff of the Faculty of CiTG.	H.V.
2014, 2015	Series of discussions (deconstructive)	Building with Nature concept and how this relates to "Engineering with Nature" and "Working with Nature".	Dr. Ir. Ronald Waterman.	J.S.
26 March 2015	Delft Infrastructures and Mobility Initiative (DIMI) Brainstorm on Building with Nature (deconstructive)	Perceptions of Building with Nature.	Key people from TU Delft and the EcoShape Foundation.	J.S.
8 October 2015	Congress: DIMI on Tour! Building with Nature (deconstructive)	Applications and teaching of Building with Nature.	Primarily engineers, ecologists, consultants, architects, urban planners, and students.	J.S. and H.V.
28 January 2016	DIMI Congress on Deltas and Ports of the Future: towards new synergy of Delta technology, Spatial Design and Governance (reconstructive)	Propositions for discussion on effective implementation and development of the Building with Nature concept and associated governance challenges.	Wide ranging set of participants, including regional policy makers, engineering consultants, urban planners, ecologists, and diverse decision-makers in water, coastal, and environmental domains.	J.S. and H.V.
2017	INTERREG 5B project Building with Nature (reconstructive)	Preparation, presentation, and discussion of business cases for Building with Nature projects, undertaken on behalf of the EcoShape Foundation.	Primarily coastal, water, and environmental authorities of the North Sea countries: Netherlands, Germany, Denmark, Sweden, Belgium, and Scotland	H.V.
2015, 2016, 2017, 2018, 2019	Masters level engineering course Building with Nature in Hydraulic Engineering (deconstructive and reconstructive phases)	Campus-based, blended education on designing Building with Nature solutions using case studies.	CiTG students from TU Delft.	J.S. and H.V.
2016, 2017, 2018, 2020	MOOC Engineering: Building with Nature 1x (deconstructive and reconstructive phases)	Online education on the Building with Nature concept and project examples.	Over 20,000 participants worldwide from a wide range of backgrounds. Primarily engineers, ecologists, and planners.	J.S. and H.V.
2017, 2018	Virtual Exchange Building with Nature (deconstructive and reconstructive phases)	Credit acquiring online education on the Building with Nature concept and project examples.	Engineering, ecology, and planning students from universities in the Netherlands and other countries, e.g., Queensland University, Rice University, University of Leuven, as well as Wageningen University and Twente University.	J.S.

Table A1. *Cont.*

Year	Activity with Critical Reflection Process (in Brackets)	Substantive Content	Participants	Author
2011–2017	NatureCoast research program (deconstructive and reconstructive phases)	Nature-based nourishment of coastal systems.	Primarily civil engineering, ecology, and policy researchers. Public policy organizations, consultants, funders, and societal organizations also participated.	J.S.
2015–2020	CoCoChannel research project (deconstructive and reconstructive phases)	Co-designing coasts using natural channel–shoal dynamics.	Primarily civil engineering, ecology, and policy researchers. Public policy organizations, consultants, and funders also participated.	J.S.
2016–2020	Sustainable Ports in Africa research project (deconstructive and reconstructive phases)	Applying Building with Nature concepts in the integrated design of port developments for enhanced sustainability.	Coastal engineers, ecologists, economists, and policy analysts. Wide range of supporting parties, such as harbor authorities in the Netherlands and Africa, port logistics and operational companies, engineering consultants, dredging companies, societal and environmental organizations, and financiers.	J.S. and H.V.
2016	Workshop: The Sand Motor, 5 years of Building with Nature (reconstructive)	Participatory international reflection on the (success of) Building with Nature concept and how this relates to "Engineering with Nature".	Invited international delegates included renowned ecologists and engineering scientists as well as acknowledged leaders of the "Engineering with Nature" concept. The majority of attendees were engineers, scientists, and ecologists.	J.S. and H.V.
2019	BeSafe research program (initiated 2015) (reconstructive)	Decision-making for nature-based solutions to flood defense.	Primarily civil engineering, ecology, and policy researchers. Public policy organizations, consultants, and funders also participated.	H.V. and J.S.
2 November 2017	PIANC NL meeting (reconstructive)	Building with Nature application to ports in Africa and elsewhere and how this relates to "Working with Nature".	Mainly engineering consultants, harbor authorities, and some scientists.	J.S.
May 2018	PIANC World Congress (reconstructive)	International applicability of Building with Nature concepts to ports and harbors, relationship to "Working with Nature" and "Engineering with Nature", governance challenges.	Wide range of international participants, primarily port and harbor engineers and scientists but including international trade and diplomatic delegates.	J.S.
2020	Beyond Engineering: Building with Nature 2x (reconstructive)	Online education on the social context of the Building with Nature concept, including project examples.	Over 800 engineers, ecologists, and planners as participants.	J.S. and H.V.

References

1. Chester, M.V.; Markolf, S.; Allenby, B. Infrastructure and the environment in the Anthropocene. *J. Ind. Ecol.* **2019**, *23*, 1006–1015. [CrossRef]
2. Daily, G. *Nature's Services: Societal Dependence on Natural Ecosystems*; Island Press: Washington, DC, USA, 1997; p. 412.
3. Environment Protection Agency. Wetlands: Protecting Life and Property from Flooding. EPA843-F-06-001, Office of Water. 2006. Available online: https://www.epa.gov/sites/production/files/2016-02/documents/flooding.pdf (accessed on 29 May 2020).
4. Smallegan, S.M.; Irish, J.L.; van Dongeren, A.R.; den Bieman, J.P. Morphological response of a sandy barrier island with a buried seawall during Hurricane Sandy. *Coast. Eng.* **2016**, *110*, 102–110. [CrossRef]
5. Holtz, M. Saved by the Mangroves? A Philippine Town Dodges Haiyan's Storm Surge. Science, Tech and Environment 2013, November 29. 2013. Available online: http://www.pri.org/stories/2013-11-29/saved-mangroves-philippine-town-dodges-haiyans-storm-surge (accessed on 29 May 2020).
6. Beattie, J.; Morgan, R. Engineering Edens on This 'Rivered Earth'? A review article on water management and hydro-resilience in the British Empire, 1860s–1940s. *Environ. Hist.* **2017**, *23*, 39–63. [CrossRef]
7. Waterman, R.E. Integrated Coastal Policy via Building with Nature. Ph.D. Thesis, Delft University of Technology, Delft, The Netherlands, 2010. Available online: http://resolver.tudelft.nl/uuid:fa9a36f9-7cf8-4893-b0fd-5e5f15492640 (accessed on 29 May 2020).
8. EcoShape. Building with Nature. Available online: www.ecoshape.nl (accessed on 29 May 2020).
9. Bridges, T.S.; Banks, C.J.; Chasten, M.A. Engineering with nature. Advancing system resilience and sustainable development. *Mil. Eng.* **2016**, *108*, 52–54. Available online: https://ewn.el.erdc.dren.mil/pub/Pub_1_EWN_TME-JanFeb2016.pdf (accessed on 29 May 2020).
10. The World Association for Waterborne Transport Infrastructure. PIANC Position Paper 'Working with Nature'. 2011. Available online: https://www.pianc.org/uploads/files/EnviCom/WwN/WwN-Position-Paper-English.pdf (accessed on 29 May 2020).
11. Mitsch, W.J. Ecological Engineering: A new paradigm for engineers and ecologists. In *Engineering within Ecological Constraints*; Schulze, P.C., Ed.; National Academy Press: Washington, DC, USA, 1996; pp. 114–132.
12. Van Wesenbeeck, B.K.; Mulder, J.P.M.; Marchand, M.; Reed, D.J.; de Vries, M.B.; de Vriend, H.J.; Herman, P.M.J. Damming deltas: A practice of the past? Towards nature-based flood defences. *Estuar. Coast. Shelf Sci.* **2014**, *140*, 1–6. [CrossRef]
13. Browder, G.; Ozment, S.; Bescos, I.R.; Gartner, T.; Lange, G.-M. *Integrating Green and Gray: Creating Next Generation Infrastructure*; World Bank and World Resources Institute: Washington, DC, USA, 2019; p. 134. ISBN 978-1-56973-955-6.
14. Vreugdenhil, H.S.I.; Slinger, J.H.; Kater, E. The influence of scale preferences on the design of a water innovation: A case in Dutch river management. *Environ. Manag.* **2010**, *46*, 29–43. [CrossRef] [PubMed]
15. Brookfield, S. The concept of critical reflection: Promises and contradictions. *Eur. J. Soc. Work.* **2009**, *12*, 293–304. [CrossRef]
16. Temmerman, S.; Meire, P.; Bouma, T.; Herman, P.M.; Ysebaert, T.; De Vriend, H. Ecosystem-based coastal defence in the face of global change. *Nature* **2013**, *504*, 79–83. [CrossRef]
17. Fook, J.; Gardner, F. *Practising Critical Reflection: A Resource Handbook*; McGraw-Hill Education: London, UK, 2007; p. 218.
18. Morley, C. *Critical Reflection as a Research Methodology, In Knowing Differently: Arts-Based and Collaborative Research*; Liamputtong, P., Rumbold, J., Eds.; Nova Science Publishers Inc.: Hauppauge, NY, USA, 2008; pp. 265–280.
19. Bontje, L. Narrative Perspectives on the Development of Coastal Pilot Projects. Ph.D. Thesis, Delft University of Technology, Delft, The Netherlands, 2017. [CrossRef]
20. Bontje, L.; Slinger, J.H. A narrative method for learning from innovative coastal projects—Biographies of the Sand Engine. *Ocean. Coast. Manag.* **2017**, *142*, 186–197. [CrossRef]
21. Slinger, J.H. *Engineering: Building with Nature 101× Video #07—Distilling Engineering Design Principles*; TU Delft: Delft, The Netherlands, 2016.
22. Slinger, J.H.; Nava Guerrero, G.d.C. *Engineering: Building with Nature 101× Video #08—Distilling Ecological Design Principles*; TU Delft: Delft, The Netherlands, 2016.

23. Van Koningsveld, M.; Slinger, J.H. *Building with Nature video #04—The Building with Nature Design Process @ TU Delft 2015*; TU Delft: Delft, The Netherlands, 2015.
24. Giddens, A. *The Constitution of Society. Outline of the Theory of Structuration*; University of California Press: Berkeley, CA, USA, 1984; p. 417.
25. Jensen, A.A.; Hoffman, L.; Møller, B.T.; Schmidt, A.; Christiansen, K.; Elkington, J.; van Dijk, F. *Life Cycle Assessment. A Guide to Approaches, Experiences and Information Sources*; Environmental Issues Series No. 6; European Environment Agency: Brussels, Belgium, 1997; p. 119. Available online: https://www.eea.europa.eu/publications/GH-07-97-595-EN-C/Issue-report-No-6.pdf (accessed on 29 May 2020).
26. Garel, E.; Camba Rey, C.; Ferreira, O.; van Koningsveld, M. Applicability of the "Frame of Reference" approach for environmental monitoring of offshore renewable energy projects. *J. Environ. Manag.* **2014**, *141*, 16–28. [CrossRef]
27. Bakker, K. Constructing 'public' water: The World Bank, urban water supply, and the biopolitics of development. *Environ. Plan. D Soc. Space* **2013**, *31*, 280–300. [CrossRef]
28. Clifford-Holmes, J.K.; Palmer, C.G.; de Wet, C.G.; Slinger, J.H. Operational manifestations of institutional dysfunction in post-apartheid South Africa. *Water Policy* **2016**, *18*, 998–1014. [CrossRef]
29. Rittel, H.W.J.; Webber, M.M. Dilemmas in a general theory of planning. *Policy Sci.* **1973**, *4*, 155–169. [CrossRef]
30. Wamsler, C.; Niven, L.; Beery, T.H.; Bramryd, T.; Ekelund, N.; Jönsson, K.L.; Osmani, A.; Palo, T.; Stålhammar, S. Operationalizing ecosystem-based adaptation: Harnessing ecosystem services to buffer communities against climate change. *Ecol. Soc.* **2016**, *21*, 31. [CrossRef]
31. Stive, M.J.F.; de Schipper, M.A.; Luijendijk, A.P.; Aarninkhof, S.G.J.; van Gelder-Maas, C.; van Thiel de Vries, J.S.M.; de Vries, S.; Henriquez, M.; Marx, S.; Ranasinghe, R. A new alternative to saving our beaches from sea-level rise: The sand engine. *J. Coast. Res.* **2013**, *29*, 1001–1008. [CrossRef]
32. Rijkswaterstaat. *Rijkswaterstaat Annual Report 2015. Rijkswaterstaat |cd0616mc74*; Ministry of Infrastructure and the Environment: Den Haag, The Netherlands, 2016.

2

A Novel Equivalent Continuum Approach for Modelling Hydraulic Fractures

Eziz Atdayev [1,*], Ron C. K. Wong [1] and David W. Eaton [2]

[1] Department of Civil Engineering, University of Calgary, Calgary, AB T2N 1N4, Canada; rckwong@ucalgary.ca

[2] Department of Geoscience, University of Calgary, Calgary, AB T2N 1N4, Canada; eatond@ucalgary.ca

* Correspondence: e.atdayev@gmail.com

Abstract: Hydraulic fracturing has transitioned into widespread use over the last few decades. There are a variety of numerical methods available to simulate hydraulic fracturing. However, most current methods require a large number of input parameters, of which the values of some parameters are poorly constrained. This paper proposes a new method of modelling the hydraulically fractured region using void-ratio dependent relation to define the permeability of the fractured region. This approach is computationally efficient and reduces the number of input parameters. By implementing this method with an equivalent continuum representation, uncertainties are reduced arising from heterogeneity and anisotropy of earth materials. The computational efficiency improves modelling performance in stress sensitive zones such as in the vicinity of the injection well or near faults.

Keywords: hydraulic fracturing; void ratio; permeability; FEM; ABAQUS

1. Introduction

Hydraulic fracturing is a fluid injection process intended to create fractures in order to increase formation permeability [1]. This process is applicable in both vertical and horizontal wells, but is now performed mainly in extended horizontal wells [2] to allows multiple hydraulically fractured zones to be completed in a single well [3]. Each hydraulically fractured zone is then referred to as a stage, and a single extended horizontal well may have more than 20 stages. Geomechanical analysis is crucial in the pre-assessment of the geological formation to ensure the safety and effectiveness of the practice [4]. Moreover, considering the high cost of drilling and completion programs, modelling is a key assessment tool for hydraulic fracturing. There are a multitude of techniques and methods available for the modelling of hydraulic fracturing [5–11].

Analytical approaches are common for modelling hydraulic fractures. These methods are primarily based on the well established Perkins–Kern–Nordgren (PKN) and Khristianovic–Geertsma–de Klerk (KGD) models [12–17]. These models remain highly useful for a quick hydraulic fracturing assessment at a site. The PKN model is better suited for cases of fractures with length equal to or greater than twice the fracture height. These conditions are generally applied to hydraulic fracturing in shale reservoirs [18]. The KGD method tends to bea better representative of cases where the fracture length is equal to or less than its height. Such fracture conditions are mainly applicable to brownfields applications where fracturing is used to revive production [19,20].

Based on these analytical solutions, there are numerous numerical methods available for hydraulic fracture modelling. Among the most common numerical modelling techniques are the cohesive zone method (CZM), extended finite element method (XFEM), and discrete element method (DEM) [9,21–26]. Hybrid techniques such as the finite element method-discrete element method (FEM-DEM) have also started to emerge. Such techniques attempt to combine the advantages of two methods to improve their performance [27,28].

Another type of modelling approach that has received less attention is a class of methods called equivalent continuum techniques. These methods attempt to represent fractured media through a generalised approach that provides descriptive results for a sufficiently large scale. Previous applications of equivalent continuum approaches have tended to largely focus on relatively shallow depths of analysis [29–32].

2. Literature Review on Hydraulic Fracturing Modelling Methods

The practice of hydraulic fracturing has resulted in an extensive body of literature focused on modelling techniques. Some of these techniques are briefly described below.

2.1. *Analytical Models*

Analytical solutions tend to be highly generalised and simplified through a set of assumptions. For example, assumptions that underlie the PKN and KGD models are: Homogeneous, isotropic, and linear elastic rock, laminar flow, no effects of additives such as proppant, and simple fracture geometry. The main distinction between PKN and KGD models is the assumed cross-sectional shape of the hydraulic fracture. In the PKN model, the hydraulic fracture is assumed to have an elliptical cross-section, while the KGD model assumes a rectangular shape. The KGD model can easily be converted to radial coordinates, which is useful to represent penny-shaped cracks. General solutions for pressure, width, and length for PKN, KGD, and penny-shaped hydraulic fracture models are widely known [17,33].

The use of analytical solutions can be extended to the formations with natural fractures using the leak-off coefficient. The introduction of leak-off coefficient to the analytical solutions improves the agreement of planar hydraulic fracture geometry with field data. However, analytical solutions are constrained by the simplified definition of approximate hydraulic fracture geometries. On the other hand, numerical models are capable of simulating complex hydraulic fracture geometries and their effects on host medium.

2.2. *Cohesive Zone Method (CZM)*

The cohesive zone method utilises cohesive elements within a continuum model to represent fractures in the rock. Cohesive elements use traction-separation relation for mechanical strength weakening [34]. In the cohesive zone, this relationship defines the energy release equation, wherein if a specific threshold strain/stress is reached, then the cohesive element is relaxed and allowed to deform more freely. In addition, the permeability relation for the cohesive zone uses a parallel plate theory, which is an effective representation for fracture permeability. As outlined below, a recent enhancement of the CZM is known as XFEM.

2.3. *Extended Finite Element Modelling (XFEM)*

Extended finite element modelling (XFEM) uses the same fundamental approach as the cohesive zone method, except that the fractured zone is defined differently. In particular, for the XFEM, heaviside, junction, and branch enrichment functions are introduced into the domain. These functions are used to calculate where the fracture tip will propagate and how the deformation of the fracture will be imposed onto surrounding rock [35].

Both CZM and XFEM attempt to model fractures individually, which complicates the computation process due to the increased number of equations in the model. Additionally, the large dimensional difference between the fractures and general modelling domain imposes challenges for model convergence. Another challenge for XFEM approaches is mesh dependency due to the use of enrichment functions, the dimensions of the element specifying the orientation and displacement of fractures become highly mesh dependent. Although the CZM approach does not suffer from the same degree of mesh dependency, it requires a pre-definition of the fracture geometry, which imposes limitations on how a fracture will propagate.

2.4. Discrete Element Method (DEM)

The discrete element method (DEM) is a modelling approach wherein each mesh element is considered as a separate continuum particle and the intervening space between is treated as fractures [36,37]. These fractures use the same type of physics as in the case of CZM, with respect to strength degradation and fracture initiation. A superiority of the DEM over the CZM is an ability for fractures to propagate in any direction in 3-dimensional space. The convergence rate is improved, compared to XFEM, due to a similar dimensionality of the fractures and continuum elements.

The improvements provided by the DEM approach are based on the treatment of fractures as contact points between the elements. However, this method also has high computational requirements, as it generally requires a large number of model elements in order for the model to be realistic and representative.

2.5. FEM-DEM

FEM-DEM is a hybrid approach that combines the advantages of DEM with finite element modelling (FEM) to reduce the computational load. The DEM component is applied within model micro-scale sections, and the stresses resultant from the DEM are then imposed within the FEM, which models the macro-scale domain [38]. This method is most applicable if the domain can be discretised into high- and low-sensitivity subdomains. It should also be mentioned that FEM-DEM codes are available that combine these two methods for different purposes [27,28].

2.6. Equivalent Continuum Methods (ECM)

Equivalent continuum methods assume that within a representative element volume (REV) the medium can be treated as homogeneous. Instead of defining fractures individually and modelling the strength degradation, this means that the properties of the domain are generalised and modelled as a continuum. The region of interest can then be discretised based on the regions of high and low fracture density, allowing for the modelling of the fractured and virgin zones of the reservoir [18,39]. The use of an ECM representation makes it simpler to apply well established numerical approaches such as the finite element method (FEM) to solve governing equations that describe the physical processes and behaviour of the medium.

Most of previous applications of ECM are focused on soils and near-surface materials [29–32], rather than for modelling hydraulic fracturing in rocks at a greater depth. This may be due to the general availability of a wide range of empirical methods to define the properties of fractured materials at shallow depths [40,41].

A novel ECM method introduced in this paper attempts to address the necessity of optimised hydraulic fracturing modelling. The introduced method in this paper uses the dependency of the permeability on void ratio to define the hydraulic fracture propagation. This method allows for an easier convergence due to uniform mesh dimensions and the exploitation of an already present coupled fluid flow and continuum relations.

3. Methodology

3.1. Average Volumetric Strain in Fractured Zone

During hydraulic fracturing, it is a common practice to record microseismicity, injection pressure, and rates. These data are then used to analyse the effectiveness of the hydraulic fracturing program. One type of analysis used in this paper is a comparison of estimates of volumetric strains based on microseismicity and injection pressure. In estimating the volumteric strain using the injection pressure, rock is assumed to be cohesion-less and fractures to open when change in injection pressure equals minimum in-situ stress. Thus, the volumetric strain using injection pressure is given by σ'_{min}/K, where σ'_{min} is the minimum in-situ effective stress and K denotes bulk modulus of the fractured domain. The volumetric strain using microseismicity is estimated using V_{inj}/SRV, where V_{inj} is the injected volume and SRV represents stimulated rock volume (SRV). SRV is defined as a three dimensional rock volume where new fractures and opened natural fractures during hydraulic fracturing are contained [42]. The uniqueness of the strain when fracturing occurs suggests that both of the volumteric strains should exhibit same solution. Thus, this uniqueness allows one to relate the hydraulic fracturing permeability to volumteric strain which then can be related to the void ratio.

3.2. Void Ratio Dependent Permeability

Permeability and porosity of a medium are often closely related. Relationships between these two quantities are widely used in reservoir engineering through semi-log porosity-permeability plots [43]. In our ECM method, a similar approach is used to estimate formation permeability. Unlike conventional porosity-permeability relations, our proposed relation uses a step-like function. This provides a good representation for hydraulic fracturing, due to steep spatial gradients of permeability in the vicinity of a hydraulic fracture [17]. Practicality and robustness of this approach arise from the availability of built-in relationships in ABAQUS FEM between permeability and void ratio [34]. This functionality is described by Equation (1).

$$\eta(e) = a * (tanh((\frac{e}{e_{thresh}})^{\alpha}))^{\alpha} \tag{1}$$

where η is permeability, a is a maximum permeability, e is the void ratio, e_{thresh} is the minimum void ratio after which fracture permeability starts to dominate, and α is a factor related to rock brittleness and fracture density.

Void ratio is a scalar quantity while permeability is a tensor. Therefore, the anisotropy of permeability, i.e., in which it varies with direction, must be handled through transformation. In our implementation, permeability anisotropy is introduced using a normalisation and denormalisation method proposed by [44]. In this method, first the microseismic data are normalised using the dimensions of the SRV using Equation (2),

$$\begin{aligned} X_{trans} &= X * \frac{L_{scal}}{L_{max}} \\ Y_{trans} &= Y * \frac{L_{scal}}{L_{min}} \\ Z_{trans} &= Z * \frac{L_{scal}}{L_{int}} \end{aligned} \tag{2}$$

where X, Y, and Z are coordinates of detected events in the microseismic cloud along the maximum, minimum, and intermediate stress directions, respectively; L_{max}, L_{min}, and L_{int} specify the SRV length parameters along the maximum, minimum, and intermediate stress directions, respectively; $L_{scal} = \sqrt[3]{L_{max} L_{min} L_{int}}$, and X_{trans}; Y_{trans}, and Z_{trans} are transformed coordinates of detected events in microsesimic cloud.

To illustrate this transformation, assume a notional SRV characterised by an elliptical microseismic cloud in a 2D plane. The elliptical shape implies that the resultant fracture propagation rate and formation permeability are both anisotropic. After normalising the microseismic data in this canonical example, the resultant microseismic cloud will be circumscribed by a circular boundary. These steps are graphically depicted in Figure 1. This approach allows for a simplification of the analysis into a more straightforward one-dimensional analysis.

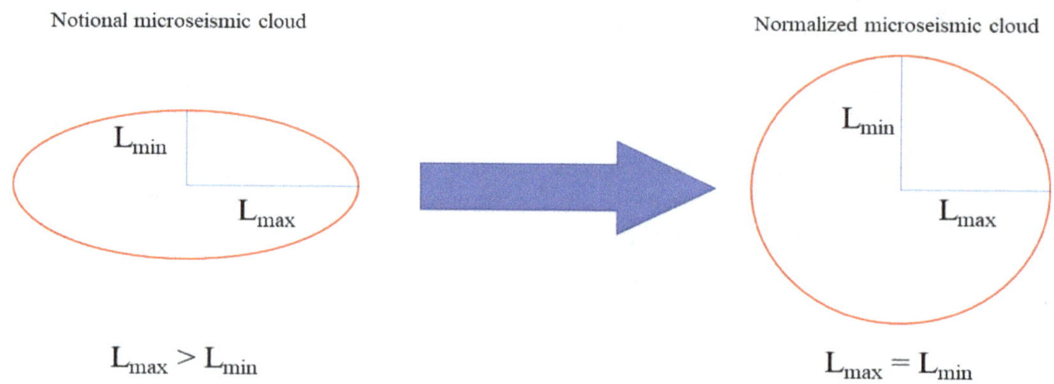

Figure 1. Applying the normalisation method by [44] on a notional stimulated rock volume (SRV) microseismic cloud. The notional microseismic cloud is assumed to be within anisotropic 2-dimensional ellipse. After normalisation, the microseismic cloud can be bounded with a circular shape reducing the analysis to a 1-dimensional scale.

The permeability function that is obtained using the resultant normalised data are denormalised using Equation (3),

$$\begin{aligned} k_x &= (\frac{L_{max}}{L_{scal}})^2 * k_0 \\ k_y &= (\frac{L_{min}}{L_{scal}})^2 * k_0 \\ k_z &= (\frac{L_{int}}{L_{scal}})^2 * k_0 \end{aligned} \tag{3}$$

where, k_0 is the permeability estimated using one-dimensional analysis and k_x, k_y, and k_z are permeabilities in X, Y, and Z directions, respectively.

To provide a case study of our ECM approach, the proposed relation was applied to the Hoadley field data to demonstrate the applicability of this method using real data.

3.3. Case Study—Hoadley Tight Sand Reservoir

The Hoadley gas field is located northwest of Red Deer and southwest of Edmonton, Alberta, Canada. The pay formation is the Glauconite tight sand member of the Cretaceous Mannville Group, which is overlain by the coal-bearing Medicine River Formation. Open-hole, multi-stage hydraulic fracturing was carried out in two horizontal injection wells and monitored using a 12-level geophone array from a nearby vertical well [45]. The configuration of the wells and directions of maximum and minimum in-situ stresses are shown in Figure 2a. In Figure 2a blue and red line and, the big blue dot indicate two injection and vertical observation wells, respectively. The scattered dots are the observed microseismicity during hydraulic fracturing. The different colour is assigned for microseismicity observed at each stage. Simplified stratigraphy of the Hoadley field, as well as a depth histogram showing the number of microseismic events at given depths, is provided in Figure 2b. This paper is focused on a single stage of the hydraulic fracturing program in well 1. The injection pressure and rate profiles for stage 12 hydraulic fracturing in well 1 are plotted in Figure 3.

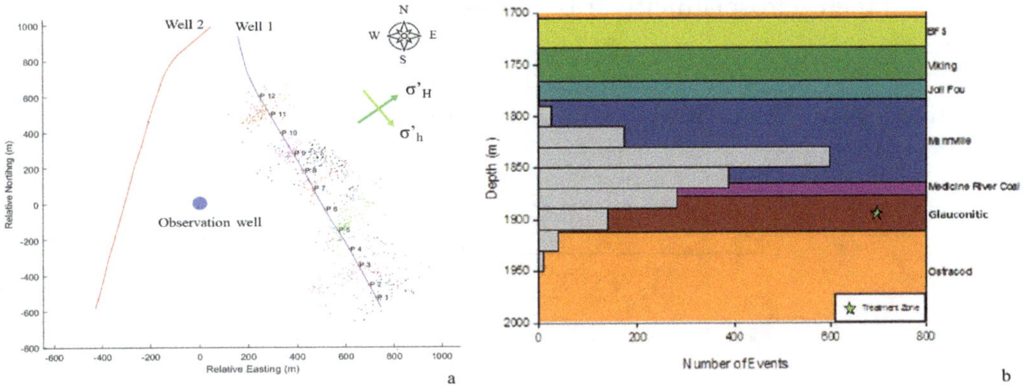

Figure 2. Hoadley field well orientation (**a**) and simplified stratigraphy (**b**). Depth histogram of microseismic events. Gray bars indicate total number of microseismic events recorded at a given depth interval during the hydraulic fracturing [18].

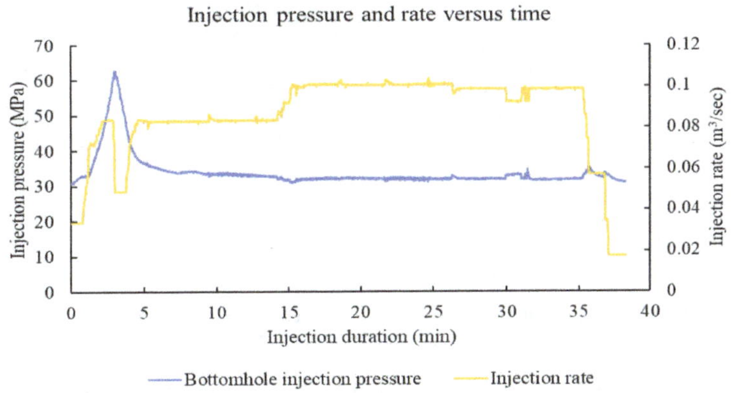

Figure 3. Field recorded injection pressure and rate for stage 12 hydraulic fracturing.

Furthermore, the volumteric strain using the injected volume and injection pressure methods are plotted in Figure 4. The other field data are also included for comparison [45–48]. The trend shown in Figure 4 suggests unique volumetric strain when fractures open, which allows relating the fractured rock permeability to volumetric strain.

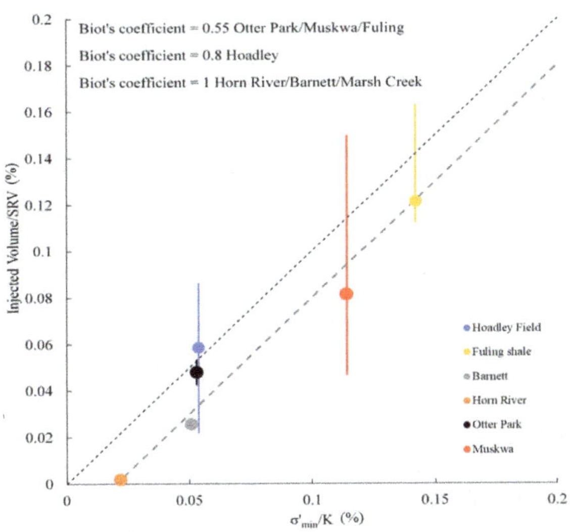

Figure 4. Volumetric strain comparison using the injected volume and minimum in-situ stress approaches. The plotted relation suggests that fracture propagates when strain due to injection equals the strain due to minimum in-situ stress.

Despite most of the microseismicity apparently concentrating above the target unit, this paper models only the Glauconite member. We focus on the in-zone microseismic response due to the abundance of geomechanical property data for Glauconite member, whereas other units are depicted through correlations only. The input parameters of the Glauconite member for ABAQUS FEM are provided in Table 1.

Table 1. The summary of Glauconite data.

Property	Value
Porosity	8–20% [49] 15% [18]
Permeability	0.001–0.01 mD [49] 0.07 mD [18]
Density	2220 kg/m^3 [50]
Depth	2075 m [49] 1900 m [18]
Pore pressure *gradient*	8.5–16.1 kPa/m [49] 4.86 kPa/m [18]
σ_v **gradient**	24.09 kPa/m [18]
σ_H **gradient**	25.79 kPa/m [18]
σ_h **gradient**	11.66 kPa/m [18]
Young's modulus	45.04 GPa [18]
Poisson's ratio	0.23 [18]

The first step in ECM simulation of the hydraulic fracturing is to normalise the microseismic data to reduce the analysis to one-dimensional form. Prior to normalisation, the microseismic data was rotated to align with maximum horizontal stress. Then the coordinates were translated to be relative to the injection port of stage 12. The planar view of rotated and translated microseismic event coordinates for stage 12 are illustrated in Figure 5a. The values for L_{max}, L_{min}, and L_{int} are 150 m, 50 m, and 75 m, respectively. The normalised microseismic cloud for stage 12 is presented in Figure 5b.

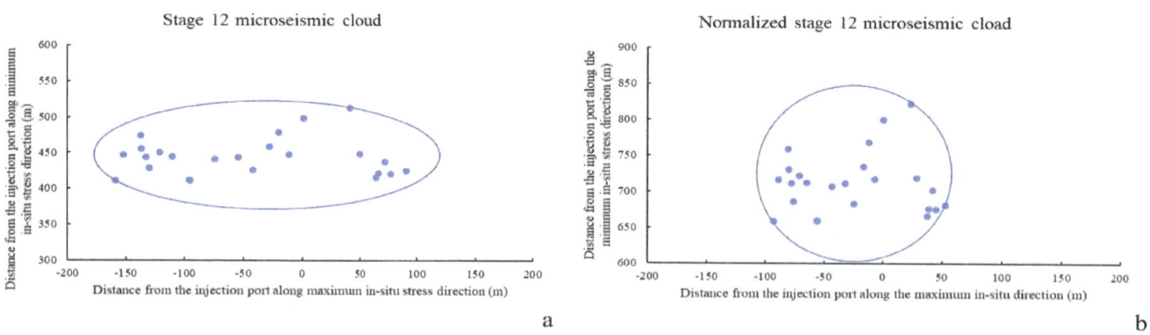

Figure 5. The planar microseismic event coordinates for stage 12 rotated to align with maximum in-situ stress direction and translated to be relative to injection port coordinates (**a**). The normalised stage 12 microseismic event coordinates using $L_{max} = 150$ m, $L_{min} = 50$ m, and $L_{int} = 75$ m (**b**).

The void ratio dependent permeability relation is constructed to match the normalised microseismic cloud. The resulting relation is shown in Figure 6a. As outlined above, this relationship is then denormalised to obtain anisotropic permeability. The resultant permeabilities along the maximum, minimum, and intermediate stress directions, respectively are plotted in Figure 6b.

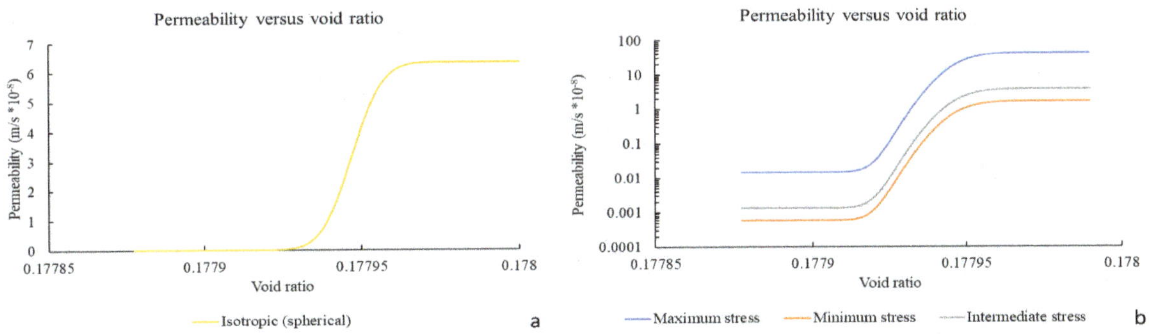

Figure 6. The best matching void ratio dependent permeability relation for normalised and isotropic microseismic cloud (**a**). The denormalised void ratio dependent permeabilities along maximum, minimum, and intermediate in-situ stress directions for stage 12 hydraulic fracturing (**b**).

The anisotropic permeability relation is then used in ABAQUS to combine with the continuity equation, including the stress change effect. The constructed model uses C3D8P three-dimensional pore pressure solid element with an approximate mesh size of 20 m. The modelled region dimensions are 1000 m by 250 m by 225 m at the maximum, minimum and vertical stress directions, respectively. The modelled region dimensions are much larger than the SRV to avoid boundary effects. At the outer and top surfaces of the model, a constant pressure boundary is applied to ensure a constant far-field effective stress effect. This injection rate is only 12.5% of the average injection rate since the constructed model uses symmetry boundary along three surfaces. The inner and bottom surfaces use a symmetry boundary. The injection is defined using concentrated fluid flow load on a single node at a constant rate of 0.015 m^3/s. In-situ stresses and pore pressure are introduced through pre-defined fields in ABAQUS. A constant value at 1900 m is applied for the in-situ stresses and pore pressure. The boundary conditions and injection port locations are indicated in Figure 7 using a simplified sketch.

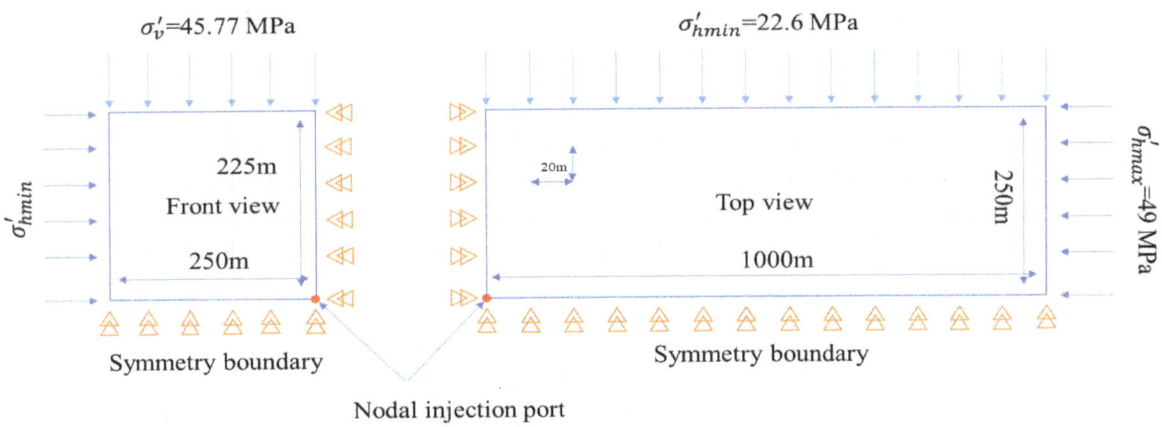

Figure 7. The boundary conditions and injection port location. The single headed arrows triangles indicate loading directions and symmetry surfaces, respectively. Red dot represents the location of injection port.

The results from the ECM are compared to field data, as well as the PKN and KGD models. The injection pressure profile obtained from three models and field observation are provided in Figure 8. The PKN and KGD models account for the intact rock cohesion of 11.55 MPa. Furthermore, to account for the natural fractures in the formation, the leak-off coefficient is introduced into analytical solutions. The leak-off coefficient is estimated based on fracture half-length of 150 m at the end of 38 min injection period. As indicated in Figure 8, the analytical models slightly overestimate injection pressure while ECM shows a good match with reported data.

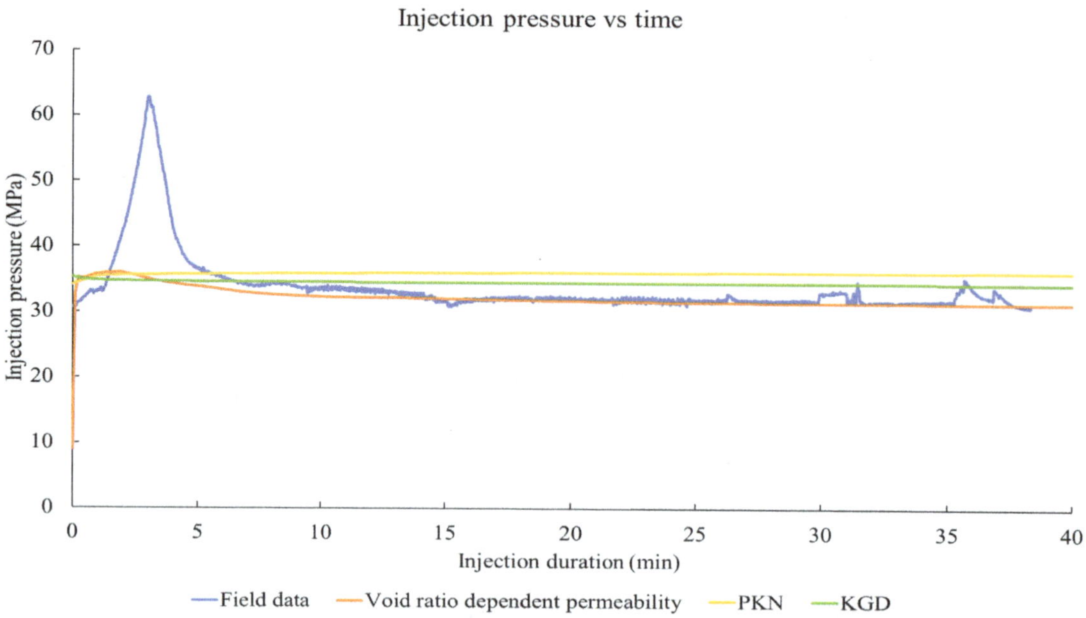

Figure 8. The injection pressure comparison between Equivalent Continuum Methods (ECM), Perkins–Kern–Nordgren (PKN), and Khristianovic–Geertsma–de Klerk (KGD) models against the field observation. As it is observed, the analytical solutions slightly overestimate the injection pressure while ECM shows a good match.

The analytical and ECM estimates for fracture length are compared against observed microseismicity. Fracture length along maximum stress comparison is provided in Figure 9. Besides the previously mentioned methods a diffusivity model estimate is included in Figure 9. The diffusivity model estimates the fracture length growth according to $r(t) = \sqrt{4\pi Dt}$ where $r(t)$ is the fracture length as a function of time in m, D is diffusivity in m^2/s and time in s. The estimated diffusivity using the microseismicity along maximum stress direction is 4 m^2/s. This indicates the case for fracture length increment with assumed constant permeability. As observed from Figure 9, both analytical solutions show the same fracture length elongation profile as the estimation of fluid leak-off coefficient based on the assumed final length of the fracture. For the ECM it is assumed that microseismicity is observed at a given node when pore pressure reaches 12.12 MPa. This satisfies the Mohr Coulomb failure criterion for Glauconite formation [18]. The ECM estimate for fracture length is lower compared to other methods. This is possible due to some of the microseismicity being not due to pore pressure change but due to total stress change around the increased pore pressure region.

Furthermore, the fracture length estimates along minimum in-situ stress direction are compared in Figure 10. As the PKN and KGD models are one-dimensional solutions, they do not show any fracture extend along minimum in-situ stress direction. In addition, the diffusivity model assumes a radial flow. Therefore, it overestimates the fracture extend along the minimum in-situ stress. The ECM approach underestimates the fracture length for the minimum in-situ stress direction, too. This further confirms the possibility of some of the microseismicity due to total stress change effects. The model can be calibrated to match better the microseismic cloud however, the injection pressure profile would differ from the field observation. Considering the higher reliability of injection data compared to microseismicity, it is more preferential to match the injection data.

Another advantage of the proposed ECM has optimised modelling efficiency. The ECM and XFEM models were run using a personal computer with Core i7 2.4 GHz quad-core processor. The ECM required 8768.1 s of CPU time to model 3600 s (or 60 min) of injection period with no divergence problem. However, XFEM required about 28,800 s CPU time before running into convergence problem after 300 s (or 5 min) of the injection period. Therefore, for given generalised data set ECM approach poses as a more efficient modelling method. The XFEM might become a more preferred option if more detailed geomechanical data will allow for a better optimisation of this method.

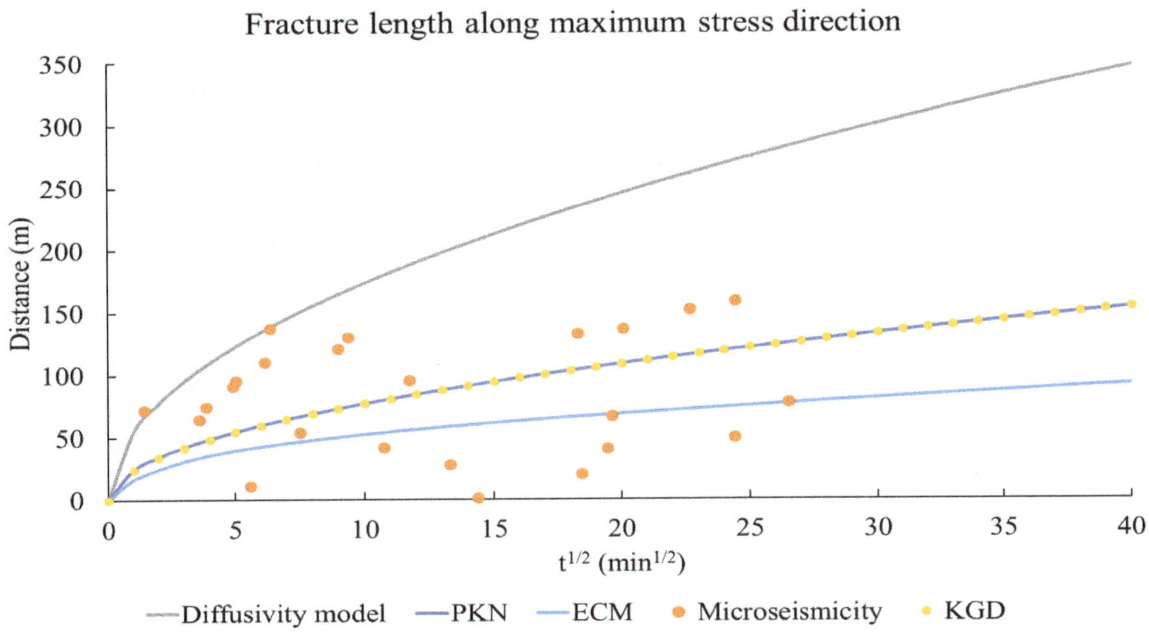

Figure 9. Comparison of fracture length increment along maximum in-situ stress direction using diffusivity, PKN, KGD, and ECM models. The orange dots indicate recorded microseismicity along the maximum in-situ stress direction.

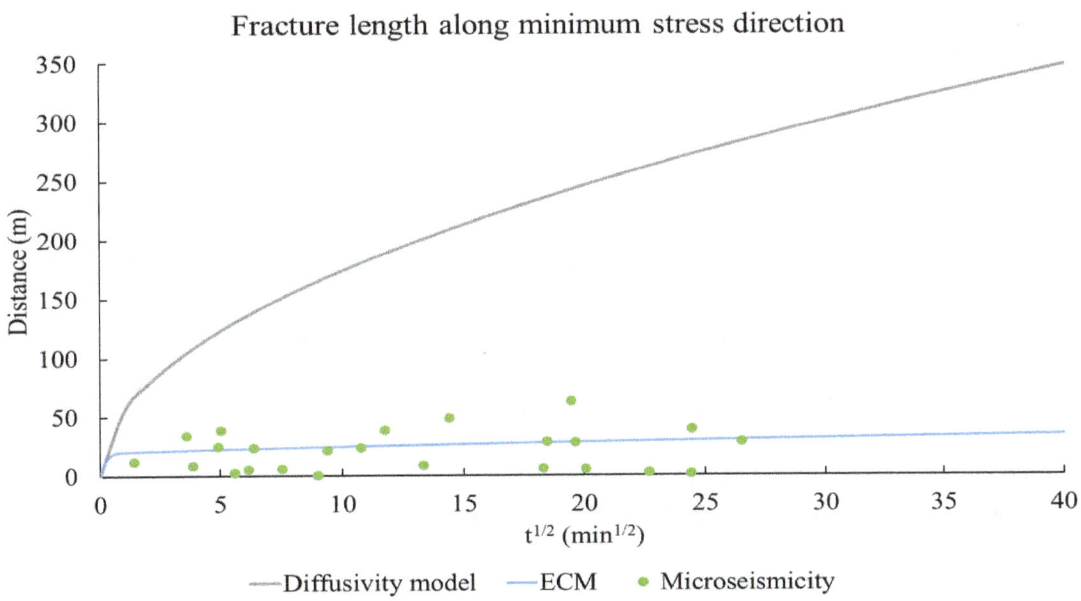

Figure 10. Comparison of fracture length increment along minimum in-situ stress direction using diffusivity and ECM models. The green dots indicate recorded microseismicity along minimum in-situ stress direction.

4. Conclusions

Based on increasing interest in the use of optimised modelling techniques to simulate hydraulic fracturing, we developed a robust and computationally efficient equivalent continuum method. Our ECM method implemented using ABAQUS applies a step-like relationship between void ratio and permeability and incorporates microseismic data to define the stimulated volume of a reservoir. Anisotropy is accounted for by applying a transformation that enables the use of a standard one-dimensional modelling approach with a uniform mesh. This method is particularly useful for cases where there is insufficient data available to define parameters for other numerical approaches, including stress/strain sensitive zones in the target formation. In addition, the use of representative element volume allows the homogenisation of rock elastic and fracture properties for a modelled region. The proposed ECM is more representative compared to PKN, KGD, and diffusivity models and more efficient compared to the XFEM approach. A case study of hydraulic fracturing of tight sand in the Hoadley gas field, Alberta, illustrates the utility of our method for characterising hydraulic fracturing.

Author Contributions: Conceptualisation, E.A.; methodology, E.A.; writing—original draft preparation, E.A.; writing—review and editing, R.C.K.W. and D.W.E.; supervision, R.C.K.W.; funding acquisition, D.W.E. All authors have read and agreed to the published version of the manuscript.

Acknowledgments: This research was funded by Natural Sciences and Engineering Research Council of Canada grant number CRDPJ/474748-2014. Sponsors of the Microseismic Industry Consortium are also thanked for their ongoing support of this initiative.

Abbreviations

The following abbreviations are used in this manuscript:

FEM	Finite Element Modelling
CZM	Cohesive Zone Modelling
XFEM	Extended Finite Element Modelling
ECM	Equivalent Continuum Method
SRV	Stimulated Reservoir Volume

References

1. Li, Q.; Xing, H.; Liu, J.; Liu, X. A review on hydraulic fracturing of unconventional reservoir. *Petroleum* **2015**, *1*, 8–15. [CrossRef]
2. Montgomery, C.T.; Smith, M.B. Hydraulic fracturing: History of an enduring technology. *JPT J. Pet. Technol.* **2010**, *62*, 26–32. [CrossRef]
3. Wu, K.; Olson, J.E. A simplified three-dimensional displacement discontinuity method for multiple fracture simulations. *Int. J. Fract.* **2015**, *193*, 191–204. [CrossRef]
4. Warpinski, N. A review of hydraulic-fracture induced microseismicity. In Proceedings of the 48th US Rock Mechanics/Geomechanics Symposium, Minneapolis, MN, USA, 1-4 June 2014; American Rock Mechanics Association. Available online: https://www.onepetro.org/conference-paper/ARMA-2014-7774 (accessed on 24 November 2020).
5. Yao, Y.; Gosavi, S.V.; Searles, K.H.; Ellison, T.K. Cohesive fracture mechanics based analysis to model ductile rock fracture. In Proceedings of the 44th US Rock Mechanics Symposium and 5th US-Canada Rock Mechanics Symposium, Salt Lake City, UT, USA, 27–30 June 2010.
6. Haddad, M.; Sepehrnoori, K. Simulation of multiple-stage fracturing in quasibrittle shale formations using pore pressure cohesive zone model. In Proceedings of the Unconventional Resources Technology Conference, Denver, CO, USA, 25–27 August 2014; pp. 1777–1792. [CrossRef]
7. Haddad, M.; Sepehrnoori, K. Integration of XFEM and CZM to model 3D multiple-stage hydraulic fracturing in quasi-brittle shale formations: Solution-dependent propagation direction. In Proceedings of the AADE National Technical Conference and Exhibition, San Antonio, TX, USA, 8–9 April 2015; pp. 8–9.
8. Haddad, M.; Sepehrnoori, K. Simulation of hydraulic fracturing in quasi-brittle shale formations using characterized cohesive layer: Stimulation controlling factors. *J. Unconv. Oil Gas Resour.* **2015**, *9*, 65–83. [CrossRef]
9. Wang, H. Numerical modeling of non-planar hydraulic fracture propagation in brittle and ductile rocks using XFEM with cohesive zone method. *J. Pet. Sci. Eng.* **2015**, *135*, 127–140. [CrossRef]

10. Zielonka, M.G.; Searles, K.H.; Ning, J.; Buechler, S.R. Development and validation of fully-coupled hydraulic fracturing simulation capabilities. In Proceedings of the SIMULIA Community Conference, Providence, RI, USA, 20–22 May 2014; pp. 19–21.
11. Arndt, S.; Zee, W.V.D.; Hoeink, T.; Hughes, B.; Nie, J. Hydraulic fracturing simulation for fracture networks. In Proceedings of the SIMULIA Community Conference, Berlin, Germany, 18–21 May 2015; Volume 4.
12. Nasirisavadkouhi, A. A comparison study of KGD. *PKN Modif. P3D Model* **2015**. [CrossRef]
13. Yu, G.; Aguilera, R. 3D analytical modeling of hydraulic fracturing stimulated reservoir volume. In *SPE Latin America and Caribbean Petroleum Engineering Conference, Mexico City, Mexico, 16–18 April, 2012*; Society of Petroleum Engineers, SPE-153486-MS, SPE. Available online: https://www.onepetro.org/conference-paper/SPE-153486-MS (accessed on 24 November 2020).
14. Bouteca, M.J. 3D analytical model for hydraulic fracturing: Theory and field test. In Proceedings of the SPE Annual Technical Conference and Exhibition, New Orleans, Louisiana, USA, 27–30 September 1984, doi:10.2523/13276-ms. [CrossRef]
15. Stalgorova, E.; Mattar, L. Analytical model for unconventional multifractured composite systems. *SPE Reserv. Eval. Eng.* **2013**, *16*, 246–256. [CrossRef]
16. Geertsma, J.; De Klerk, F. A rapid method of predicting width and extent of hydraulically induced fractures. *J. Pet. Technol.* **2007**, *21*, 1571–1581. [CrossRef]
17. Nordgren, R. Propagation of a vertical hydraulic fracture. *Soc. Pet. Eng. J.* **1972**, *12*, 306–314. [CrossRef]
18. Maulianda, B. On Hydraulic Fracturing of Tight Gas Reservoir Rock. Ph.D. Thesis, University of Calgary, Calgary, AB, Canada, 2016. [CrossRef]
19. Gonzalez, D.; Holman, R.; Richard, R.; Xue, H.; Morales, A.; Kwok, C.K.; Judd, T. Accounting for production shadow in infill/DUC well hydraulic fracturing modeling and calibration. In *SPE Liquids-Rich Basins Conference-North America, Midland, Texas, USA, 13–14 September, 2017*; Society of Petroleum Engineers. Available online: https://www.onepetro.org/conference-paper/SPE-187479-MS (accessed on 24 November 2020).
20. Valiullin, A.; Makienko, V.; Yudin, A.; Overin, A.; Gromovenko, A. Channel fracturing technique helps to revitalize brown fields in langepas area. In *Society of Petroleum Engineers—SPE Oil and Gas India Conference and Exhibition, Mumbai, India, 24–26 November 2015*; Society of Petroleum Engineers. Available online: https://www.onepetro.org/conference-paper/SPE-178131-MS (accessed on 24 November 2020). [CrossRef]
21. Chen, Z.; Bunger, A.P.; Zhang, X.; Jeffrey, R.G. Cohesive zone finite element-based modeling of hydraulic fractures. *Acta Mech. Solida Sin.* **2009**, *22*, 443–452. [CrossRef]
22. Wang, X.L.; Liu, C.; Wang, H.; Liu, H.; Wu, H.A. Comparison of consecutive and alternate hydraulic fracturing in horizontal wells using XFEM-based cohesive zone method. *J. Pet. Sci. Eng.* **2016**, *143*, 14–25. [CrossRef]
23. Guo, J.; Luo, B.; Lu, C.; Lai, J.; Ren, J. Numerical investigation of hydraulic fracture propagation in a layered reservoir using the cohesive zone method. *Eng. Fract. Mech.* **2017**, *186*, 195–207. [CrossRef]
24. Yao, Y.; Liu, L.; Keer, L.M. Pore pressure cohesive zone modeling of hydraulic fracture in quasi-brittle rocks. *Mech. Mater.* **2015**, *83*, 17–29. [CrossRef]
25. Carrier, B.; Granet, S. Numerical modeling of hydraulic fracture problem in permeable medium using cohesive zone model. *Eng. Fract. Mech.* **2012**, *79*, 312–328. [CrossRef]
26. Yao, Y. Linear elastic and cohesive fracture analysis to model hydraulic fracture in brittle and ductile rocks. *Rock Mech. Rock Eng.* **2012**, *45*, 375–387. [CrossRef]
27. Lavrov, A.; Larsen, I.; Holt, R.M.; Bauer, A.; Pradhan, S. Hybrid FEM/DEM simulation of hydraulic fracturing in naturally-fractured reservoirs. In Proceedingd of the 48th US Rock Mechanics/Geomechanics Symposium, Minneapolis, MN, USA, 1–4 June 2014; pp. 626–633.
28. Munjiza, A.; John, N.W. Mesh size sensitivity of the combined FEM/DEM fracture and fragmentation algorithms. *Eng. Fract. Mech.* **2001**, *69*, 281–295. [CrossRef]
29. Min, K.B. Fractured Rock Masses as Equivalent Continua—A Numerical Study. Ph.D. Thesis, Mark Och Vatten, Stockholm, Sweden, 2004.
30. Oda, M. An equivalent continuum model for coupled stress and fluid flow analysis in jointed rock masses. *Water Resour. Res.* **1986**, *22*, 1845–1856. [CrossRef]
31. Villeneuve, M.C.; Heap, M.J.; Kushnir, A.R.; Qin, T.; Baud, P.; Zhou, G.; Xu, T. Estimating in situ rock mass strength and elastic modulus of granite from the Soultz-sous-Forêts geothermal reservoir (France). *Geotherm. Energy* **2018**, *6*, 1–29. [CrossRef]

32. JianPing, Y.; WeiZhong, C.; DianSen, Y.; JingQiang, Y. Numerical determination of strength and deformability of fractured rock mass by FEM modeling. *Comput. Geotech.* **2015**, *64*, 20–31. [CrossRef]
33. Mack, M.G.; Warpinski, N.R. Mechanics of hydraulic fracturing. In *Reservoir Stimulation*; Wiley: New York, NY, USA, 2000; pp. 1–6.
34. SIMULIA. ABAQUS 6.16-1. 2016. Available online: http://130.149.89.49:2080/v2016/index.html (accessed on 24 November 2020).
35. Youn, D.J.; Griffiths, D.V. Hydro-Mechanical Coupled Simulation of Hydraulic Fracturing Using the Extended Finite Element Method (XFEM). Ph.D Thesis, Colorado School of Mines, Golden, GO, USA, 2016.
36. Kang, Y.; Yu, M.; Miska, S.; Takach, N.E. Wellbore stability: A critical review and introduction to DEM. In Proceedings of the SPE Annual Technical Conference and Exhibition, New Orleans, LA, USA, 4–7 October 2009; Volume 4, pp. 2689–2712. [CrossRef]
37. Zhou, J.; Huang, H.; Mattson, E.; Wang, H.F.; Haimson, B.C.; Doe, T.W.; Oldenburg, C.M.; Dobson, P.F. *Modeling of Hydraulic Fracture Propagation at the kISMET Site Using a Fully Coupled 3D Network-Flow and Quasi-Static Discrete Element Model*; Technical Report; Idaho National Lab (INL): Idaho Falls, ID, USA, 2017.
38. Nguyen, T.K.; Combe, G.; Caillerie, D.; Desrues, J. FEM x DEM modelling of cohesive granular materials: Numerical homogenisation and multi-scale simulations. *Acta Geophys.* **2014**, *62*, 1109–1126. [CrossRef]
39. Maulianda, B.; Savitri, C.D.; Prakasan, A.; Atdayev, E.; Yan, T.W.; Yong, Y.K.; Elrais, K.A.; Barati, R. Recent comprehensive review for extended finite element method (XFEM) based on hydraulic fracturing models for unconventional hydrocarbon reservoirs. *J. Pet. Explor. Prod. Technol.* **2020**. [CrossRef]
40. Hoek, E.; Diederichs, M.S. Empirical estimation of rock mass modulus. *Int. J. Rock Mech. Min. Sci.* **2006**, *43*, 203–215. [CrossRef]
41. Oda, M. Similarity rule of crack geometry in statistically homogeneous rock masses. *Mech. Mater.* **1984**, *3*, 119–129. [CrossRef]
42. Mayerhofer, M.; Lolon, E.; Warpinski, N.; Cipolla, C.; Walser, D.; Rightmire, C. What Is Stimulated Rock Volume? In Proceedings of the SPE Shale Gas Production Conference, Fort Worth, TX, USA, 16–18 November 2009.
43. Ahmed, T. *Reservoir Engineering Handbook*; Gulf Professional Publishing: Oxford, UK, 2018.
44. Shapiro, S.; Hummel, N. Nonlinear diffusion-based interpretation of hydraulic fracturing induced seismicity. In Proceedings of the 75th EAGE Conference & Exhibition-Workshops, London, UK, 10–13 June 2013.
45. Eaton, D.; Caffagni, E.; van der Baan, M.; Matthews, L. Passive seismic monitoring and integrated geomechanical analysis of a tight-sand reservoir during hydraulic-fracture treatment, flowback and production. In Proceedings of the Unconventional Resources Technology Conference, Denver, CO, USA, 25–27 August 2014; pp. 1537–1545.
46. Ren, L.; Lin, R.; Zhao, J. Stimulated reservoir volume estimation and analysis of hydraulic fracturing in shale gas reservoir. *Arab. J. Sci. Eng.* **2018**, *43*, 6429–6444. [CrossRef]
47. Li, Q.; Aguilera, R. Parametric probabilistic models for fluid diffusivity inversion and forward microseismic generation using seismicity rates. *SPE West. Reg. Meet.* **2018**. [CrossRef]
48. Xu, Y.; Adefidipe, O.; Dehghanpour, H. Estimating fracture volume using flowback data from the Horn River Basin: A material balance approach. *J. Nat. Gas Sci. Eng.* **2015**, *25*, 253–270. [CrossRef]
49. Reynolds, M.; Thomson, S.; Peyman, F.; Hung, A.; Quirk, D.; Chen, S. A Direct Comparison of Hydraulic Fracture Geometry and Well Performance between Cemented Liner and Openhole Packer Completed Horizontal Wells in a Tight Gas Reservoir. In Proceedings of the PSPE Hydraulic Fracturing Technology Conference, Woodlands, TX, USA, 6–8 February 2012; pp. 1–19. [CrossRef]
50. Broger, E.K.; Syhlonyk, G.E.; Zaitlin, B.A. Glauconite Sandstone Exploration: A Case Study from the Lake Newell Project, Southern Alberta. In *Petroleum Geology of the Cretaceous Mannville Group Western Canada—Memoir 18*; American Association of Petroleum Geologists, AAPG: Tulsa, OK, USA, 1997; pp. 140–168. Available online: http://archives.datapages.com/data/cspg_sp/data/018/018001/140_cspgsp0180140.htm (accessed on 24 November 2020).

3

Three-Dimensional Turbulence Numerical Simulation of Flow in a Stepped Dropshaft

Yongfei Qi [1,2], Yurong Wang [1,*] and Jianmin Zhang [1,*]

1 State Key Laboratory of Hydraulics and Mountain River Engineering, Sichuan University, Chengdu 610065, China; xjqiyongfei@163.com
2 Department of Foundation Studies, Xinjiang Institute of Engineering, Urumqi 830023, China
* Correspondence: wangyurong@scu.edu.cn (Y.W.); zhangjianmin@scu.edu.cn (J.Z.).

Abstract: The dropshaft structure is usually applied in an urban drainage system to connect the shallow pipe network and the deep tunnel. By using the renormalization group (RNG) k~ε turbulence model with a volume of fluid method, the flow pattern and the maximum relative water depth over a stepped dropshaft with a different central angle of step were numerically investigated. The calculated results suggested that the flow in the stepped dropshaft was highly turbulent and characterized by deflection during the jet caused by the curvature of the sidewall. According to the pressure distribution on the horizontal step and the flow pattern above the step, the flow field was partitioned into the recirculating region, the wall-impinging region and the mixing region. In addition, with the increase in the central angle of step, the scope of the wall-impinging region and the mixing region increased and the scope of the recirculating region remained nearly unchanged. The maximum water depth increased with the increase in discharge. In the present work we have shown that, as the value of the central angle of step increased, the maximum water depth decreased initially and increased subsequently.

Keywords: stepped dropshaft; numerical simulation; flow region; central angle of step

---

1. Introduction

A dropshaft is a hydraulic structure that is installed in drainage systems and tunnel schemes to convey extreme rainfall so that it will not trigger urban flooding. The energy dissipation of a dropshaft is very notable. Studies by Rajaratnam et al., Chanson, and Camino et al. have shown that the energy dissipation rate of the dropshaft is normally more than 75% [1–4].

According to the flow features of a vertical dropshaft, the general dropshaft structure can be divided into four categories: plunge-flow dropshaft [5,6], vortex-flow dropshaft [7,8], baffle-flow dropshaft [9,10], and helicoidal-ramp dropshaft [11,12]. In recent years, vortex-flow dropshafts have become popular due to the excellent energy dissipation and air removal in the construction. The flow characteristic has been investigated by engineers and scientists through laboratory experimentation and numerical simulation. Jain and Kennedy [13] reported the application of vortex dropshafts in the Milwaukee Metropolitan Sewerage District in 1984. In 2010, Hager [14], combined with previous engineering experience, put forward a complete design plan for a vortex dropshaft. Zhao et al. [15] investigated the performance of a vortex drop structure with a relatively small height-to-diameter ratio and confirmed that the energy dissipation rate and air entrainment rate were very significant. Del Giudice and Gisonni [16] optimized the intake of a vortex drop structure in Naples and solved the problem of supercritical approach flow. Natarius [17] installed the Vortex Flow Insert (VFI) in a vortex drop which can effectively eliminate the public odor problem in that drop. Yu and Lee [18] investigated the hydraulics of tangential vortex intake and presented the design guideline that can convey the flow steadily without being influenced by the hydraulic jump.

In order to improve energy dissipation and reduce safety hazards, an innovative design of vortex-flow structures has been developed [19]. A series of steps was deployed on the dropshaft. The flow hit each step as an impinging jet, with the energy being dissipated through jet breakup in the air, collision and swirling on the step, and increased friction between jet and sidewall [20,21]. From the perspective of safety, the steps will significantly increase the air concentration and reduce the risks of cavitation damage [22]. The features of the standing waves concerning the relative height, the location, and the extent were studied theoretically and experimentally by Wu et al. [23].

Although researchers of existing studies have gained a comprehensive understanding of this particular type of dropshaft, little is known about the intricacies of the flow and influence of central angle of step on hydraulic properties. Therefore, in order to study the effect of central angle of step on the flow pattern and maximum water depth on the sidewall, numerical investigations of a stepped dropshaft with different central angles were carried out.

2. Numerical Simulation

2.1. Volume of Fluid Method

The volume of fluid [24] model, which relies on the fact that two or three phases are not interpenetrating, provides an efficient and economical way to track the volume fraction of each of the fluids throughout the domain. Because different fluid components are solved by a single set of momentum equations, the volume fraction of unit phase is defined in order to track the free surface of each computational cell. For an air–water two-phase flow model, in each computational cell, the sum of the volume fractions of the air and water is unity. Specifically, computational cells without water have a value of zero. Full cells are assigned a value of 1 and partially filled cells have a value between 0 and 1˙ Therefore, the volume fraction of air or water is defined as α_a and α_w, respectively. The relationship between α_a and α_w can be given as follows:

$$\alpha_a + \alpha_w = 1 \tag{1}$$

The variables and their attributes are shared by air and water and represent volume-averaged values if the volume fraction of air and water is obtained. Therefore, at any given control volume, the variables and their attributes represent either air or water, or a mixture of them. The tracking of the interface between air and water is accomplished by the solution of the continuity equation with the following form:

$$\frac{\partial \alpha_w}{\partial t} + u_i \frac{\partial \alpha_w}{\partial x_i} = 0 \tag{2}$$

where α_w is the volume fraction of water and u_i and x_i are the velocity components and coordinates (i = 1, 2, 3), respectively.

The density ρ and molecular viscosity μ, which are the volume-fraction-averaged properties and not constants can be expressed as follows:

$$\rho = \alpha_w \rho_w + (1 - \alpha_w)\rho_a \tag{3}$$

$$\mu = \alpha_w \mu_w + (1 - \alpha_w)\mu_a \tag{4}$$

where ρ_w and ρ_a are the density of water and air and μ_w and μ_a are the molecular viscosity of water and air. By the iterating solution of the volume fraction of water α_w, ρ and μ can be calculated.

2.2. Turbulence Model

With the rapid development of computer technology, numerical methods have been applied more and more to complex flows of industrial relevance [25–30]. The most widely used approach in both academia and industry for the modeling of turbulent flows is Reynolds-averaged Navier–Stokes (RANS) methods that solve additional transport equations for turbulence and introduce the turbulence eddy viscosity to simulations to mimic the effect of turbulence. In the RANS modeling framework, two distinct turbulence models were used, namely, k~ε and k~ω. The k~ε turbulent models have historically been used in flow simulations because of the better convergence and lower memory [31,32]. Desirable results can be achieved in various contexts by the k~ω turbulent models, such as boundary layers with adverse pressure gradients and flow separation [33,34]. There are many papers which simulated the complicated hydraulic properties in stepwise and vortex shaft spillway structures using the k~ε model [35–38], yielding reliable results compared with laboratory experiments. As an improvement of the standard k~ε model, the RNG k~ε model [39] is more accurate in simulating flows with high strained rates and streamline bending [40–42], such as the swirling flow. Therefore, the RNG k~ε turbulent model was adopted to model the intricate flow in the stepped dropshaft. The computational fluid dynamics module of the ANSYS 16.0 software (ANSYS®, Canonsburg, PA, USA), Fluent [43], was utilized to investigate the flow over the stepped dropshaft. The equations of the turbulent kinetic energy, k, and its dissipation rate, ε, are as follows:

k equation:

$$\frac{\partial}{\partial t}(\rho k) + \frac{\partial}{\partial x_i}(\rho k u_i) = \frac{\partial}{\partial x_j}\left[(\mu + \frac{\mu_t}{\sigma_k})\frac{\partial k}{\partial x_j}\right] + G_k - \rho\varepsilon, \tag{5}$$

ε equation:

$$\frac{\partial}{\partial t}(\rho\varepsilon) + \frac{\partial}{\partial x_i}(\rho\varepsilon u_i) = \frac{\partial}{\partial x_j}[(\mu + \frac{\mu_t}{\sigma_\varepsilon})\frac{\partial \varepsilon}{\partial x_j}] + C_{1\varepsilon}^*\frac{\varepsilon}{k}G_k - C_{2\varepsilon}\rho\frac{\varepsilon^2}{k}, \tag{6}$$

$$\mu_t = \rho C_\mu \frac{k^2}{\varepsilon} \tag{7}$$

$$G_k = \mu_t(\frac{\partial u_i}{\partial x_j} + \frac{\partial u_j}{\partial x_i})\frac{\partial u_i}{\partial x_j}, \tag{8}$$

$$C_{1\varepsilon}^* = C_{1\varepsilon} - \frac{\eta(1-\eta/\eta_0)}{1+\beta\eta^3}, \tag{9}$$

$$\eta = \sqrt{\frac{G_k}{\rho C_{\mu\varepsilon}}}, \tag{10}$$

where ρ is the corresponding density, μ is the dynamic viscosity, μ_t is the turbulence viscosity, u_i is the velocity component in the ith direction, t is the time, G_k is the generation of turbulent energy caused by the average velocity gradient, and C_μ, $C_{1\varepsilon}$, $C_{2\varepsilon}$, β, and η_0 are empirical constants. In the RNG k~ε model, the empirical constants are given as $C_\mu = 0.0845$, $\sigma_k = \sigma_\varepsilon = 0.7179$, $C_{1\varepsilon} = 1.42$, $C_{2\varepsilon} = 1.68$, $\beta = 0.012$, and $\eta_0 = 4.38$.

2.3. Numerical Algorithm

The Pressure-Implicit with Splitting of Operators (PISO) algorithm, based on the higher degree of the approximate relation between the corrections for pressure and velocity, was applied to implicitly couple the velocity and pressure. The implicit finite volume method was used to solve the iteration. The calculated domain was divided into discrete control volumes by the unstructured grid that has a high adaptability and is self-adjusting to the complicated geometry and boundary. The detailed location of the free surface is determined by the geometric reconstruction scheme, which is accurate and applicable for general unstructured meshes in Fluent. This scheme assumes that the interface between two fluids has a linear slope within each cell, and then uses this linear shape for calculation of the advective flux. Ultimately, the volume fraction of each cell can be calculated according to the advective flux balance of previous calculations. Refining the grid on interface between water and air is important for transient VOF calculations using the Geo-Reconstruct formulation. It was unfortunate that the water level in the dropshaft could not be measured beforehand because of the complexity and randomness in the multi-phase flow process. Consequently, in order to get the high-precision results, all the three calculated configurations were refined with total grid numbers of approximately 0.30 million, 0.46 million, and 0.65 million, respectively (the specific content will be introduced later).

The appropriate setting of time-dependent solution parameters is the magnitude guarantee of robustness and efficiency for calculation. For transient volume of fluid calculations that use the implicit scheme of VOF, the fixed times stepping method was selected. The max iterations per time step, a maximum for the number of iterations per time step, was set as 50. Since the ANSYS Fluent formulation is fully implicit, there is no stability criterion that must be met in determining the time step size. A good way to adjust the size of time step is to satisfy the ideal number of iterations per time step, which ranges from 5 to 10. To be specific, if the calculation needs only a few iterations per time step, the time step size should be increased; otherwise, it should be decreased. In this work, the range of the time step size was adjusted from 0.0001 s to 0.001 s and the iteration number was always less than 5.

2.4. Geometric Model

The calculated domain is shown in Figure 1. It has the same size as the physical model, given in the literature [19]. The numerical model consists of three main parts, namely, the inlet section, the vertical shaft section, and the outlet section. The inlet channel cross-section was rectangular, 0.15 m wide and 0.25 m deep. The length of the inlet channel was 1 m. For the vertical shaft section, the height was 1.834 m, the external radius (R) was 0.25 m, and the internal radius (r) was 0.1 m. The shaft vertical section consisted of 14 steps, which connected the inlet channel and the outlet channel. The width, b, and the height, h, of each step were 0.15 m and 0.131 m, respectively. They were numbered from S1 to S14. The central angle of step, one important structure feature of dropshaft, was defined as θ, and configurations with $\theta = 120°$, $150°$, or $180°$ were investigated. The dimensions of the cross-section of the outlet channel were $2 \times 0.15 \times 0.131$ m (length $\times$ width $\times$ height).

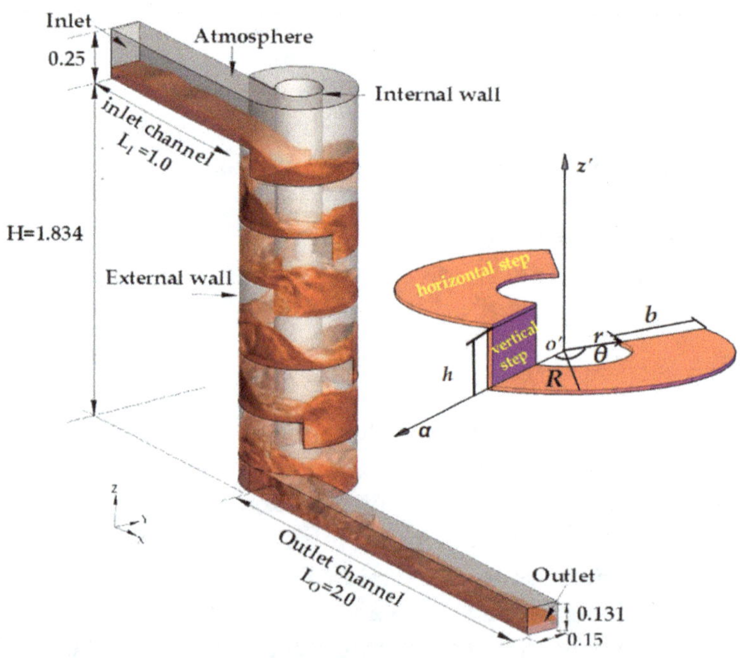

Figure 1. Computational model of the dropshaft.

In order to facilitate analysis, the cylindrical coordinate systems were established. The z-axis was the center axis of shaft. The relative polar coordinates were established on each step, in which the center of the circle was defined as the pole and the starting edge of the step was defined as the polar axis.

2.5. Boundary Conditions and Cases

The boundary conditions were as follows:

(1) Inlet boundary: the velocity inlet was used for the intake, which was set at 0.89–2.69 m/s;
(2) Outlet boundary: the outlet boundary was set as pressure outlet and the normal gradient of all variables was equal to 0;
(3) Free surface: the free surface of water was assumed to be the pressure inlet and the pressure value was $P = 0$; and
(4) Wall boundary: no-slip velocity boundary condition; the near-wall regions of the flow were analyzed using the method of standard wall function.

In this study, nine cases with different flow rates and different central angles of step were investigated. The calculation program is shown in Table 1. The design condition of verified cases was set up the same way as the experiment from the literature [19]: h is the step height; D is the width-radius ratio, which is the ratio of step width to external radius; i is the slope coefficient, which is spreading out from the external wall; and Fr is the Froude number of approach flow.

Table 1. Summary of the operating conditions for the simulations.

h (m)	D	θ (°)	i	Q (m^3/s)	Fr	Case
0.131	0.6	150	0.20	12.75	0.58	test1
				41.50	1.90	test2
		120	0.25	80.00	3.67	1
				48.00	2.20	2
				26.50	1.22	3
		150	0.20	80.00	3.67	4
				48.00	2.20	5
				26.50	1.22	6
		180	0.17	80.00	3.67	7
				48.00	2.20	8
				26.50	1.22	9

2.6. Verification

2.6.1. Grid Testing

It is very important to use the grid reasonably for the accuracy of calculation results. Therefore, the effects of different grid densities on the uncertainty of the calculation results were tested using the grid convergence index (GCI) [44] with grid numbers of approximately 0.65 million (grid 1), 0.46 million (grid 2), and 0.30 million (grid 3). The GCI is defined as follows:

$$\mathrm{GCI} = \frac{1.25|(\phi_1 - \phi_2)/\phi_1|}{(k_2 - k_1)^p - 1}, \tag{11}$$

in which ϕ_i represents the variable of the ith calculation, k_i represents the representative grid size for the ith calculation, and let $k_1 < k_2 < k_3$. For three-dimensional calculations, k is given by the following:

$$k = \left(\left[\frac{1}{N}\sum_{i=1}^{N}(\Delta V_i)\right]\right)^{1/3} \tag{12}$$

The order of p in Equation (11) can be estimated by the following:

$$p = \frac{1}{\ln(k_2/k_1)} \left| \ln|(\phi_3 - \phi_2)/(\phi_2 - \phi_1)| + \left| \ln\left(\frac{(k_2/k_1)^P - \operatorname{sgn}[(\phi_3 - \phi_2)/(\phi_2 - \phi_1)]}{(k_3/k_2)^P - \operatorname{sgn}[(\phi_3 - \phi_2)/(\phi_2 - \phi_1)]} \right) \right| \right|. \tag{13}$$

In this paper, the pressure and tangential velocity that, respectively, were written in dimensionless form as p/h and $V_\tau/V_{\tau\max}$, were variables. Figure 2b,d presents pressure and tangential velocity along a radial line that was in the yoz plane and paralleled to the y-axis (as depicted in Figure 2a) for different grid densities with Q = 41.25 m^3/s. The data obtained from grid 2 were close to the data obtained with the grid 1 and the computational efficiency decreased by 20%. In Figure 2c,e, the maximum uncertainties in pressure and tangential velocity were approximately 3.12% and 4.17%, respectively, implying that the discretization uncertainties were little in most locations. For the consideration of computational efficiency and accuracy, the grid number of 0.46 million (grid 2) was adopted in the present study.

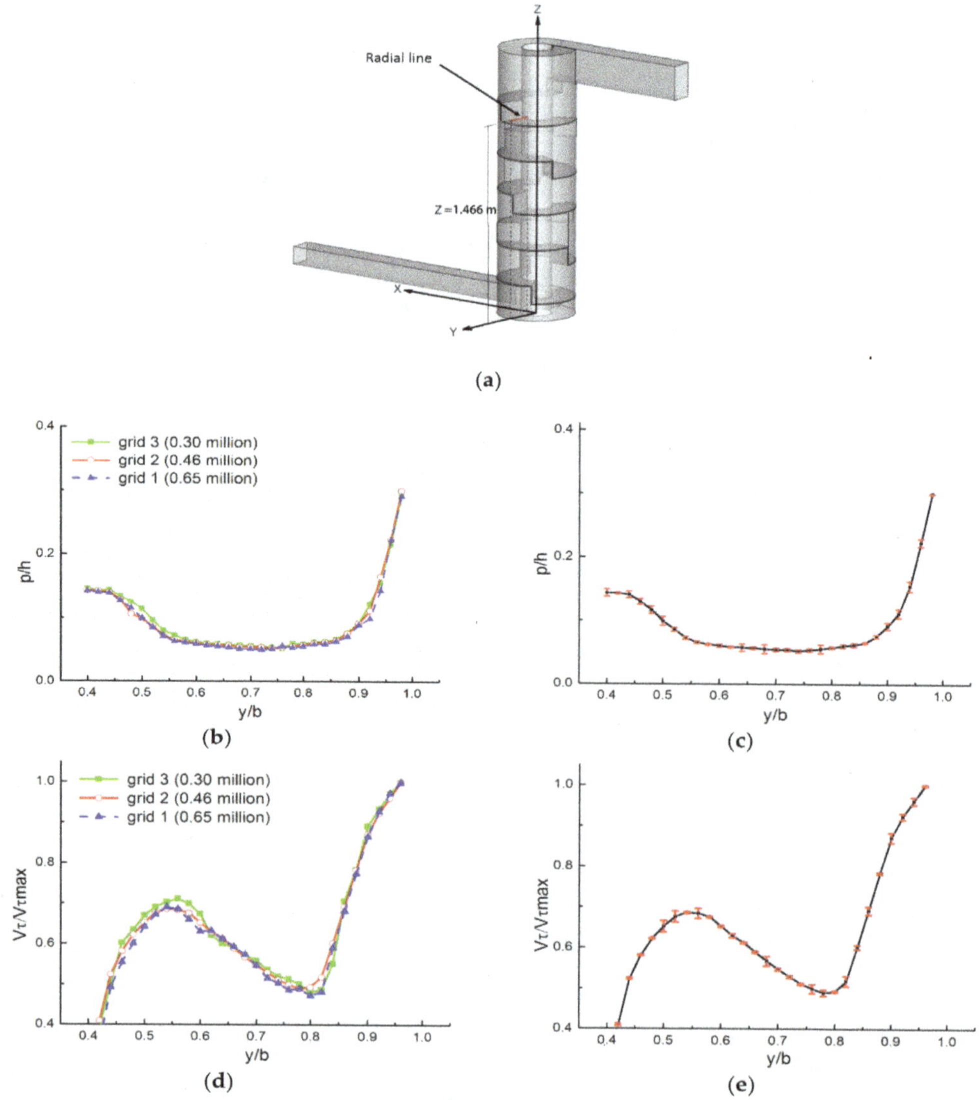

Figure 2. Grid convergence index value for different grid densities: (**a**) location of radial line; (**b**) pressure profile in models with different mesh numbers; (**c**) fine-grid solution, with discretization error bars computed using Equation (11); (**d**) tangential velocity profile in models with different mesh numbers; (**e**) fine-grid solution, with discretization error bars computed using Equation (11).

2.6.2. Model Verification

The calculated flow pattern was fairly consistent with the experimental results under different discharges, as is shown in Figure 3. When transferring from the upper steps to the lower steps, and owing to the collision between the flow and the steps, part of the flow changed direction with the mainstream travelling downstream. The flow that changed direction impacted the vertical surface of the step and formed a reflux. In addition, the cavities near the vertical step were obvious. As shown in Figure 3a,c, the main differences are as follows: (1) the recirculation zone for Q = 41.5 m^3/s was smaller than that of Q = 12.75 m^3/s; and (2) for Q = 12.75 m^3/s, there was an increase in the water level when flow entered the next step, but for Q = 41.5 m^3/s, the increase in water level occurred after the nappe flow impacted the horizontal step. As can be seen from Figure 3b,d, the simulated flow patterns were in good agreement with experimental results.

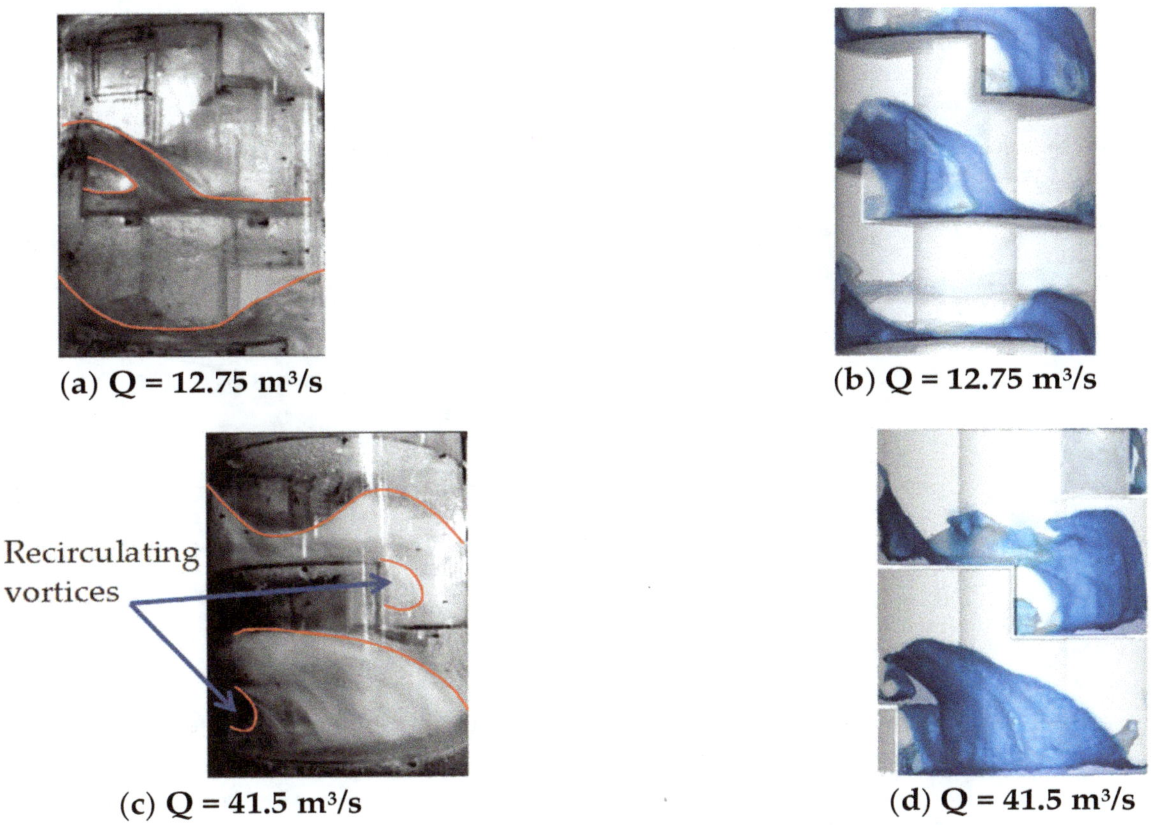

Figure 3. Comparison of experimental and calculated flow pattern: (**a**,**c**) flow pattern in experiment; (**b**,**d**) flow pattern in simulation.

In addition, experimental data obtained from Wu et al. [19] were used to verify the numerical results by comparing the stream-wise and radial pressure at S9, as is shown in Figure 4. In Figure 4a, four measuring points (P1~P4) were set on the center line of the bottom step along the flow direction. P3, and the other three measuring points (P5, P6, P7) were on a straight line that was paralleled to the y-axis in the radial direction. It can be seen from Figure 4b,c that, at most locations, the calculated results of pressure distribution were fairly consistent with that of laboratory tests. The maximums of the relative error for Q = 12.75 m^3/s and Q = 41.5 m^3/s were 7.5% and 10%, respectively, indicating that the numerical simulations produced reliable and acceptable results.

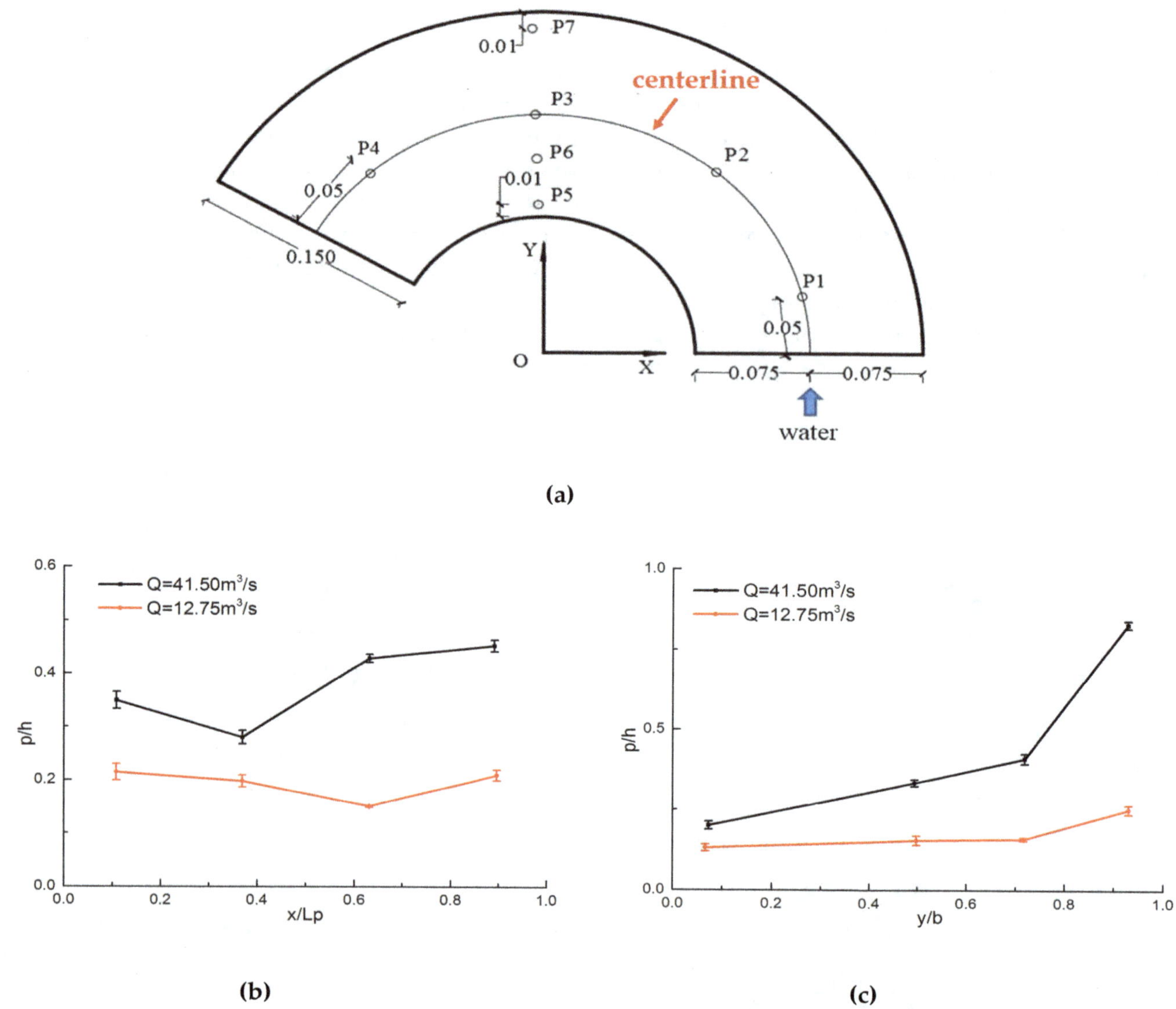

Figure 4. Comparison of numerical and experimental data: (**a**) location of the measuring points (m); (**b**) pressure distribution along the flow direction; (**c**) pressure distribution along the radical direction. Notes: L_p represents the arc length of the center line; b is the width of the horizontal step.

2.6.3. Fluctuation of Calculation Results

There existed instabilities and uncertainties in the computed results because of the complexity of the flow. In order to improve the reliability of results, the pressure for R = 0.25 m at S6 and the water surface on outer wall at S6 were monitored to analyze the fluctuation of pressure and water level in the calculation process. After a long time calculation, when the average relative errors of the mass flow rate were below 10%, the flow variables were considered relatively constant in this work. Five relative locations, namely, 10°, 30°, 60°, 90°, and 120°, were selected from each of the nine cases for determining the dispersion of pressure and the water surface at the same time period. The data in a duration of 10 s with an interval of 1 s were selected for analysis. The fluctuation of the data was calculated in terms of the standard deviation ($\sigma_{p/h}$ and $\sigma_{h_w/h}$). Table 2 shows the standard deviation of the variables for each case. In Table 2, the standard deviations of pressure and water surface were small, ranging from 0.0565 to 0.1008, indicating the reliability of the simulated results.

Table 2. Summary of the standard deviations of pressure and water surface in different locations at the same time period, where Er_{ave} represents the average relative errors of the mass flow rate of inlet and outlet; $\sigma_{p/h}$ and $\sigma_{h_w/h}$ are the standard deviation of pressure and water surface.

	Er_{ave} (%)		10°	30°	60°	90°	120°
case1	8.25	$\sigma_{p/h}$	0.0707	0.0692	0.0723	0.0818	0.0726
		$\sigma_{h_w/h}$	0.0852	0.0912	0.0823	0.0797	0.0906
case2	7.18	$\sigma_{p/h}$	0.0675	0.0704	0.0681	0.0619	0.0621
		$\sigma_{h_w/h}$	0.0823	0.0834	0.0885	0.0818	0.0911
case3	4.58	$\sigma_{p/h}$	0.0754	0.0652	0.0801	0.0781	0.0702
		$\sigma_{h_w/h}$	0.0925	0.0922	0.0879	0.0921	0.0942
case4	7.32	$\sigma_{p/h}$	0.0583	0.0621	0.0612	0.0565	0.0669
		$\sigma_{h_w/h}$	0.0725	0.0822	0.0850	0.0818	0.0861
case5	7.07	$\sigma_{p/h}$	0.0754	0.0689	0.0692	0.0717	0.0722
		$\sigma_{h_w/h}$	0.0628	0.0587	0.0603	0.0614	0.0592
case6	5.68	$\sigma_{p/h}$	0.0718	0.0782	0.0811	0.0777	0.0798
		$\sigma_{h_w/h}$	0.0823	0.0898	0.0912	0.0884	0.0879
case7	8.12	$\sigma_{p/h}$	0.0905	0.0972	0.0883	0.0878	0.0928
		$\sigma_{h_w/h}$	0.0923	0.0985	0.0984	0.1008	0.0954
case8	7.22	$\sigma_{p/h}$	0.0661	0.0622	0.0704	0.0683	0.0688
		$\sigma_{h_w/h}$	0.0921	0.0918	0.0856	0.0885	0.0892
case9	6.64	$\sigma_{p/h}$	0.0775	0.0721	0.0605	0.0644	0.0786
		$\sigma_{h_w/h}$	0.0858	0.0805	0.0734	0.0713	0.0728

All verifications for computational models, including the grid sensitivity study, the model verification, and the fluctuation of the calculated results, showed some errors and uncertainties due to the use of the eddy viscosity-based turbulence models. The errors of such eddy viscosity-based turbulence models in flow separation and recirculation zones have been documented in prior studies [45,46]. Nevertheless, using RNG k~ε to reveal the general flow patterns could satisfy the research requirements at the present stage.

3. Results and Analysis

3.1. Region Division in the Flow

As the simulation reveals, the flow transfers from one step to another, in which the flow regime has similar features on different steps. First, the successive impinging produces a bifurcation, where part of the flow moves backwards and forms a recirculation zone, while the main flow rotates along the sidewall with the vortex intensity decaying. Second, the flow undergoes a significant mixing and jumping process at the end of step. In order to present the complicated flow pattern of the stepped dropshaft accurately, the flow field is divided into the recirculating region (I), the wall-impinging region (II) and the mixing region (III), according to the flow characteristics. The relative ranges of these regions are expressed in terms of angles, namely, α_1, α_2, and α_3, respectively. In the subsequent sections, flow regions are discussed qualitatively according to individual streamlines and analyzed quantitatively on the basis of the pressure distribution for R = 0.25 m.

Figure 5 highlights the streamlines on the first step in each configuration. Although the central angle of each configuration was different, they shared similar flow features. In particular, the flow pattern had similar characteristics as well on other steps having the same central angle. Hence, only the flow on S6 in the different configuration under Q = 80 m^3/s was analyzed.

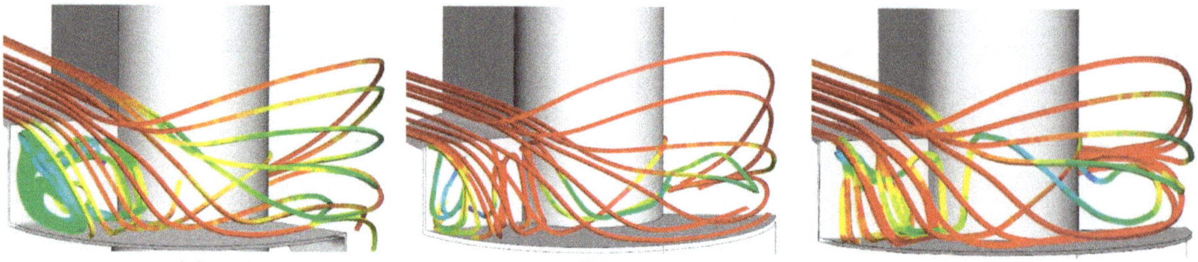

(a) (b) (c)

Figure 5. Streamlines above the S2 in different configurations: (**a**) 120°; (**b**) 150°; and (**c**) 180°.

(1) Recirculating region

The recirculating region was the zone where there was insignificant change of pressure gradient in the radial direction and streamline deflection in the opposite direction. Figure 6 illustrates the flow pattern along the flow direction in the cross-section for different configurations. Recirculating vortices can be found on the step due to the pressure difference between the upper and lower surfaces of the nappe flow during the falling process. Combined with the pressure difference and resistance caused by the nappe flow colliding with the horizontal step surface, the nappe flow departed from its original direction and formed a reflux. In addition, the location of recirculating moved from the external wall to the internal wall because of the curvature of the sidewall along the downstream direction. Figure 7 presents the pressure distributions on the horizontal step surfaces and the range of flow regimes. Most of the falling points were located at the intersection line of the horizontal step and the external wall. The most remarkable feature was that it was the minimum value of relative pressure that could be detected in the recirculating region.

(2) Wall-impinging region

The definition for wall-impinging region was the same as the recirculating region, which was mainly based on pressure variations and streamline features. Therefore, the range of the wall-impinging region was from the location where the minimum relative pressure occurred to the location where the local minimum pressure took place.

In the wall-impinging region, the flow was subjected to resistance and centrifugal force. However, the location and time that the inner and outer fluid were impacting the step or sidewall were different because of the curvature. First, the tangential velocity changed abruptly because of the impact of the surface flow near the external wall on that of the lateral one, leading it to increase rapidly and then to decline gradually on the water surface. The flow near the internal wall jetted along the tangent of that under pressure difference and gravity, and the maximum distance of the falling point of flow was in the circumferential direction, whereas the minimum was in the radial direction. For the lower level fluid, the flow quickly dived to the bottom of the step and traveled forward along the sidewall under the influence of gravity. However, the falling points of inside flow were close to the axis of the horizontal step surface. Deflected by the step, the flow deviated to the internal wall, resulting in it hitting the internal wall and attaching to it.

Therefore, the water surface was higher on the external wall than it was on the internal wall (Figure 5). Moreover, maximum pressure occurred and the pressure gradient showed drastic change at the start of the wall-impinging region (Figure 6).

(3) Mixing region

The range of the mixing region was from the location where the local minimum pressure occurred to the end of the step. The possibility of a relatively steady flow in the mixing region may be determined by the declining intensity of the swirling flow. The flow near the external wall moved inward from the bottom and the one close to the internal wall moved to the opposite from the surface due to the inward pressure gradient near the horizontal step which prevailed over the centrifugal force. As a result, there were less variations in the water surface in the transverse direction. Flow from the internal and the external walls mixed and interacted with each other near the axis of the step, causing the fluctuation of water surface along the circumference. At the end of the step, the elevation of the water surface caused by the transverse gradient in the outer wall was higher than that in the inner wall. The pressure distribution in the mixing region exhibited no obvious change.

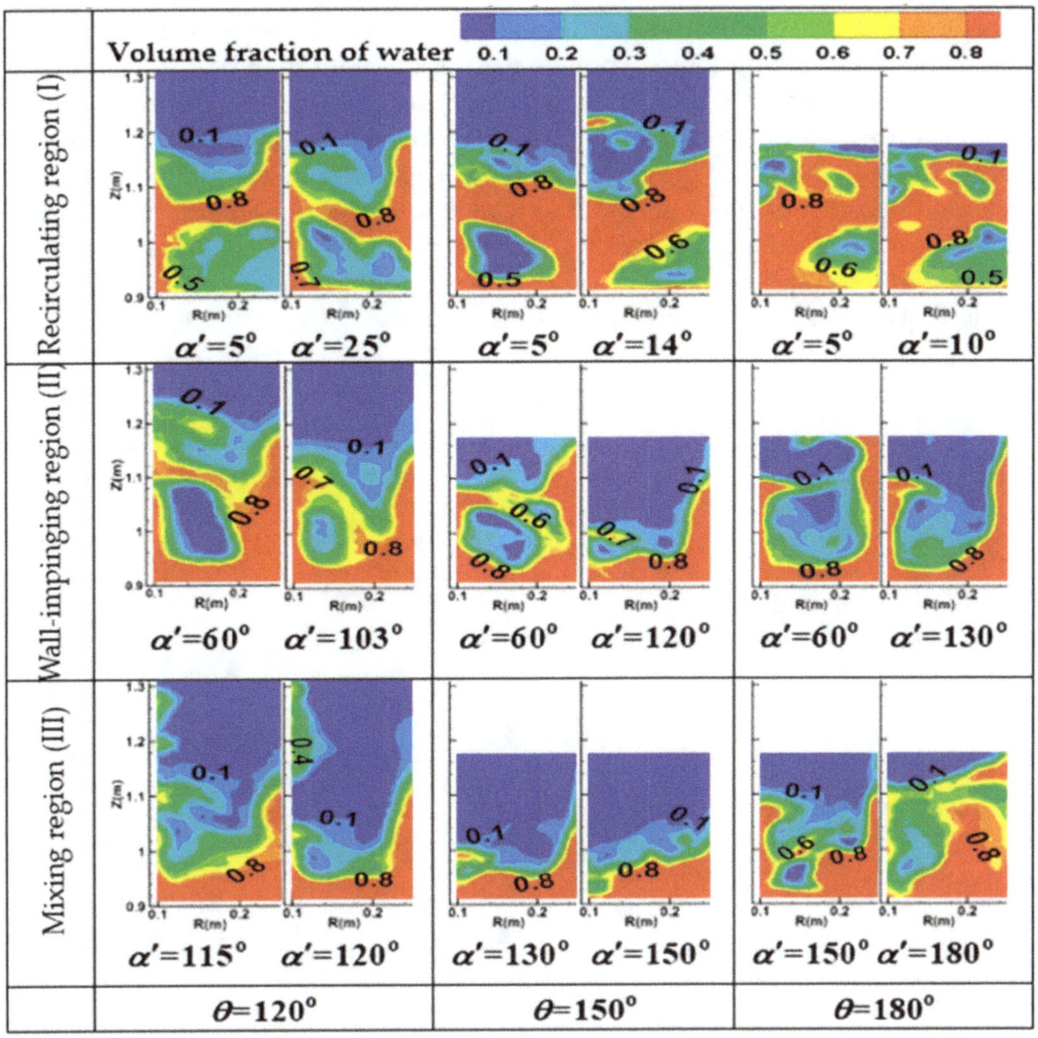

Figure 6. Flow pattern in the cross-section for different configurations, where α' is the relative location of the cross-section.

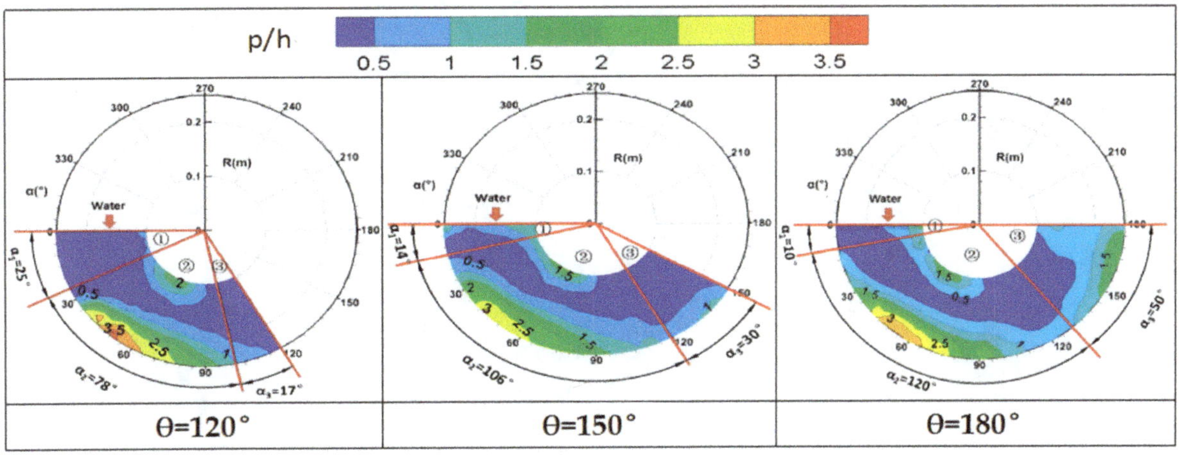

Figure 7. Pressure distribution and flow regions at S6.

3.2. Regional Scope

Figure 8 highlights the pressure distributions on the horizontal step for $R = 0.25$ m. As can be seen, the impact pressure fluctuated greatly with the maximum and minimum occurring in the recirculating region and the wall-impinging region, respectively, under all conditions. In addition, the pressure in the wall-impinging region showed greater gradient than that in the mixing region and the recirculating region. As the θ increased, α_1 decreased, while α_2 and α_3 increased. The cause of this phenomenon was attributed to the increase in the circumferential length of the step. These changes were able to bring the falling point of the jet flow to the vertical step closer and to make the recirculating region smaller. On the contrary, the ranges of the wall-impinging region and the mixing region were enhanced. Flow regime had little impact on the spectrum of flow regions in most cases with $\theta = 120°$ and $\theta = 150°$, but a different phenomenon could be observed when $\theta = 180°$. When flow discharge increased, α_2 increased, whereas α_3 decreased.

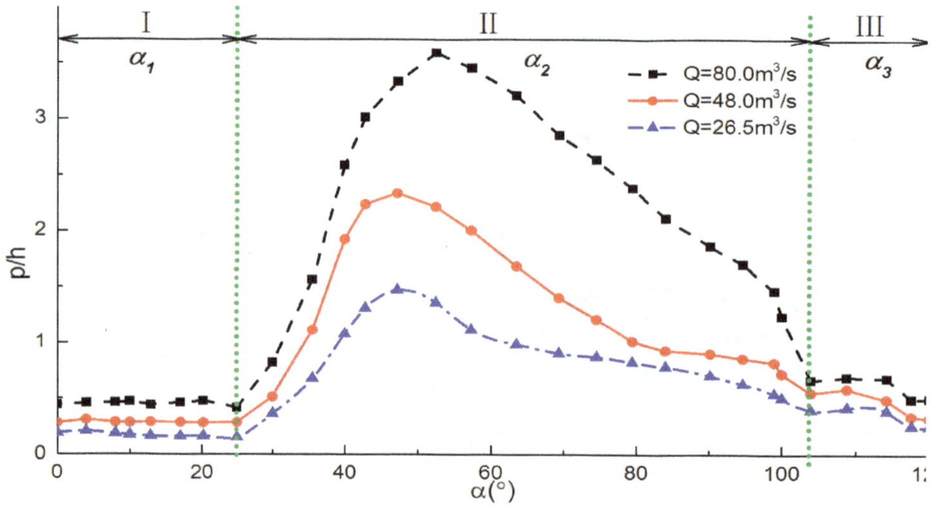

(a)

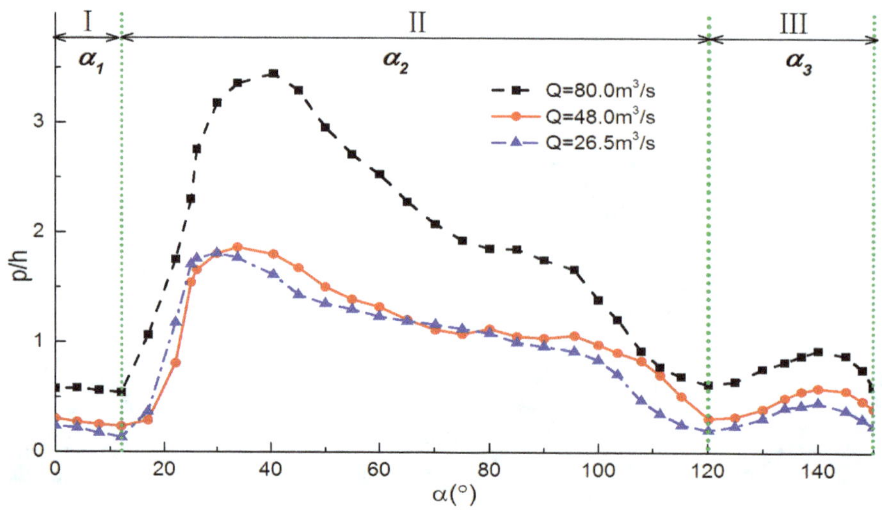

(b)

Figure 8. *Cont.*

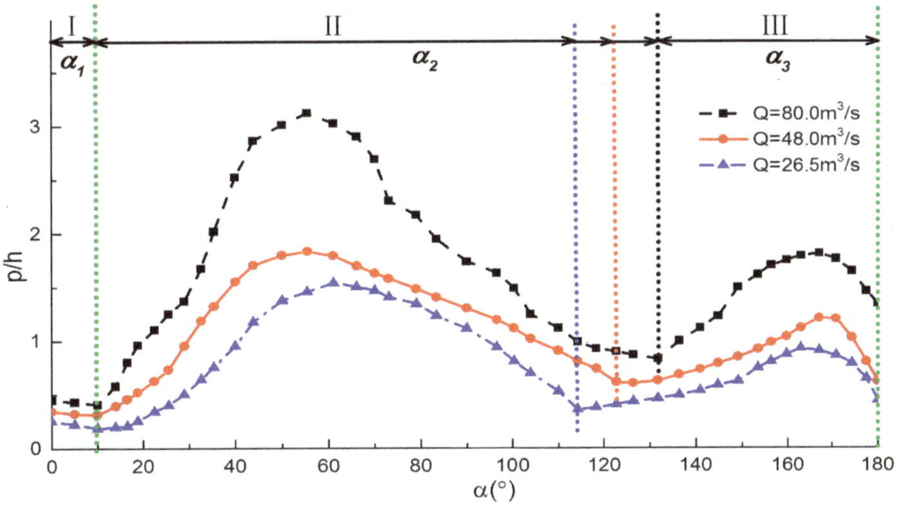

(c)

Figure 8. Pressure distribution for R = 0.25 m in different θ: (**a**) 120°; (**b**) 150°; and (**c**) 180°.

Figure 9a displays the average range of flow region on each step. Although some slight variations can be found, it is noticeable that the ranges of the regions were approximately the same under different flow conditions, indicating that the scope was mainly affected by the step angle rather than the flow regime. The chart compares the average range of different flow regions in all configurations.

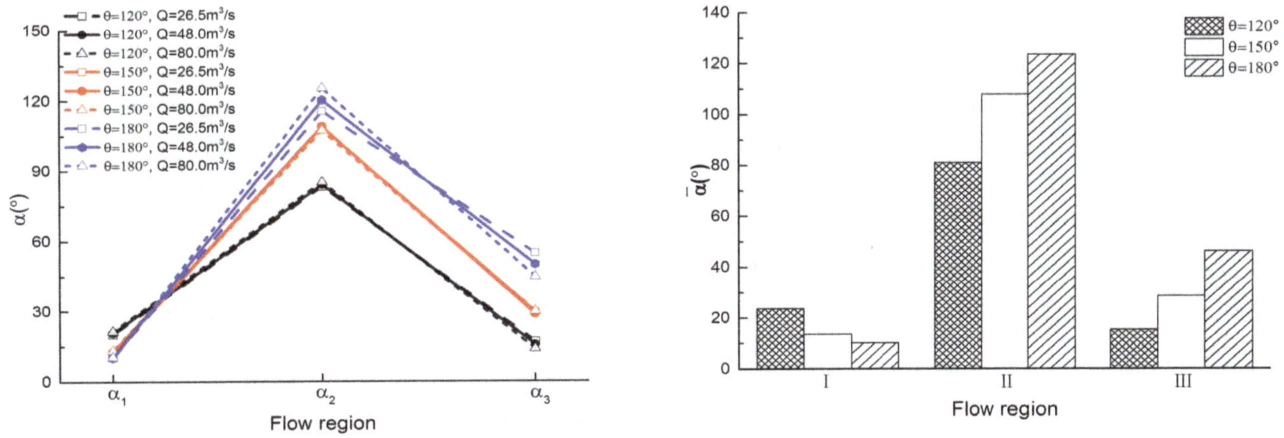

(a) (b)

Figure 9. Flow region: (**a**) range of flow regions and (**b**) range of flow regions under different configurations.

3.3. Water Depth

For the traditional dropshaft, standing waves can only occur in a supercritical approach flow [47–49] channel due to sidewall deflection or curvature; therefore, the height of sidewalls must be carefully designed in a prototype environment. In addition, these standing waves can spread to downstream channels, causing fluctuations in the water surface. Nevertheless, for the dropshaft in the present work, sidewall standing waves were generated on each step, although the standing wave was quite insignificant under a small flow rate. Therefore, research studies should be undertaken to gain insights into the typical properties of standing waves. A sketch of the wave along the external wall is shown in Figure 10. h_{max} is the maximum water depth and α_{max} is the relative position of the maximum water depth. The distance between the steps is h_s, and the relative position of the vertical plane of the upper step is α_s. A dimensionless height, h^s/h, for $\theta = 120°$ and $\theta = 180°$ is, respectively,

3 and 2, and the location, α_s, for both $\theta = 120°$ and $\theta = 180°$ is 0. However, for $\theta = 150°$, h_s/h and α_s vary significantly with h_s/h ranging from 3 to 2 and α_s ranging from 0° to 60°.

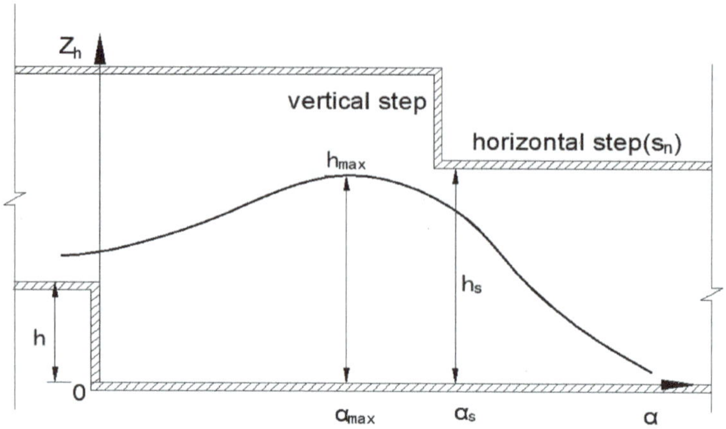

Figure 10. Water depth along the external wall.

Figure 11 demonstrates the flow profile along the external wall. The flow-depth curves indicated that the water surface along the outer wall presented a similar wavy pattern under different conditions.

(1) With the increase in the central angles of step, drastic fluctuation began to occur. For $\theta = 120°$, the water-surface profile was steady and the water depth decreased gradually along the flow direction in which the maximum relative water depth occurred in the recirculating region. For $\theta = 150°$, the variation of free surface became remarkable. When the central angle of step increased to 180°, the oscillation of water surface was more violent than that of the other cases.

For the small center angle, as the rotational flow approached the mixing region, the flow fell into the next step suddenly owing to the small range of the mixing region, which could lead to a dramatic change in axial velocity gradient and water depth. However, the larger the angle, the greater the diffusion in the mixing region. Water surface fluctuated greatly due to the resistance of the sidewall and centrifugal force.

(2) The influence of flow rate on water depth was relatively noteworthy: the water level rose with the increase in discharge under different conditions. For $\theta = 120°$ and $\theta = 150°$, the free surface changed without distinct oscillation under all circumstances. For $\theta = 180°$, the water level oscillated more vigorously with a further increase in discharge, causing an increase in the water surface, and hit the vertical surface of the next step in the mixing region.

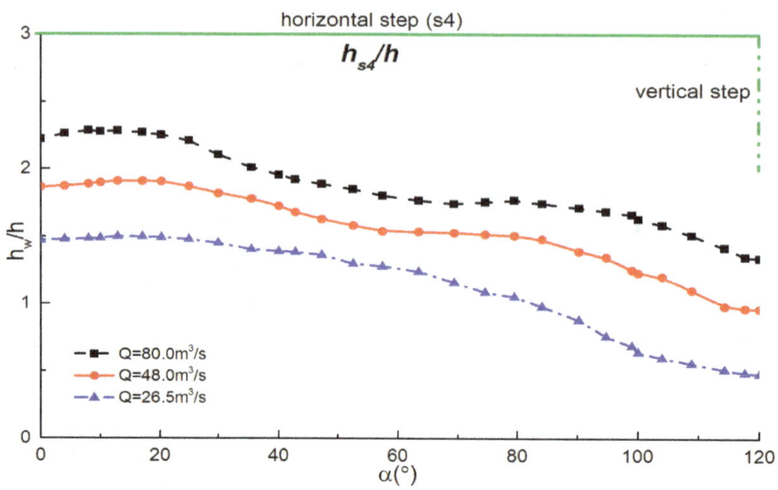

(a)

Figure 11. *Cont.*

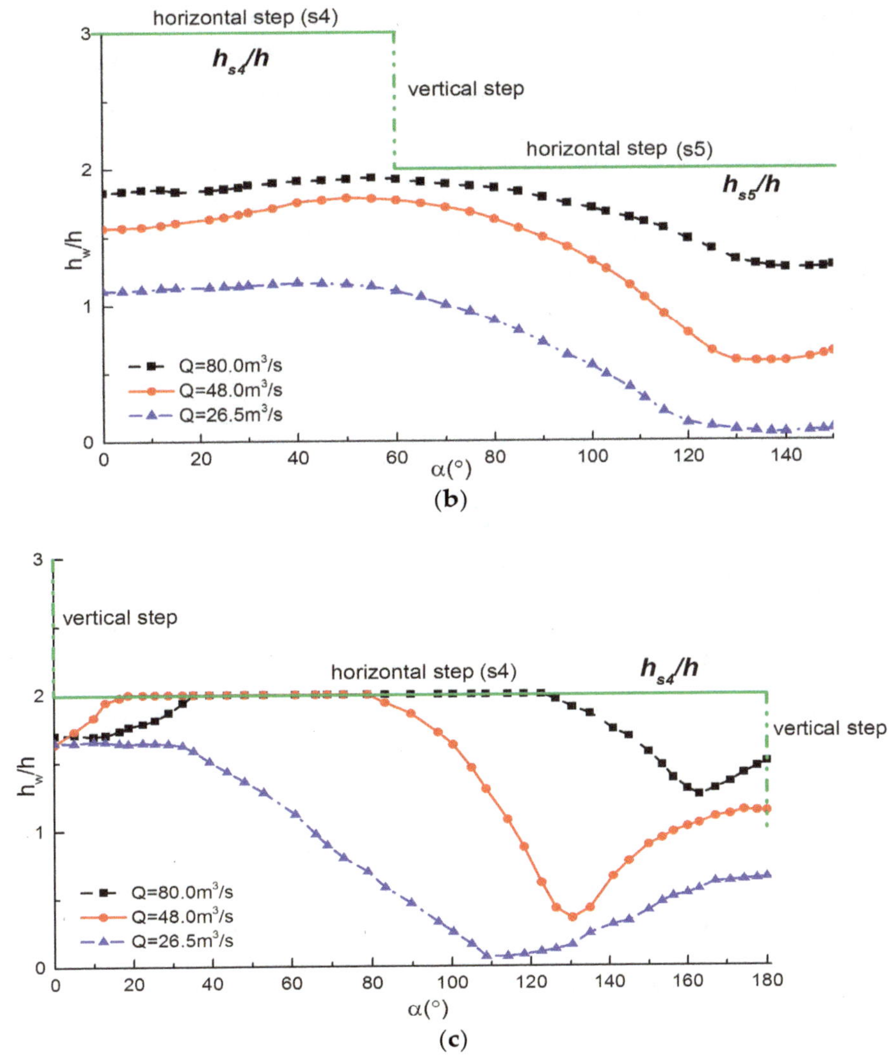

Figure 11. Water surface curves: (**a**) 120°; (**b**) 150°; and (**c**) 180°.

Figure 12 illustrates the change of relative maximum water depth in all cases and yields the following conclusions:

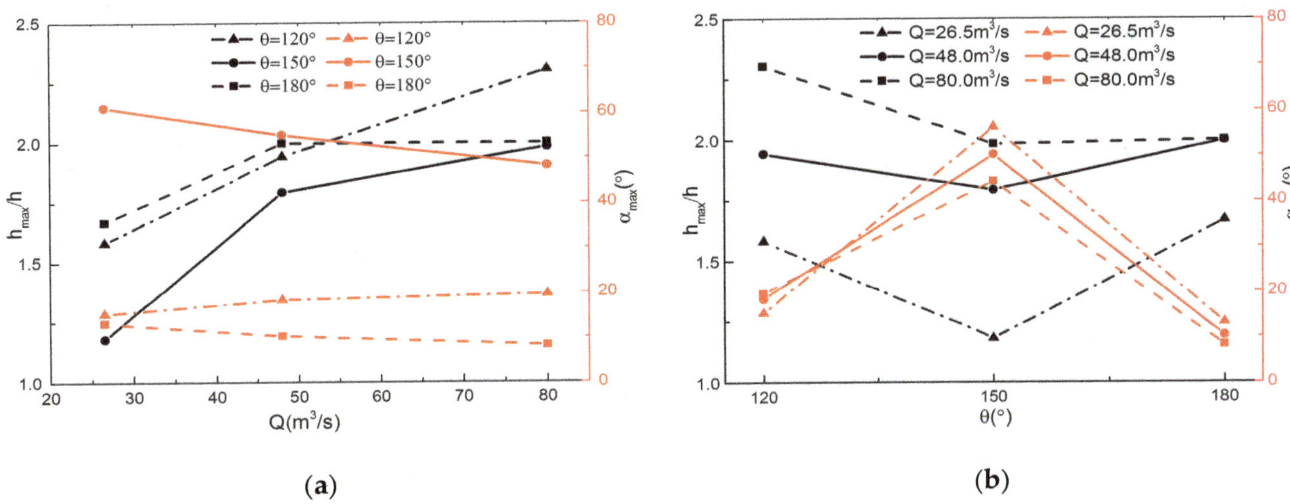

Figure 12. Water depth characteristics: (**a**) the relationship between h_{max}/h, α_{max} and Q and (**b**) the relationship between h_{max}/h, α_{max} and θ.

(1) For a fixed central angle of step θ, $h_{\max}/h$ and $\alpha_{\max}$ increased as Q increased.

(2) For a fixed discharge Q, $h_{\max}/h$ and α_{max} showed no obvious trend as θ increased.

(3) With the increment of θ, $h_{\max}/h$ decreased initially and then increased, whereas $\alpha_{\max}$ exhibited the opposite phenomenon.

(4) With the increase in Q, $h_{\max}/h$ increased and $\alpha_{\max}$ presented no evident change.

Therefore, it seemed that the range of $h_{\max}/h$ depended on the discharge and the central angle of step, whereas the range of $\alpha_{\max}$ was associated with the central angle of step. For $\theta = 180°$, when $h_{\max}/h = 2$ ($Q = 48$ m^3/s, 80 m^3/s), the flow impacted the bottom of the upper step, which can only occur on some steps when $\theta = 150°$ for $Q = 80$ m^3/s. For $\theta = 120°$, $h_{\max}/h$ was considerably lower than the height of the upper step in all discharge.

4. Conclusions

For all the cases considered above in this paper, the RNG k~ε turbulence flow model is capable of predicting the main characteristics of the flow. This included the definition of the flow regime and the determination of the maximum water depth. The following conclusions can be drawn:

1. Because of the complexity of the flow regime, the flow was partitioned into the recirculating region, the wall-impinging region, and the mixing region.

2. The range of the three regions was affected by the central angle of step. With the increase in θ, the range of the wall-impinging region and the mixing region increased, whereas the range of the recirculating region had no significant change.

3. The maximum water depth increased with the increase in the flow rate, whereas the flow rate showed little effect on $\alpha_{\max}$. Moreover, with the increase in θ, the maximum water depth initially decreased and then increased; however, $\alpha_{\max}$ displayed the opposite trend.

Although the results of this study are encouraging, same solution processes can be adopted in future work to model the energy dissipation performance and air entertainment characteristics of stepped dropshafts.

Author Contributions: Conceptualization and Methodology, Y.W. and J.Z.; Numerical Simulation, Y.Q.; Investigation, Y.Q.; Writing—Original Draft Preparation, Y.Q.; Writing—Review and Editing, Y.W. and J.Z.; Supervision, Y.W. and J.Z.; Project Administration, Y.W. and J.Z.; Funding Acquisition, Y.W. and J.Z.

References

1. Rajaratnam, N.; Mainali, A.; Hsung, C. Observations on flow in vertical dropshafts in urban drainage systems. *J. Environ. Eng.* **1997**, *123*, 486–491. [CrossRef]
2. Chanson, H. Hydraulics of rectangular dropshafts. *J. Irrig. Drain. Eng.* **2004**, *130*, 523–529. [CrossRef]
3. Chanson, H. Understanding air–water mass transfer in rectangular dropshafts. *J. Environ. Eng. Sci.* **2004**, *3*, 319–330. [CrossRef]
4. Adriana Camino, G.; Zhu, D.Z.; Rajaratnam, N. Flow observations in tall plunging flow dropshafts. *J. Hydraul. Eng.* **2014**, *141*, 06014020. [CrossRef]
5. Granata, F.; de Marinis, G.; Gargano, R.; Hager, W.H. Hydraulics of circular drop manholes. *J. Irrig. Drain. Eng.* **2010**, *137*, 102–111. [CrossRef]
6. Ma, Y.; Zhu, D.Z.; Rajaratnam, N.; van Duin, B. Energy dissipation in circular drop manholes. *J. Irrig. Drain. Eng.* **2017**, *143*, 04017047. [CrossRef]
7. Jain, S.C. Air transport in vortex-flow drop shafts. *J. Hydraul. Eng.* **1988**, *114*, 1485–1497. [CrossRef]
8. Vischer, D.; Hager, W. Vortex drops. In *Energy Dissipators*; Routledge: Abingdon, UK, 2018; pp. 167–181. ISBN 9781351451345.

9. Odgaard, A.J.; Lyons, T.C.; Craig, A.J. Baffle-drop structure design relationships. *J. Hydraul. Eng.* **2013**, *139*, 995–1002. [CrossRef]
10. Stirrup, M.; Margevicius, T.; Hrkac, T.; Baca, A. A baffling solution to Sewage Conveyance In York Region, Ontario. *Proc. Water Environ. Fed.* **2012**, *2012*, 74–90. [CrossRef]
11. Kennedy, J.F.; Jain, S.C.; Quinones, R.R. Helicoidal-ramp dropshaft. *J. Hydraul. Eng.* **1988**, *114*, 315–325. [CrossRef]
12. Tamura, S.; Matsushima, O.; Yoshikawa, S. Helicoidal-ramp type drop shaft to deal with high head drop works in manholes. *Proc. Water Environ. Fed.* **2010**, *2010*, 4991–5002. [CrossRef]
13. Jain, S.C.; Kennedy, J.F. *Vortex-Flow Drop Structures for the Milwaukee Metropolitan Sewerage District Inline Storage System*; Iowa Institute of Hydraulic Research, The University of Iowa: Iowa City, IA, USA, 1983.
14. Hager, W.H. *Wastewater Hydraulics: Theory and Practice*; Springer Science & Business Media: Berlin, Germany, 2010.
15. Zhao, C.H.; Zhu, D.Z.; Sun, S.K.; Liu, Z.P. Experimental study of flow in a vortex drop shaft. *J. Hydraul. Eng.* **2006**, *132*, 61–68. [CrossRef]
16. Del Giudice, G.; Gisonni, C. Vortex dropshaft retrofitting: Case of Naples city (Italy). *J. Hydraul. Res.* **2011**, *49*, 804–808. [CrossRef]
17. Natarius, E.M. Aeration performance of vortex flow insert assemblies in sewer drop structures. *Proc. Water Environ. Fed.* **2008**, *2008*, 842–851. [CrossRef]
18. Yu, D.; Lee, J.H. Hydraulics of tangential vortex intake for urban drainage. *J. Hydraul. Eng.* **2009**, *135*, 164–174. [CrossRef]
19. Wu, J.H.; Yang, T.; Sheng, J.Y.; Ren, W.C.; Fei, M.A. Hydraulic characteristics of stepped spillway dropshafts with large angle. *Chin. J. Hydrodyn.* **2018**, *33*, 176–180. (In Chinese) [CrossRef]
20. Christodoulou, G.C. Energy dissipation on stepped spillways. *J. Hydraul. Eng.* **1993**, *119*, 644–650. [CrossRef]
21. Sorensen, R.M. Stepped spillway hydraulic model investigation. *J. Hydraul. Eng.* **1985**, *111*, 1461–1472. [CrossRef]
22. Frizell, K.W.; Renna, F.M.; Matos, J. Cavitation potential of flow on stepped spillways. *J. Hydraul. Eng.* **2012**, *139*, 630–636. [CrossRef]
23. Wu, J.H.; Ren, W.C.; Ma, F. Standing wave at dropshaft inlets. *J. Hydrodyn. Ser. B* **2017**, *29*, 524–527. [CrossRef]
24. Hirt, C.W.; Nichols, B.D. Volume of fluid (VOF) method for the dynamics of free boundaries. *J. Comput. Phys.* **1981**, *39*, 201–225. [CrossRef]
25. Peng, Y.; Zhou, J.G.; Burrows, R. Modelling the free surface flow in rectangular shallow basins by lattice Boltzmann method. *J. Hydrau. Eng.* **2011**, *137*, 1680–1685. [CrossRef]
26. Peng, Y.; Zhou, J.G.; Burrows, R. Modelling solute transport in shallow water with the lattice Boltzmann method. *Comput. Fluids.* **2011**, *50*, 181–188. [CrossRef]
27. Peng, Y.; Zhang, J.M.; Zhou, J.G. Lattice Boltzmann Model Using Two-Relaxation-Time for Shallow Water Equations. *J. Hydrau. Eng.* **2016**, *142*, 06015017. [CrossRef]
28. Peng, Y.; Zhang, J.M.; Meng, J.P. Second order force scheme for lattice Boltzmann model of shallow water flows. *J. Hydraul. Res.* **2017**, *55*, 592–597. [CrossRef]
29. Peng, Y.; Mao, Y.F.; Wang, B.; Xie, B. Study on C-S and P-R EOS in pseudo-potential lattice Boltzmann model for two-phase flows. *Int. J. Mod. Phys. C* **2017**, *28*, 1750120. [CrossRef]
30. Peng, Y.; Wang, B.; Mao, Y.F. Study on force schemes in pseudopotential lattice Boltzmann model for two-phase flows. *Math. Probl. Eng.* **2018**. [CrossRef]
31. Galván, S.; Reggio, M.; Guibault, F. Assessment study of k-ε turbulence models and near-wall modeling for steady state swirling flow analysis in draft tube using fluent. *Eng. Appl. Comput. Fluid* **2011**, *5*, 459–478. [CrossRef]
32. Morovati, K.; Eghbalzadeh, A.; Javan, M. Numerical investigation of the configuration of the pools on the flow pattern passing over pooled stepped spillway in skimming flow regime. *Acta Mech.* **2015**, *227*, 1–14. [CrossRef]
33. Devolder, B.; Troch, P.; Rauwoens, P. Performance of a buoyancy-modified k-ω and k-ω SST turbulence model for simulating wave breaking under regular waves using OpenFOAM®. *Coast. Eng.* **2018**, *138*, 49–65. [CrossRef]
34. Fuhrman, D.R.; Dixen, M.; Jacobsen, N.G. Physically-consistent wall boundary conditions for the k-ω turbulence model. *J. Hydraul. Res.* **2010**, *48*, 793–800. [CrossRef]

35. Bai, Z.L.; Zhang, J.M. Comparison of different turbulence models for numerical simulation of pressure distribution in V-shaped stepped spillway. *Math. Probl. Eng.* **2017**, *2017*, 1–9. [CrossRef]
36. Li, S.; Zhang, J. Numerical investigation on the hydraulic properties of the skimming flow over pooled stepped spillway. *Water* **2018**, *10*, 1478. [CrossRef]
37. Chan, S.; Qiao, Q.; Lee, J.H. On the three-dimensional flow of a stable tangential vortex intake. *J. Hydro-Environ. Res.* **2018**, *21*, 29–42. [CrossRef]
38. Gao, X.P.; Zhang, H.; Liu, J.J.; Sun, B.; Tian, Y. Numerical investigation of flow in a vertical pipe inlet/outlet with a horizontal anti-vortex plate: Effect of diversion orifices height and divergence angle. *Eng. Appl. Comput. Fluid Mech.* **2018**, *12*, 182–194. [CrossRef]
39. Yakhot, V.; Orszag, S.A. Renormalization group analysis of turbulence. I. Basic theory. *J. Sci. Comput.* **1986**, *57*, 1722. [CrossRef]
40. Liu, Z.P.; Guo, X.L.; Xia, Q.F.; Fu, H.; Wang, T.; Dong, X.L. Experimental and numerical investigation of flow in a newly developed vortex drop shaft spillway. *J. Hydraul. Eng.* **2018**, *144*, 04018014. [CrossRef]
41. Zhang, W.; Wang, J.; Zhou, C.; Dong, Z.; Zhou, Z. Numerical simulation of hydraulic characteristics in a vortex drop shaft. *Water* **2018**, *10*, 1393. [CrossRef]
42. Guo, X.L.; Xia, Q.F.; Fu, H.; Yang, K.L.; Li, S.J. Numerical study on flow of newly vortex drop shaft spillway. *J. Hydraul. Eng.* **2016**, *47*, 733–741. (In Chinese) [CrossRef]
43. *ANSYS Fluent Theory Guide*; Release 16.0; ANSYS Inc.: Canonsburg, PA, USA, 2015.
44. Celik, I.B.; Ghia, U.; Roache, P.J. Procedure for estimation and reporting of uncertainty due to discretization in CFD applications. *J. Fluids Eng.* **2008**, *130*. [CrossRef]
45. Iaccarino, G.; Mishra, A.A.; Ghili, S. Eigenspace perturbations for uncertainty estimation of single-point turbulence closures. *Phys. Rev. Fluids* **2017**, *2*, 024605. [CrossRef]
46. Mishra, A.A.; Iaccarino, G. Uncertainty Estimation for Reynolds-Averaged Navier–Stokes Predictions of High-Speed Aircraft Nozzle Jets. *AIAA J.* **2017**, 3999–4004. [CrossRef]
47. Del Giudice, G.; Gisonni, C.; Rasulo, G. Design of a scroll vortex inlet for supercritical approach flow. *J. Hydraul. Eng.* **2010**, *136*, 837–841. [CrossRef]
48. Hager, W.H. Vortex drop inlet for supercritical approaching flow. *J. Hydraul. Eng.* **1990**, *116*, 1048–1054. [CrossRef]
49. Kawagoshi, N.; Hager, W. Wave type flow at abrupt drops Wave type flow at abrupt drops: I. Flow geometry. *J. Hydraul. Res.* **1990**, *28*, 235–252. [CrossRef]

4

Collapsing Mechanisms of the Typical Cohesive Riverbank along the Ningxia–Inner Mongolia Catchment

Guosheng Duan [1], Anping Shu [1,*], Matteo Rubinato [2], Shu Wang [1] and Fuyang Zhu [1]

1 School of Environment, Key Laboratory of Water and Sediment Sciences of MOE, Beijing Normal University, Beijing 100875, China; duanguosheng2004@163.com (G.D.); wangshu861217@gmail.com (S.W.); fuyangzhu1017@gmail.com (F.Z.)
2 Department of Civil and Structural Engineering, The University of Sheffield, Sir Frederick Mappin Building, Mappin Street, Sheffield S1 3JD, UK; m.rubinato@sheffield.ac.uk
* Correspondence: shuap@bnu.edu.cn.

Abstract: As one of the major sediment sources in rivers, bank collapse often occurs in the Ningxia–Inner Mongolia catchment and, to date, it caused substantial social, economic and environmental problems in both local areas and downstream locations. To provide a better understanding of this phenomenon, this study consisted of modifying the existing Bank Stability and Toe Erosion Model (BSTEM), commonly used to investigate similar phenomena, introducing new assumptions and demonstrating its applicability by comparing numerical results obtained against field data recorded at six gauging stations (Qingtongxia, Shizuishan, Bayan Gol, Sanhuhekou, Zhaojunfen, and Toudaoguai). Furthermore, the impact of multiple factors typical of flood and dry seasons on the collapse rate was investigated, and insights obtained should be taken into consideration when completing future projects of river adaptation and river restoration.

Keywords: Ningxia–Inner Mongolia; Yellow River; riverbank collapse; BSTEM model; flood & dry season; sediment transport

1. Introduction

Riverbank collapse is a phenomenon caused by several natural factors (e.g., intense rainfall, site topography, properties of the river bed and hydraulic conditions of the river flow) [1] and artificial factors (e.g., man-made bank undercutting, basal clean-out) [2]. The Yellow River is the China's second-longest river and previous studies [3] confirmed that 518.38 km^2 of the riverbank along the Ningxia–Inner Mongolia part of the Yellow River was eroded between 1958 and 2008. This elevated rate of riverbank erosion is due to continuous variations of flow levels, causing the deposition of loose riverbed material in alluvial plains, especially after heavier rainfall events typical of climate change [4,5]. To date, riverbank erosion has caused substantial social (e.g., traffic disruption due to flooding), economic (e.g., loss of farm land) and environmental problems (e.g., sediment dynamics and water quality) in both local areas and downstream locations [6,7].

Previous studies were completed to tackle this challenge and to identify what can induce riverbanks to collapse in catchments in both developing and developed countries. Grabowski [8] found that the soil texture, the soil structure, the unit weight and water composition [9,10] all play a principal role on the banks erosion. While non-cohesive banks are eroded through discrete particle entrainment that can be quantified using the magnitude of the shear stress and the particle size [11,12], cohesive banks are eroded through entrainment of aggregates [13] and this mechanism is very challenging to diagnose when considering the electrochemical forces acting between them [14,15]. The detachment and erosion phenomena of cohesive material by gravity and/or flowing water are typically controlled by a variety of physical, electrical and chemical forces, such as cohesion, electrochemical forces, pore-water pressure and matric suction [15]. These forces within and between aggregates cause their erosion to be complex. Furthermore, cohesive riverbanks are usually poorly drained due to their silt and clay composition, and thus may experience excess pore water pressures, one of the main agents of subaerial erosion. Clayey banks are also susceptible to desiccation cracking and slaking from wetting–drying cycles [16]. Numerous studies have examined the impact of water content on the riverbank erosion [17–22] while other studies have investigated the subaerial erosion due to the weakening and weathering of bank materials [23–25]. Additionally, Yu [26] completed a series of experimental tests to quantify the influence and the impact of the water infiltration on bank collapse and consequent sediment deposition. They identified that partial erosion can be due to fluvial hydraulic forces or gravitational forces. Moreover, total erosion of the bank can be generated if both forces are combined at their maximum magnitude, with a consequent disintegration of the bank and the transportation of fallen blocks within the flow, changing the bed elevation. Focusing on the Ningxia–Inner Mongolia banks, the erosion process has been found to be influenced by three specific factors: (i) the sand blown by the wind from the desert [27–30]; (ii) the sediment transported by the river from the upstream catchment [31,32]; and (iii) the material falling from the bank due to natural collapse [33]. It is fundamental to replicate all these governing processes within numerical models when assessing stability of the riverbanks. A large amount of literature is available and it covers all the numerical modeling aspects investigated to date [34–38]. One of the numerical tools most commonly used is the Bank Stability and Toe Erosion Model (BSTEM), developed by the National Sedimentation Laboratory in Oxford, Mississippi, USA [39], which has been continually modified and improved by the authors since its creation to improve its performance. To model the riverbanks' stability, BSTEM calculates a factor of safety (F_s) for multi-layers combining three equilibrium-method models: (i) the layers simulated are horizontal; (ii) vertical slices simulate tension crack; and (iii) the common failure mechanisms are cantilever failures. BSTEM also assumes that all the collapsed material is totally removed from the bank and does not accumulate on the toe [40]. BSTEM is frequently used to simulate: (i) the banks' stability and the consequent sediment loading inside the river [41]; (ii) stream rehabilitation projects [42]; and (iii) erosion and failure mechanisms [43,44]. Despite several findings that have resulted in a better understanding of this phenomenon, to the authors' knowledge there is a lack of studies which involve the evaluation of BSTEM for long-term streambank erosion and consequent long-term failure.

Therefore, to fill this gap, this work comprises a collection of field datasets in the Ningxia–Inner Mongolia area that were implemented on the BSTEM model to obtain a new solution to the mechanisms of long-term bank erosion. To achieve that, three hypothesis different from those applied on the original BSTEM model were executed: (i) when the soil collapses from the bank, part of it is washed away by the water, and the remaining part accumulates on the toe; (ii) the shape of the material that accumulates on the toe is considered to have a triangle form considering the cross-sectional view and the angle away from the river bank is the angle of repose of the sediment, which is taken as 30 degrees; and (iii) the downstream area of a bank is considered the first one to be affected by the collapse process.

The paper is organized as follows. In the next sections the study area, the procedure to collect field data and the numerical analysis applied to replicate the river bank collapse recorded (Section 2) are described. Section 3 presents the comparison between numerical and experimental datasets and the factors that could influence the rate of collapse are discussed. Finally, Section 4 reports the conclusions.

2. Materials and Methods

2.1. Case Study

The Ningxia–Inner Mongolia catchment (Figure 1) is located in the lower part of the upper Yellow River and has a length of 913.5 km, starting from Qingtongxia (Ningxia Hui autonomous region) and ending at Hekouzhen (Inner Mongolia autonomous region). This site includes six gauging stations (Qingtongxia, Shizuishan, Bayan Gol, Sanhuhekou, Zhaojunfen, and Toudaoguai) and it is characterized by four desert regions (HedongSandyland, Ulan Buh Desert, Tengger Desert and Hobq Desert) and two alluvial plains (Yinchuan Plain and Hetao Plain).

The climate of this region can be classified as temperate continental climate, which is characterized by long cold winters, and warm to hot summers [45]. The annual rainfall (maximum and minimum peaks commonly reached in summer and winter, respectively) ranges from 100 to 288 mm.

The case study selected is located in Dengkou which is in the Inner Mongolia Province. The gauging station Bayan Gol is adjacent to it. Being part of the Hetao plain, the Dengkou catchment is characterized by a low gradient (≈0.17‰) and loose riverbed materials. This site was carefully chosen based on three specific criteria. Firstly, preliminary monitoring datasets of this site have confirmed that the riverbank was actively retreating. Secondly, the site was sufficiently close to a gauging station (about 5 km), hence detailed rainfall readings including time-series, frequencies and water surface elevations associated with each flow condition flowing next to the bank site were available and could be collected. Last but not least, the morphology and the soil properties of this area are representative of the cohesive streambank along all the Ningxia–Inner Mongolia catchment which has been drastically eroded from 1958 to 2008 [3]. Therefore, it is fundamental to provide a better understanding of this phenomenon, especially in this zone, to provide mitigation measures and plan actions to reduce the effects due to erosion of the riverbanks.

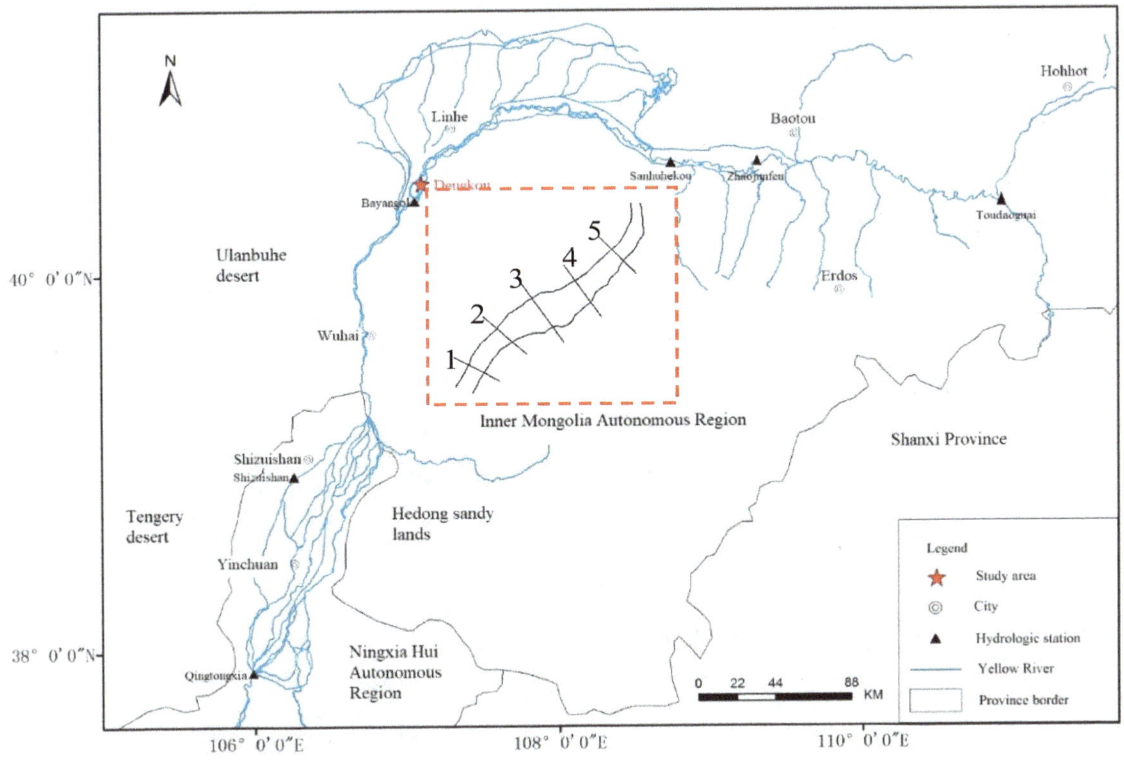

Figure 1. Location of the case study considered for this work. 1, 2, 3, 4 and 5 are the five surveying cross sections (500 m far from each other) selected for the collection of field-data.

2.2. Field Data

To analyze the complex processes of riverbank collapse affecting Dengkou, it was necessary to inspect the nature of the bank and the typical flow conditions in the parts of the Yellow river crossing this area. Five cross sections (1, 2, 3, 4 and 5, Figure 1), spaced 500 m from each other, were identified along the banks for the acquisition of field datasets. Parameters and methods selected by the authors for the monitoring and collection of datasets from the field study are displayed in Table 1.

Table 1. Indexes and methods for in-site measurement.

	Bank Shape	Soil Composition	Mechanical Properties
Parameter monitored	Bank height (m) Bank gradient (°) Length of bank toe (m)	Median diameter (μm) Unit weight (g/cm^3) Moisture content (%)	Critical shear stress (kPa) Internal friction angle (°) /
Equipment utilized for the monitoring	Gradometer (±1°), Tape (±1 cm)	Scale, Laser particle sizer (±5 μm)	Direct shear apparatus

Additionally, to measure the magnitude of bank collapse along the area of study, five stakes were positioned vertically to the flow direction at each section selected for the monitoring (Figure 2). Five measurements were conducted respectively on 21 May, 13 July, 9 August, 23 September and 30 September in 2011 to quantify the collapse distance. The collapse distances between each section recorded are displayed in Table 2.

Table 2. Collapse distances (m) recorded in 2011.

Time Scale	Section 1	Section 2	Section 3	Section 4	Section 5
21 May–13 July	4.60	4.50	5.80	5.30	5.60
13 July–9 August	3.05	2.80	3.10	3.95	3.60
9 August–23 September	5.25	5.08	6.80	6.20	6.50
23 September–30 September	1.68	1.45	1.10	2.40	1.45
Total	14.58	13.81	16.80	17.85	17.15

Figure 2. Collapse distance measurement for surveying Section 1.

2.2.1. Hydrologic Data

Monthly average hydrologic datasets (2011, year of site inspection) obtained from the gauging station Bayan Gol used for this study are showed in Table 3.

Table 3. Monthly average hydrologic data (2011).

Month	Flow (m^3/s)	Velocity (m/s)	Water Depth (m)	Sediment Concentration (kg/m^3)	Sediment Transport Rate (kg/s)
1	494.23	0.68	3.09	0.21	102
2	568.11	0.74	3.02	0.22	123
3	495.58	1.06	1.94	1.23	615
4	684.77	1.08	1.34	2.16	1480
5	444.00	0.89	1.03	0.64	283
6	369.60	0.81	0.85	0.59	217
7	368.35	0.87	0.83	0.77	283
8	528.00	1.32	0.94	3.20	1690
9	817.30	1.42	1.40	1.95	1590
10	612.48	1.37	0.99	1.27	779
11	495.90	1.18	0.62	1.17	580
12	592.52	0.95	2.00	2.58	1530

2.2.2. Bank Characteristics of Each Section

Riverbank properties at each section (1, 2, 3, 4 and 5, Figure 1) were obtained through five field inspections in 2011 (1. 21 May, 2. 13 July, 3. 9 August, 4. 23 September and 5. 30 September), and details are showed in Table 4. During the characteristic long cold winters in this region, the Yellow River typically freezes, hence the summer season has been selected as sampling period to avoid the quantification of additional effects on river bed changes [46,47] which could have made the already complex comparison between experimental datasets and numerical results even more challenging.

Table 4. Characteristics of the sections selected for this study.

Section	Bank Height (m)	Bank Gradient (°)	Toe Length (m)	Median Diameter (μm)	Wet Unit Weight (g/cm^3)	Critical Shear Stress (kPa)	Internal Friction Angle (°)
1	1.9	82	0.85	33	1.55	7.2	32
2	2.1	77	0.9	35	1.48	7.5	31
3	2.3	81	0.95	32	1.50	7.3	32
4	1.7	84	0.8	37	1.46	7.5	30
5	1.6	86	0.75	30	1.52	7.2	31

2.3. *Riverbank Collapse Characterisation*

When collecting field data, observations were made to get more insights about the erosion that typically affects the riverbanks of the Ningxia–Inner Mongolia catchment. Figure 3 shows an example of the real case study under investigation prior and after the erosion. Observations confirmed that the bank material is cohesive, the riverbank slope is steep (77–86°), and the pattern of bank collapse is considered to be planar, in conformity with previous studies conducted [48,49]. Additionally, while monitoring the site, it was possible to identify different stages of the continuous collapse process characterized as follows: (i) the bank toe erosion initiates (Figure 4a); (ii) tension cracks develop (Figure 4b); (iii) shearing starts and parts of the bank breaks and detaches from the main body (Figure 4c); (iv) bank failure occurs (Figure 4d). When the soil collapses from the bank, part of it is washed away by the water and the remaining part accumulates at the bank toe. The new shape of the bank including the additional material eroded then becomes the initial form of the next bank collapse.

Figure 3. Example of the existing bank slope (**left**) and an example of riverbank collapse along Dengkou reach (**right**).

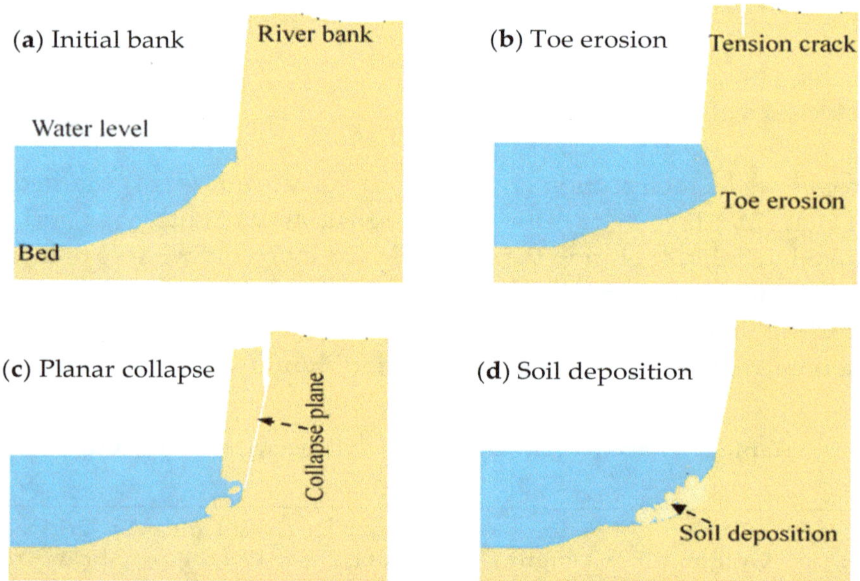

Figure 4. Typical riverbank processes observed at the case study of Ningxia–Inner Mongolia catchment.

2.4. Numerical Modeling

In this study, the BSTEM model, developed by the National Sedimentation Laboratory in Oxford, MS, USA, was used and modified according to the assumptions presented in the introduction.

2.4.1. BSTEM Method

There are two separate modules in the conventional BSTEM model: (i) the toe erosion module and (ii) the stability of the bank module [39]. Equation (1) is commonly applied to predict the width of the bank toe eroded due to the hydraulic conditions impacting on the cohesive riverbank [16,50–52]:

$$B = \kappa(\tau - \tau_c)^a \Delta t \tag{1}$$

where B is the toe erosion width (m); κ is the erodibility coefficient (m^3 N^{-1} s^{-1}); τ is the average shear stress (Pa); τ_c is soil's critical shear stress (Pa); a is an exponent usually assumed to be unity, Δt is the time interval (s). κ and τ_c are functions of the soil properties and they are characterized by the following relationship, $\kappa = 0.2\tau_c^{-0.5}$. For non-cohesive soils, τ_c is typically estimated based on the median particle diameter of the soil [53]. Rinaldi [54] noted the difficulty to accurately estimate κ but despite that, other methods provided solutions via a variety of methods to calculate κ and τ_c. One of these methods was developed by Hanson [55] using an in situ jet-test device and his experimental results are currently used universally. In Equation (1), τ can be calculated using Equation (2):

$$\tau = \gamma_w RS \tag{2}$$

where γ_w is the weight of water (N/m^3); R is the hydraulic radius (m); S is the channel slope.

For the specific scenario of planar failure assumed for this study, the riverbank stability analysis involved the computation of a safety factor, identified as the ratio between the resisting and the driving forces applied to the failure zone, which is typically calculated by using Equation (3) [56–59]:

$$F_s = \frac{S_R}{S_D} \tag{3}$$

where F_s is the safety factor with respect to bank failure ($F_s < 1$ indicates the possibility of an eventual collapse); S_R is the shear strength of the soil (kPa); S_D is the driving stress (Pa). S_R can be calculated using Equation (4):

$$S_R = c' + S\tan\phi^b + \Psi\tan\phi' \tag{4}$$

where c' is the effective cohesion (the effective cohesion of the soil is the cohesion of the soil in the anti-shear process) (Pa); S is the normal stress (Pa); ϕ^b is an angle that describes the relationship between the shear strength and the matric suction (°); Ψ is the matric suction which contrasts the pore-water pressure (Pa); ϕ' is the effective internal friction angle (°). S_D can be calculated using Equation (5):

$$S_D = W\sin\beta \tag{5}$$

where W is the weight of the wet soil per unit area of the failure plane (Nm^{-2}); β is the angle of the failure plane (°) [39]. Note that the BSTEM model takes c' into account to represent the cohesion effect. Nevertheless, the influence of cohesive materials on erosion would be considered through other approaches, as well. For example, Dodaro et al. [60,61] modified Shields parameter to predict erosion phenomena at a cohesive sediment mixture.

2.4.2. BSTEM Method Modified

As defined in Section 2.3, the phenomenon of bank collapse is characterized by four different stages. Despite multiple factors influencing the rate of the riverbank collapse, as previously mentioned, continuous collapses are also affected by the intensity of the antecedent collapse magnitude. When the collapse happens, the riverbank affected typically divides into two sections within a moment-frame: one part collapsed is naturally washed away by the river and the remaining part usually deposits at the bank toe. Before the next bank collapse takes place, the additional deposited soil is eroded. Therefore, it is very important to characterize the continuous changing shape of the riverbank to correctly predict the effects of the continuous erosion. Although the original BSTEM model provides some insights into this process and is very helpful for the analysis of the riverbank stability, it should be noted that BSTEM assumes that the falling part which has collapsed from the riverbank does not accumulate on the toe and is immediately removed. Hence it does not accurately represent the reality of the entire erosion process typical on natural rivers. Authors observed multiple shapes and forms during the erosion process following the collapse of material and have tried to characterize each of them, but the literature published to date lacks formulae to accurately estimate and quantify the amount of material deposited. Therefore, to address this gap, three changes were made in the original BSTEM model as follows.

(i) The quantity of soil falling into the river and carried away by the natural streamflow in the form of suspended material can be calculated using Equation (6):

$$M_0 = (S_* - S_v)Hd \tag{6}$$

where M_0 is the weight of soil carried away per unit length (kg/m); S_v is the sediment concentration (kg/m^3), typically obtained by field monitoring; H is the height of the bank (m); d is the width of the flow in the proximity of the bank (m). Because there was a main current which was about 20 m away from the bank, d is assumed as a constant (d = 20). S_* is the capacity of the flow to transport suspended material (kg/m^3) and can be calculated using Equation (7) [62]:

$$S_* = K\left(U_L^3/gh\omega\right)^m \tag{7}$$

where K and m are parameters that can be obtained from the literature [60]; U_L is flow velocity (m/s); g is gravitational acceleration (m/s^2); h is water depth (m); and ω is the velocity of the particles setting (m/s) which can be calculated by using Equation (8) [62]:

$$\omega = (\gamma_s - \gamma_w)gD^2/25.6\gamma_w v \tag{8}$$

where D is the grain size of the material (m); v is kinematic viscosity coefficient (m^2/s).

(ii) The quantity of deposition soil can be calculated by using Equation (9):

$$M_d = (W_0 - M_0)x \tag{9}$$

where M_d is the quantity of material deposited (kg); W_0 is the quantity of material collapsed (excluding the part suspended in the flow) (kg); x is a ratio between the material deposited and the material collapsed.

(iii) The new shape generated by the deposition of material on the bank toe is assumed to be triangular and the angle away from the riverbank is the angle of repose of sediment. In general, loose sediment grains in water accumulate with an angle close to the sediment angle of repose [63,64].

In the calculation process, the time step is selected as the hour. This decision is justified by the fact that since the flow rate, water depth, flow rate and other data are used as a daily average, the 24 h cycle was considered more appropriate. After the original boundary has passed 24 h, the river bank stability was tested: if the river bank was still stable, then the next set of water depth and flow rate conditions was inserted and the river bank toe erosion module continued to run. When the river bank stability coefficient was less than 1, the collapse occurred and consequently the collapse rate should have been recorded. For this scenario, the amount of collapse was calculated according to the shape of the collapse area identified, and the amount of sediment deposited at the foot of the slope was calculated by using Equations (6)–(9). The shape of the material deposited on the slope was considered to be triangular, and the angle away from the river bank was the angle of repose of the sediment, which was taken as 30 degrees. Based on the collapse rate, the new shape of the river bank was obtained and taken as a new initial condition for the next collapse.

The scheme utilized by the modified BSTEM model can be summarized with the flow chart displayed in Figure 5.

It is complex to quantify in real time the collapse processes along the entire riverbank of the study area due to flow conditions under continuous change, but it is possible to provide an approximation of the riverbank collapse processes by using averaged parameters such as monthly average water depths, flowrates, sediment concentration and average grain size. In addition, there are also parameters that require an accurate calibration during the entire calculation process, such as x. τ_c can be obtained by using values furnished in the literature [65]. By applying trial and error techniques, x = 0.39 corresponds to the optimal numerical value to replicate the realistic distance of collapse measured for this case study and discrepancies obtained between numerical and field measurements are listed in Tables 5–9.

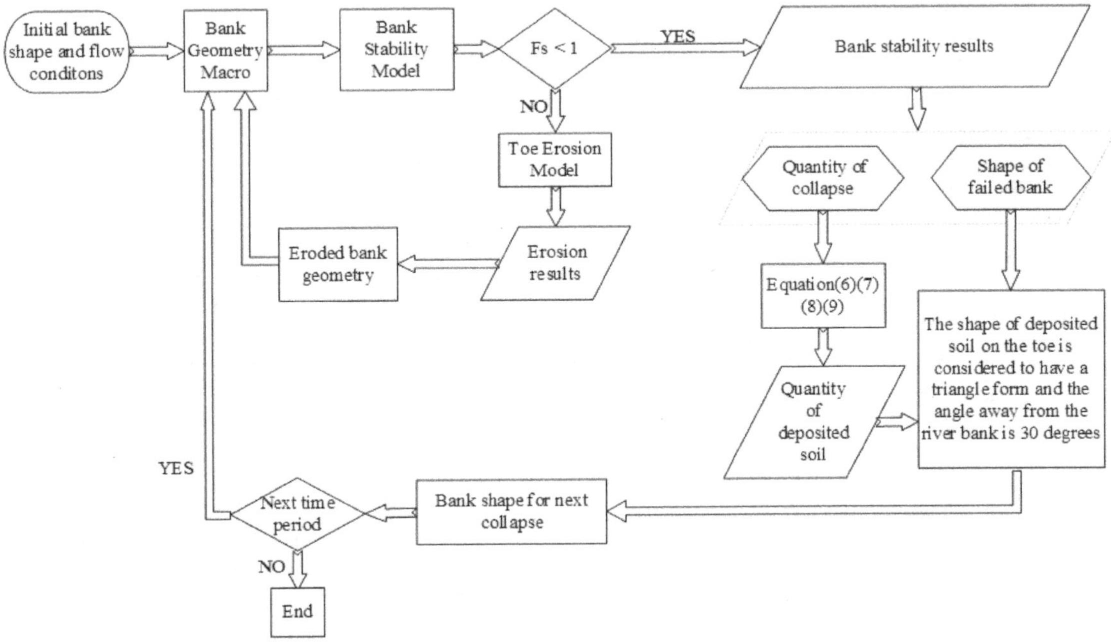

Figure 5. Flow chart of the established model.

Table 5. Numerical results, field measurements and errors for surveying Section 1 of Figure 1.

Time Scale	Calculating Collapse Distance (m)	Monitoring Collapse Distance (m)	Error (%)
21 May–13 July	4.96	4.60	7.83
14 July–9 August	3.24	3.05	6.23
10 August–23 September	5.55	5.25	5.71
23 September–30 September	1.50	1.68	10.71
Total collapse	15.25	14.58	4.60

Table 6. Numerical results, field measurements and errors for surveying Section 2 of Figure 1.

Time Scale	Calculating Collapse Distance (m)	Monitoring Collapse Distance (m)	Error (%)
21 May–13 July	4.99	4.50	10.89
14 July–9 August	3.01	2.80	7.5
10 August–23 September	5.38	5.08	5.91
23 September–30 September	1.13	1.45	22.06
Total collapse	14.51	13.81	5.07

Table 7. Numerical results, field measurements and errors for surveying Section 3 of Figure 1.

Time Scale	Calculating Collapse Distance (m)	Monitoring Collapse Distance (m)	Error (%)
21 May–13 July	5.46	5.80	5.9
14 July–9 August	3.30	3.10	6.5
10 August–23 September	7.15	6.80	5.1
23 September–30 September	0.94	1.10	14.5
Total collapse	16.85	16.80	0.3

Table 8. Numerical results, field measurements and errors for surveying Section 4 of Figure 1.

Time Scale	Calculating Collapse Distance (m)	Monitoring Collapse Distance (m)	Error (%)
21 May–13 July	5.75	5.3	8.5
14 July–9 August	3.72	3.95	5.8
10 August–23 September	6.41	6.2	3.4
23 September–30 September	2.73	2.4	13.8
Total collapse	18.61	17.85	4.3

Table 9. Numerical results, field measurements and errors for surveying Section 5 of Figure 1.

Time Scale	Calculating Collapse Distance (m)	Monitoring Collapse Distance (m)	Error (%)
21 May–13 July	5.93	5.6	5.9
14 July–9 August	3.82	3.6	6.1
10 August–23 September	6.24	6.5	4
23 September–30 September	1.75	1.45	20.7
Total collapse	17.74	17.15	3.4

By analyzing the results displayed in Tables 5–9, it is possible to confirm that the error between numerical and experimental values is between 6% and 20%. There are multiple influencing factors that can affect this comparison and the shape of the bank toe, always changing under the action of water flow which provides the higher impact. As water and sand conditions are continuously varying, the error is considerably higher considering measurements vs numerical results for the smallest time frame monitored (23 September–30 September). On the other hand, for longer time scales, the total error is about 5%. Overall, the numerical results are fairly consistent with the data recorded during the monitoring campaign, considering the significant modifications applied within the numerical model.

3. Results and Discussion

This section provides the further results obtained and their interpretation.

3.1. Collapse Processes Obtained through Numerical Simulations with the Modified BSTEM Model

The riverbank collapses simulated with the modified BSTEM model for the period 21 May–30 September in 2011 are illustrated in Figures 6–10 for surveying river Sections 1–5 respectively. Collapse rates of study sections in different months are listed in Table 10.

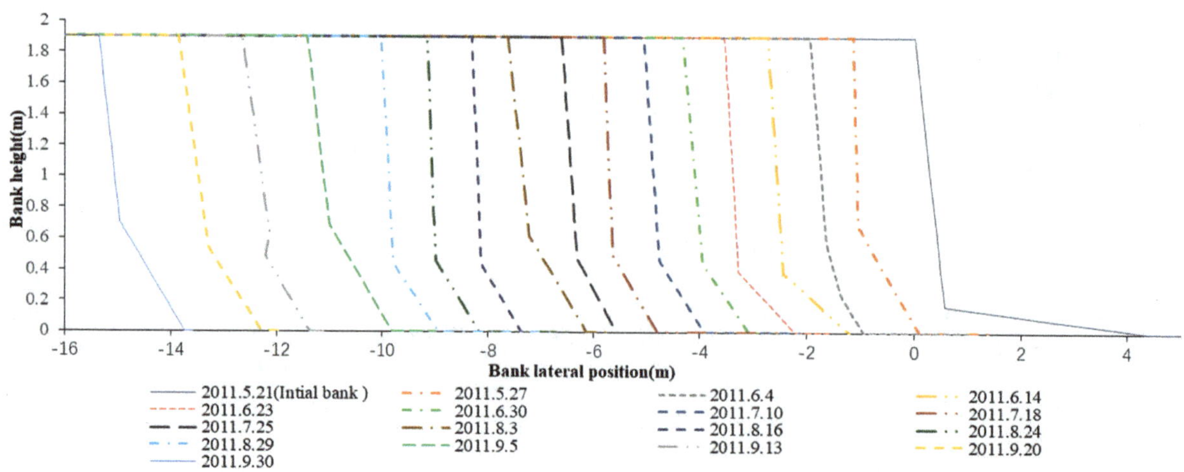

Figure 6. Collapse phases for surveying Section 1.

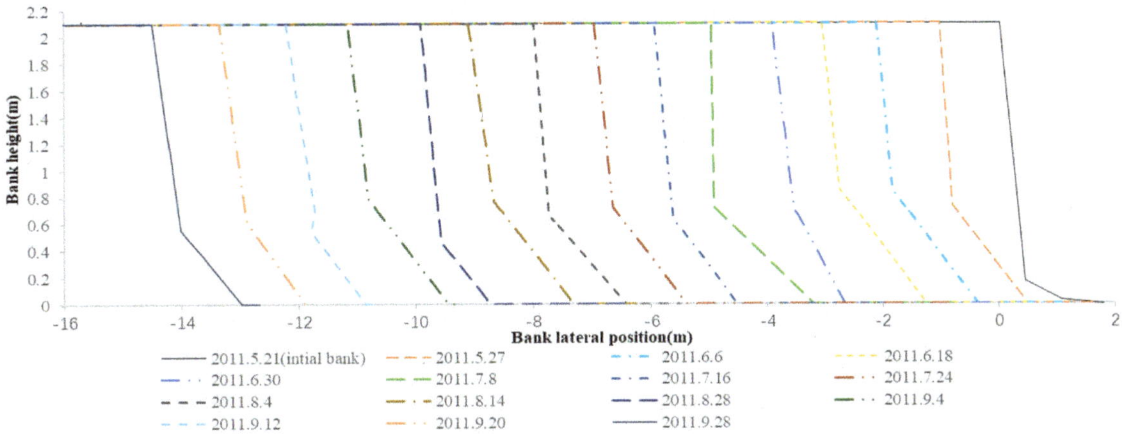

Figure 7. Collapse phases for Section 2.

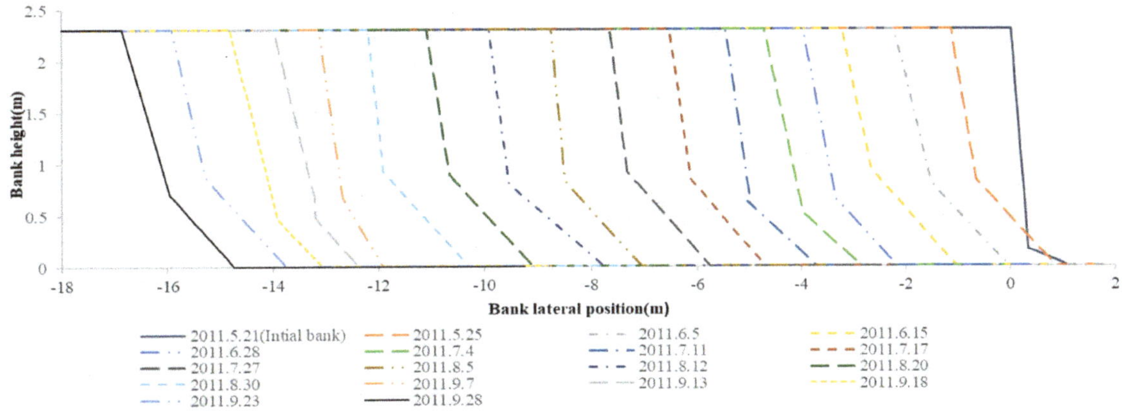

Figure 8. Collapse phases for Section 3.

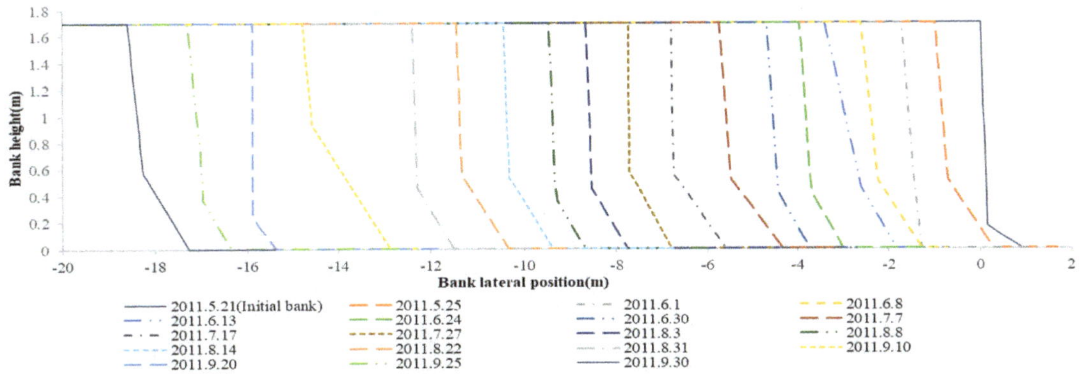

Figure 9. Collapse phases for Section 4.

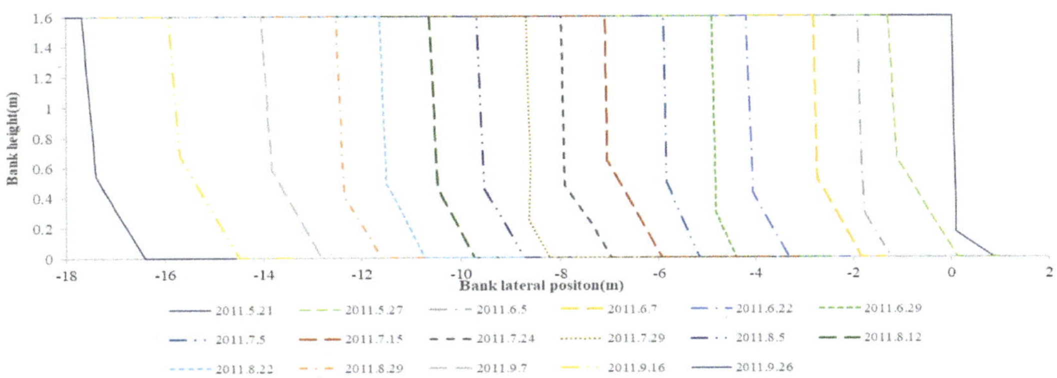

Figure 10. Collapse phases for Section 5.

Table 10. Collapse rates of study sections in different months (Unit: m/d).

Month	Section 1	Section 2	Section 3	Section 4	Section 5
5	0.095	0.094	0.104	0.091	0.118
6	0.106	0.097	0.094	0.123	0.122
7	0.074	0.098	0.119	0.098	0.123
8	0.109	0.095	0.147	0.150	0.124
9	0.178	0.153	0.155	0.207	0.172

To identify hypothetical relationships between hydraulic conditions and collapse rates, different indicators recorded were plotted versus the collapse rates estimated, and factors obtained are displayed in Figure 11.

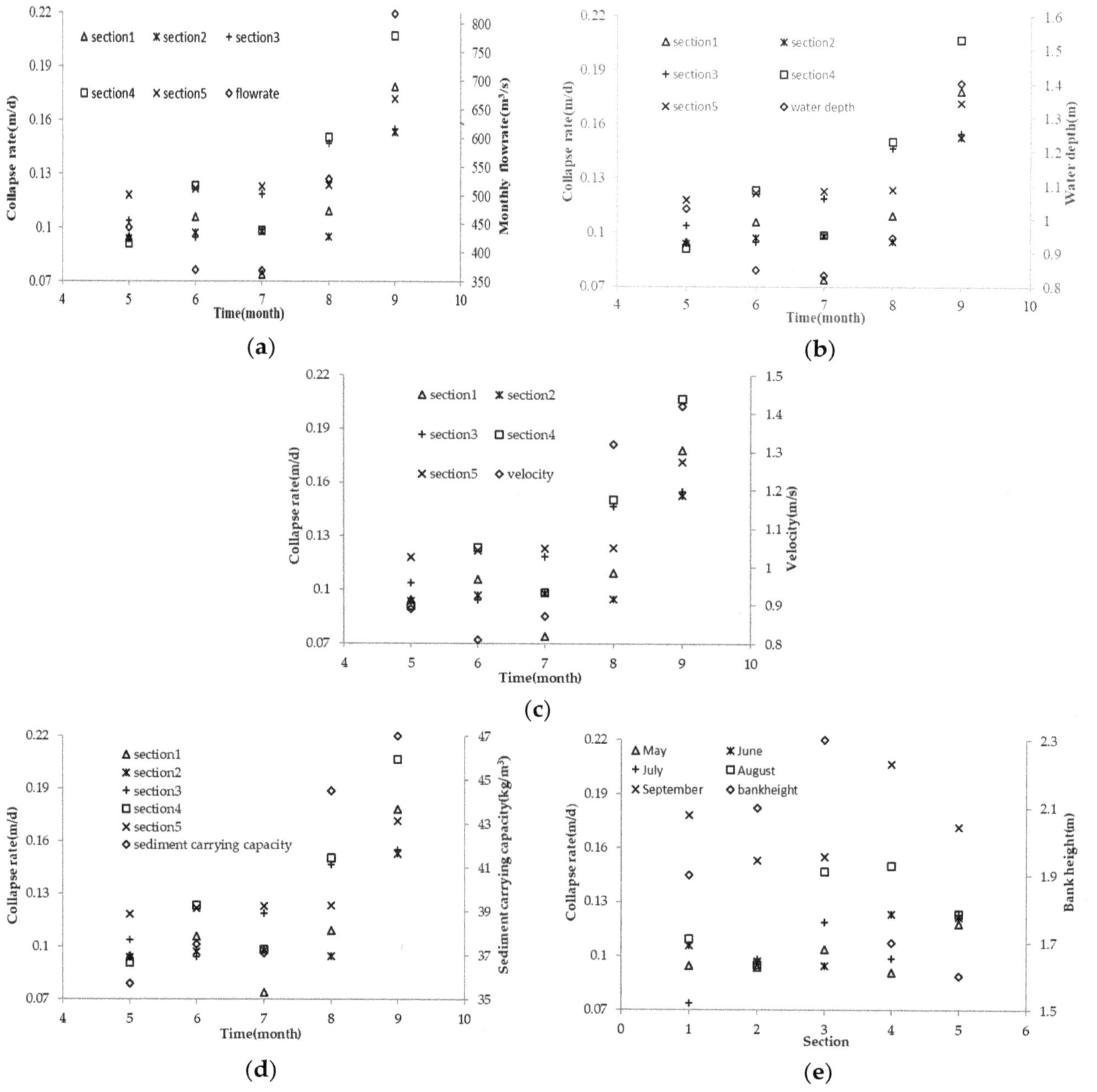

Figure 11. The relationship between flow conditions and collapse rates, the effects of (**a**) flowrate; (**b**) water depth; (**c**) velocity; (**d**) sediment carrying capacity; and (**e**) bank height.

Because the flowrate, water depth, velocity and sediment carrying capacity were of similar magnitudes from June to July, the average collapse rate was used to discuss the effects on collapse rates by influencing factors previously highlighted, and their impacts can be observed in Figure 11.

The authors have attempted to characterize the continuous collapses influenced by multiple factors that can be divided into two categories: autologous and external. The autologous factor considered is the bank height, while external factors include flow rate, water depth, velocity and sediment carrying capacity. Different factors play dissimilar roles on influencing continuous collapse processes. For example, when flow rate, velocity and sediment carrying capacity rise, bank toe erosion can consequently increase due to higher shear stresses. Additionally, when water depth increases, soil shear strength diminishes due to the soil saturation, however the rise of water depth could also improve the bank stability due to the water pressure acting on the bank.

Even if bank collapses are continuously affected by these factors, there are still some interesting regular features observed in this study. For example, in Figure 11a–d, it can be observed that the collapse rate is positively correlated with the flow rate, water depth, velocity and capacity of carrying materials. Although the bank heights are all different in altered sections, collapse rates for the locations considered are higher in the flood season (July to September) than in the dry season (May to June), due to the increase of flow rates, water depth and velocities. Due to the strength of higher flow rates, the flow shear stress is typically greater than the soil shear stress, causing continuous erosion of the bank toe. It should also be noticed then that flow conditions are playing a leading role in the collapse processes and are the main dynamic factors for the quantity of material deposited.

Another interesting phenomenon which can be noticed within the results is that continuous collapse rates are bigger in May than in June, except for Section 3 (Figure 11). During this period, which belongs to the dry season, flow rates, water depth, velocities and sediment carrying capacities are all smaller (Table 3). This is due to the associated phenomena when water depths reduce, with a consequent loss in river bank stability and more intense bank collapses. However, for Section 3, the collapse distance decreases, and the authors believe that this is due to the bank height (2.3 m, the highest among the sections).

3.2. Non Applicability of the Traditional BSTEM Method

In order to further demonstrate that the original BSTEM model is not applicable for the representation of continuous long term bank collapses, numerical and experimental results are listed in Tables 11–15, together with the errors obtained.

Table 11. BSTEM results, field measurements and errors for surveying Section 1.

Time Scale	BSTEM Method Results (m)	Monitoring Collapse Distance (m)	Error (%)
21 May–13 July	7.80	4.60	69.6
14 July–9 August	7.18	3.05	135.4
10 August–23 September	9.98	5.25	89.3
23 September–30 September	2.08	1.68	23.8
Total collapse	27	14.58	85.2

Table 12. BSTEM results, field measurements and errors for surveying Section 2.

Time Scale	BSTEM Method Results (m)	Monitoring Collapse Distance(m)	Error (%)
21 May–13 July	7.70	4.50	71.1
14 July–9 August	5.94	2.80	112.1
10 August–23 September	8.48	5.08	67.2
23 September–30 September	1.73	1.45	19.1
Total collapse	23.84	13.81	72.6

Table 13. BSTEM results, field measurements and errors for surveying Section 3.

Time Scale	BSTEM Method Results (m)	Monitoring Collapse Distance (m)	Error (%)
21 May–13 July	9.63	5.80	66.1
14 July–9 August	6.85	3.10	121.1
10 August–23 September	11.83	6.80	73.9
23 September–30 September	1.36	1.10	24.1
Total collapse	29.68	16.80	76.6

Table 14. BSTEM results, field measurements and errors for surveying Section 4.

Time Scale	BSTEM Method Results(m)	Monitoring Collapse Distance (m)	Error (%)
21 May–13 July	8.9	5.3	68.1
14 July–9 August	9.13	3.95	131.1
10 August–23 September	11.35	6.2	83.2
23 September–30 September	2.9	2.4	21
Total collapse	32.28	17.85	80.8

Table 15. BSTEM results, field measurements and errors for surveying Section 5.

Time Scale	BSTEM Method Results (m)	Monitoring Collapse Distance (m)	Error (%)
21 May–13 July	9.63	5.6	71.9
14 July–9 August	7.85	3.6	118.1
10 August–23 September	12.02	6.5	85
23 September–30 September	1.77	1.45	22
Total collapse	31.27	17.15	82.4

It can be concluded that the errors are all very large for longer monitoring periods. That is due to the hypothesis implemented within the BSTEM method, commonly used to calculate collapse distance, considering no soil deposition at the bank toe after the collapse—an unrealistic condition.

3.3. Discussions

Comparing simulation results using the modified BSTEM model with field data collected, as showed in Tables 5–9, it can be observed that errors range from 0.3% to 5.07%, hence are all acceptable and it is possible to confirm the reliability and accuracy of the modified BSTEM model to calculate continuous riverbank collapse.

Furthermore, additional outcomes have to be highlighted to increase the performance of the numerical model. Firstly, as the initial boundary conditions can have an impact on river bank collapse, supplementary studies have to be conducted to investigate this phenomenon. Secondly, due to the lack of advanced experimental measurement techniques, it is very difficult to get the river bed profile at every time step, and for this study a simplified bank shape was used in the calculations presented. This may be one more reason for the errors calculated. To solve this problem, sonar and electromagnetic wave measurement technologies may be adopted in future studies. Thirdly, in natural rivers, flow conditions such as flow rate and water depth vary continuously. To obtain bank collapse distances, monthly average flow conditions were used in this study, hence more targeted and accurate flow conditions should be used in future studies to improve the results. Finally, riverbed erosion and bank erosion typically occur at the same time and both are influenced by each other. For example, riverbed erosion influences the bank collapse by increasing the bank height and when the bank collapse happens, falling soil directly settles along the area where riverbed erosion happened before depositing at the bank toe.

As all these aspects are crucial, by implementing them within the numerical model it is predicted that results will be more accurate.

4. Conclusions

(1) As confirmed by the observations made during the monitoring of the area, a single bank collapse can be divided into four different stages that can be summarized as: (i) bank toe erosion initiation; (ii) tension crack development; (iii) shearing initiation; and (iv) bank failure occurrence. When a single collapse happens, part of the collapsed soil is washed away by the water and the remaining part accumulates at the bank toe. The new shape of the bank formed after this process along the river becomes the initial form of the next bank collapse. Bank collapse processes in a longer time scale are continuous collapses that are made from single bank collapse during short time scales.

(2) Based on the original BSTEM model, a new modified version was implemented, applying several new assumptions. The original BSTEM assumed that the falling part which had collapsed from the riverbank did not accumulate on the toe and was immediately removed. This does not accurately represent the reality of the entire erosion process typical in natural rivers, and comparisons between experimental and numerical results have shown that, by using the original approach, significant errors will affect the numerical results, as listed in Tables 11–15.

Furthermore, to address this gap, a modified BSTEM was developed to obtain the quantity of collapsed sediment accumulating on the toe. Several parameters were introduced, such as sediment concentration, sediment carrying capacity and sediment setting velocity. The quantity of collapsed sediment accumulated on the toe could have been obtained using Formulas (6)–(9).

Additionally, another main goal modifying the BSTEM model was to obtain the new shape generated by the deposition of material on the bank toe. It should be noted that the boundary conditions play an important role in the calculations of continuous collapse processes for natural rivers. The new shape generated by the deposition of material on the bank toe was the initial boundary condition of next collapse process. Based on this, after observing multiple shapes during the erosion process following the collapse of material, the shape generated by the deposition of material on the bank toe was assumed to be triangular and the angle away from the riverbank was the angle of repose of sediment. After a single collapse, the new shape of the riverbank could have been obtained from the shape of the deposited sediment and the collapsed plain.

(3) Adopting all the previous modifications within the calculation of every single collapse process, the numerical results obtained using the modified BSTEM are consistent with the monitored datasets for continuous processes, as listed in Tables 5–9.

(4) The relationship between collapse rates and influencing factors was discussed and influencing factors were divided into two categories: autologous and external (flow conditions). Different factors play different roles in continuous collapse processes. For example, the flow rate, velocity and sediment carrying capacity rise can increase bank toe erosion by increasing flow shear stress. Water depth rise can reduce soil shear strength by promoting soil saturation, and water depth rise can also increase river bank stability by providing water pressure. Soil characteristics determine the shear strength of each material, and bank height plays an important role in the bank stability analysis. For similar flow conditions, a higher bank with a steeper slope is more unstable. These factors should be considered together to better characterize the collapse mechanisms.

(5) Aspects such as precisely determining the bank shape and quantifying the more singular flow conditions are crucial, and by implementing them within the numerical model it is predicted that results will become more accurate.

Author Contributions: All the authors jointly contributed to this research. A.S. was responsible for the proposition and design of the field monitoring and the modified BSTEM method; M.R. and G.D. analysed the experimental datasets and wrote the paper; and S.W. and F.Z. participated in the experiments.

Acknowledgments: The research reported in this manuscript is supported by the National Basic Research Program of China (Grant No. 2011CB403304) and National Natural Science Foundation of China (Grant No. 11372048).

References

1. Nanson, G.C.; Hickin, E.J. A statistical analysis of bank erosion and channel migration in western Canada. *Geol. Soc. Am. Bull.* **1986**, *97*, 497–504. [CrossRef]
2. Shu, A.P.; Li, F.H.; Liu, H.F.; Duan, G.S.; Zhou, X. Characteristics of particle size distributions for the collapsed riverbank along the desert reach of the upper Yellow River. *Int. J. Sediment. Res.* **2016**, *31*, 291–298. [CrossRef]
3. Yao, Z.Y.; Ta, W.Q.; Jia, X.P.; Xiao, J.H. Bank erosion and accretion along the Ningxia–Inner Mongolia reaches of the Yellow River from 1958 to 2008. *Geomorphology* **2011**, *127*, 99–106. [CrossRef]
4. Hou, S. Analysis on recent channel evolution characteristics on Inner-Mongolia reaches of the Yellow River. *Yellow River* **1996**, *18*, 43–44.
5. Hou, S.; Chang, W.; Wang, P. Characteristics and cause of formation of channel atrophy at Inner Mongolia section of the Yellow River. *Yellow River* **2007**, *29*, 25–29.
6. Wu, J.M.; Wang, R.S.; Yao, J.H. Remote sensing monitoring and study on the change of the Yellow River course in Yinchuan plain sector. *Remote Land Resour.* **2006**, *18*, 36–40.
7. Ta, W.Q.; Xiao, H.L.; Dong, Z.B. Long-term morphodynamic changes of a desert reach of the Yellow River following upstream large reservoirs' operation. *Geomorphology* **2008**, *7*, 249–259. [CrossRef]
8. Grabowski, R.C.; Droppo, I.G.; Wharton, G. Erodibility of cohesive sediment: The importance of sediment properties. *Earth-Sci. Rev.* **2011**, *105*, 101–120. [CrossRef]
9. Thorne, C.R. Processes and mechanisms of river bank erosion. In *Gravel-Bed Rivers*; Bathurst, J.C., Thorne, C.R., Eds.; Wiley and Sons: Chichester, UK, 1982; pp. 227–259.
10. ASCE Task Committee on Hydraulics; Bank Mechanics; Modeling of River Width Adjustment. River width adjustment. I: Processes and mechanisms. *J. Hydraul. Eng.* **1998**, *124*, 881–902. [CrossRef]
11. Shields, A.; Ott, W.P.; Uchelen, J.C.V. *Application of Similarity Principles and Turbulence Research to Bed-Load Movement*; Soil Conservation Service Cooperative Laboratory California Institute of Technology: Pasadena, CA, USA, 1936.
12. Buffington, J.M.; Montgomery, D.R. A systematic analysis of eight decades of incipient motion studies, with special reference to gravel-bedded rivers. *Water Resour. Res.* **1997**, *33*, 1993–2029. [CrossRef]
13. Hooke, J.M. An analysis of the processes of riverbank erosion. *J. Hydrol.* **1979**, *42*, 39–62. [CrossRef]
14. Grissinger, E.H. Bank erosion of cohesive materials. In *Gravel-Bed Rivers*; Thorne, C.R., Bathurst, J.C., Eds.; Wiley and Sons: Chichester, UK, 1982; pp. 273–287.
15. Simon, A.; Collison, A.J.C. Pore-water pressure effects on the detachment of cohesive streambeds: Seepage forces and matric suction. *Earth Surf. Process. Landf.* **2001**, *26*, 1421–1442. [CrossRef]
16. Julian, J.P.; Torres, R. Hydraulic erosion of cohesive riverbanks. *Geomorphology* **2006**, *76*, 193–206. [CrossRef]
17. Twidale, C.R. Erosion of an alluvial bank at Birdwood, South Australia. *Z. Geomorphol.* **1964**, *8*, 189–211.
18. Lawler, D.M. The measurement of river bank erosion and lateral channel change: A review. *Earth Surf. Process. Landf.* **1993**, *18*, 777–821. [CrossRef]
19. Rinaldi, M.; Casagli, N. Stability of streambanks formed in partially saturated soils and effects of negative pore water pressures: The Sieve River (Italy). *Geomorphology* **1999**, *26*, 253–277. [CrossRef]
20. Singer, M.J.; Munns, D.N. *Soils: An Introduction*; Prentice Hall Inc.: Upper Saddle River, NJ, USA, 1999; pp. 14–46. ISBN 9780131190191.
21. Couper, P. Effects of silt–clay content on the susceptibility of riverbanks to subaerial erosion. *Geomorphology* **2003**, *56*, 95–108. [CrossRef]
22. Musa, J.J.; Abdulwaheed, S.; Saidu, M. Effect of surface runoff on Nigerian rural roads (a case study of Offa local government area). *AU J. Technol.* **2010**, *13*, 242–248.
23. Wolman, M.G. Factors influencing erosion of a cohesive river bank. *Am. J. Sci.* **1959**, *257*, 204–216. [CrossRef]
24. Knighton, A.D. Riverbank erosion in relation to streamflow conditions, River Bollin-Dean, Cheshire. *East Midl. Geogr.* **1973**, *5*, 416–426.
25. Albidin, R.Z.; Sulaiman, M.S.; Yusoff, N. Erosion risk assessment: A case study of the Langat Riverbank in Malaysia. *Int. Soil Water Conserv. Res.* **2017**, *5*, 26–35. [CrossRef]
26. Yu, M.; Wei, H.; Wu, S. Experimental study on the bank erosion and interaction with near-bank bed evolution due to fluvial hydraulic force. *Int. J. Sediment. Res.* **2015**, *1*, 81–89. [CrossRef]

27. Yang, G.S.; Tuo, W.Q.; Dai, F.N.; Liu, Y.X.; Jing, K.; Li, B.Y.; Zhang, O.Y.; Lu, R.; Hu, L.F.; Tao, Y. Contribution of sand sources to the silting of riverbed in Inner Mongolia section of Huanghe River. *J. Desert Res.* **2003**, *23*, 152–159.
28. Feng, G. The coarse sediment of the Yellow River mainly from wind sand. *J. Soil Water Conserv.* **1992**, *3*, 45–47.
29. Jing, K.; Li, J.; Li, F. Erosion yield from the middle Yellow River basin and tendency prediction. *Acta Geogr. Sin.* **1998**, *53*, 107–115.
30. Jing, K.; Li, J.; Li, F. Study on delimitation of coarse sand area in the middle Yellow River. *J. Soil Water Conserv.* **1997**, *3*, 10–15.
31. Xu, J.H.; Lv, G.C.; Zhang, S.L.; Gan, Z.M. *Definition of the Source Area of Centralized Coarse Sediment and Sediment Yield and Transportation in the Middle Yellow River*; Yellow River Press: Zhengzhou, China, 2000; pp. 11–36, ISBN 9787806214527.
32. Bai, T.; Ma, R.; Ma, X.; Ha, Y.; Huang, Q. Threshold of water and sediment in desert wide valley reaches upper the Yellow River. *J. Desert Res.* **2018**, *38*, 645–650.
33. Yu, M.; Shen, K.; Zhang, J.H. Preliminary investigation on bank properties and sediment source in Yellow River Ningmeng River Reach. *Int. J. Sediment. Res* **2014**, *4*, 39–42.
34. Rijn, L.C.V. *Principles of Sediment Transport in Rivers, Estuaries and Coastal Seas*; Aqua Publications: Delft, The Netherlands, 1993; ISBN 90-800356-2-9.
35. Catano, Y.; Passoni, G.; Pacheco, R.G.; Toro, F.M.; Naranjo, B.J. Comparison between different numerical solutions for the hydraulic dam-break wave. In Proceedings of the Iche 98-International Conference on Hydro-sciences and Engineering, Berlin, Germany, 31 August–3 September 1998.
36. Toro, E.F. *Shock-Capturing Methods for Free-Surface Shallow Flows*; Wiley: New York, NY, USA, 2001; pp. 566–571, ISBN 9780471987666.
37. Cao, Z.; Day, R.; Egashira, S. Coupled and decoupled numerical modeling of flow and morphological evolution in alluvial rivers. *J. Hydraul. Eng.* **2002**, *128*, 306–321. [CrossRef]
38. Hudson, J.; Sweby, P.K. A high-resolution scheme for the equations governing 2D bed-load sediment transport. *Int. J. Numer. Meth.* **2005**, *47*, 1085–1091. [CrossRef]
39. Simon, A.; Curini, A.; Darby, S.E.; Langendoen, E.J. Bank and near-bank processes in an incised channel. *Geomorphology* **2000**, *35*, 193–217. [CrossRef]
40. Zong, Q.; Xia, J.; Deng, C.; Xu, Q.X. Modeling of the composite bank failure process using BSTEM. *J. Sichuan Univ.* **2013**, *45*, 69–78.
41. Simon, A.; Pollen-Bankhead, N.; Mahacek, V.; Langendoen, E. Quantifying reductions of mass-failure frequency and sediment loadings from streambanks using toe protection and other means: Lake Tahoe, United States. *J. Am. Water Resour. Assoc.* **2009**, *45*, 170–186. [CrossRef]
42. Lindow, N.; Fox, G.A.; Evans, R.O. Seepage erosion in layered stream bank material. *Earth Surf. Process. Landf.* **2009**, *34*, 1693–1701. [CrossRef]
43. Wilson, G.V.; Perketi, R.K.; Fox, G.A.; Dabney, S.M.; Shields, F.D.; Cullum, R.F. Soil properties controlling seepage erosion contributions to streambank failure. *Earth Surf. Process. Landf.* **2007**, *32*, 447–459. [CrossRef]
44. Cancienne, R.M.; Fox, G.A.; Simon, A. Influence of seepage undercutting on the stability of root-reinforced streambanks. *Earth Surf. Process. Landf.* **2008**, *33*, 1769–1786. [CrossRef]
45. Xiao, C.; Hao, Y.; Liu, F. Analysis of climate change for Dengkou in recent 52 years. *J. Arid Land Resour. Environ.* **2008**, *22*, 90–93.
46. Kolerski, T.; Shen, H.T. Possible effects of the 1984 St Clair River Ice jam on bed changes. *Can. J. Civ. Eng.* **2015**, *42*, 696–703. [CrossRef]
47. Kolerski, T.; Shen, H.T. St. Clair River ice jam dynamics and possible effect on bed changes. In Proceedings of the 20th IAHR International Symposium on Ice, Lahti, Finland, 14–18 June 2010.
48. Thorne, C.R.; Tovey, N.K. Stability of composite riverbanks. *Earth Surf. Process. Landf.* **1981**, *6*, 469–484. [CrossRef]
49. Shu, A.P.; Gao, J.; Li, F.H. Lateral variation properties in river channels due to bank failure along the desert valley reach of upper Yellow River. *Adv. Water Sci.* **2014**, *25*, 77–82.
50. Osman, A.M.; Thorne, C.R. Riverbank stability analysis. I: Theory. *J. Hydraul. Eng.* **1988**, *114*, 134–150. [CrossRef]
51. Darby, S.E.; Thorne, C.R. Numerical simulation of widening and bed deformation of straight sand-bed rivers. I: Model development. *J. Hydraul. Eng.* **1996**, *122*, 184–193. [CrossRef]

52. Midgley, T.L.; Fox, G.A.; Heeren, D.M. Evaluation of the bank stability and toe erosion model (BSTEM) for predicting lateral retreat on composite streambanks. *Geomorphology* **2012**, *145*, 107–114. [CrossRef]
53. Garcia, M.H. *Sedimentation Engineering*; American Society of Civil Engineers: Reston, VA, USA, 2008; pp. 21–163, ISBN 9780784408148.
54. Rinaldi, M.; Mengoni, B.; Luppi, L.; Darby, S.E.; Mosselman, E. Numerical simulation of hydrodynamics and bank erosion in a river bend. *Water Resour. Res.* **2008**, *44*, 303–312. [CrossRef]
55. Hanson, G.J. Surface erodibility of earthen channels at high stresses. Part II—Developing an in-situ testing device. *Trans. ASAE* **1990**, *33*, 127–131. [CrossRef]
56. Amiritokaldany, E.; Darby, S.E.; Tosswell, P. Bank stability for predicting reach-scale land loss and sediment yield. *J. Am. Water Resour. Assoc.* **2003**, *39*, 897–910. [CrossRef]
57. Langendoen, E.J.; Simon, A. Modeling the evolution of incised streams. II: Streambank erosion. *J. Hydraul. Eng.* **2008**, *134*, 905–915. [CrossRef]
58. Samadi, A.; Amiritokaldany, E.; Darby, S.E. Identifying the effects of parameter uncertainty on the reliability of riverbank stability modelling. *Geomorphology* **2009**, *106*, 219–230. [CrossRef]
59. Taghavi, M.; Dovoudi, M.H.; Amiritokaldany, E.; Darby, S.E. An analytical method to estimate failure plane angle and tension crack depth for use in riverbank stability analyses. *Geomorphology* **2010**, *123*, 74–83. [CrossRef]
60. Dodaro, G.; Tafarojnoruz, A.; Stefanucci, F.; Adduce, C.; Calomino, F.; Gaudio, R.; Sciortino, G. An experimental and numerical study on the spatial and temporal evolution of a scour hole downstream of a rigid bed. In Proceedings of the International Conference on Fluvial Hydraulics, Lausanne, Switzerland, 3–5 September 2014; pp. 1415–1422.
61. Dodaro, G.; Tafarojnoruz, A.; Sciortino, G.; Adduce, C.; Calomion, F.; Gaudio, R. Modified Einstein sediment transport method to simulate the local scour evolution downstream of a rigid bed. *J. Hydraul. Eng.* **2016**, *142*. [CrossRef]
62. Wang, X.K.; Shao, X.J.; Li, D.X. *Fundamental River Mechanics*; China Water and Power Press: Beijing, China, 2002; pp. 54–98, ISBN 7-5084-1093-9.
63. Gaudio, R.; Tafarojnoruz, A.; Calomino, F. Combined flow-altering countermeasures against bridge pier scour. *J. Hydraul. Res.* **2012**, *50*, 35–43. [CrossRef]
64. Tafarojnoruz, A.; Gaudio, R.; Calomino, F. Bridge pier scour mitigation under steady and unsteady flow conditions. *Acta Geophys.* **2012**, *60*, 1076–1097. [CrossRef]
65. Tang, C.B. Laws of sediment incipient motion. *J. Hydraul. Eng.* **1963**, *2*, 3–14.

5

Numerical Analysis of the Impact Factors on the Flow Fields in a Large Shallow Lake

Haifei Liu [1], Zhexian Zhu [2,*], Jingling Liu [2,*] and Qiang Liu [2]

1 The Key Laboratory of Water and Sediment Sciences of Ministry of Education, Beijing Normal University, Beijing 100875, China; haifei.liu@bnu.edu.cn

2 School of Environment, Beijing Normal University, Beijing 100875, China; qiang.liu@bnu.edu.cn

* Correspondence: 201621180007@mail.bnu.edu.cn (Z.Z.); jingling@bnu.edu.cn (J.L.);

Abstract: Wetland acts as an important part of climatic regulation, water purification, and biodiversity maintenance. As an integral part of wetlands, large shallow lakes play an essential role in protecting ecosystem diversity and providing water sources. Baihe Lake in the Momoge Wetland is one such example, so it is necessary to study the flow pattern characteristics of this lake under different conditions. A new model, based on the lattice Boltzmann method, was used to investigate the effects of different impact factors on flow fields, such as water discharge from surrounding farmland, rainfall, wind speed, and aquatic vegetation. Importantly, this study provides a hydrodynamic basis for local ecological protection and restoration work.

Keywords: wetland; lattice Boltzmann method; shallow lake; drag force

---

1. Introduction

There are many wetlands in China that play an important role in water resource conservation and ecological diversity maintenance. Wetlands have irreplaceable ecological significance for human survival [1]. Although China has emphasized wetland protection over the past few decades, there remains a sharp decline in the area of wetlands [2]. According to official statistics of the China Forestry Administration, China's wetland area fell by 8.83% in 2013, compared with that of 2003. The shrinking of wetland area is closely related to the decrease of water resources in wetlands, and the atrophy of lake area is directly impacted by the shrinking of wetland area [3]. Therefore, it is of vital importance to study the water flow in lakes to find out the feasible measures for wetland conservation [1,4].

A large number of microbes, animals, and plants live in lakes. Therefore, improving purification ability is essential to maintaining the wetland ecological environment [5]. In China, the area of lakes in wetlands has reduced greatly, and as a result, finding ways to slow down the shrinkage of lakes has become an increasingly important topic [6]. The reduction of lake area is bound to the flow of water, and there is no doubt that the characteristics of the water flow is of great importance [5–7]. In general, gravity, bed friction, rainfall, wind speed, and topography may influence the hydrodynamic conditions of lakes of different levels [6]. In addition, vegetation in lakes can also have an important impact on water flow with its drag effect [8,9].

Physical modelling and numerical simulation are the two main approaches to studying flow characteristics [10]. With the development of computer science, numerical modeling has become prevalent [11]. The MIKE series model, developed by the Danish Institute of Water Resources and Water Environment (DHI, Hørsholm, Denmark), is a general commercial package with a wide range of applications for water simulation [12]. The Environmental Fluid Dynamics Code (EFDC) model, created by the Virginia Institute of Marine Science at the College of William and Mary (Williamsburg,

VA, USA), has many functions for water quality modelling and its partially open-source code offers good flexibility [12–14]. However, these commercial models are only based on the traditional finite difference method or the finite volume method, and further development is inhibited given that these models do not have a flexible external force term.

The lattice Boltzmann method (LBM) is a relatively new numerical method for fluid flows [15]. The related theoretical basis of LBM was carried out in the 1980s [16]. Salmon (1999) [17] developed a lattice Boltzmann model for shallow water. Compared with traditional computational methods based on the direct approach of the flow equation, it proposes a new solution for the flow equation [18]. The method is characterized by simple calculation and easy handling of boundary conditions [19]. In recent years, it has become a promising approach in computational fluid dynamics [20]. The use of the LBM to simulate different kinds of flow (e.g., open channel flows, tidal flows, and dam-break flows) has become popular. Ottolenghi et al. (2018) [21] used lattice Boltzmann method to investigate the properties of graphene oxide for environmental applications. O'Brien et al. (2002) [22] developed a lattice Boltzmann scheme to study reactive transport in porous media. Zhou (2007) [23] developed a lattice Boltzmann model for groundwater flow. Tubbs and Tsai (2009) [24] developed the parallel computation for multi-layer shallow water flows. Liu et al. (2012) [25] worked out a large eddy simulation of turbulent shallow water flows using the LBM. Prestininzi (2016) [26] presented a 2D multi-layer shallow water lattice Boltzmann model able to predict the salt wedge intrusion in river estuaries. Furthermore, Yang et al. (2017) [8] developed a rigid vegetation model with 2D shallow water equations using the LBM.

In this paper, the LBM is used to study the hydrodynamic characteristics of Baihe Lake. This is also the first numerical simulation study of Baihe Lake in China, which is helpful for local ecological and hydrological restoration. Through the numerical study of different scenarios, the model provides a quantitative investigation of different impact factors and the results could provide a theoretical basis for local wetland restoration and hydraulic engineering construction.

2. Methodology

2.1. Study Area

The study area was Baihe Lake in the Momoge Wetland of Zhenlai County, Jilin Province, China. The Momoge Wetland (45°45′~46°10′ N, 122°27′~124°04′ E) is one of the most important wetlands in Northeastern China, as shown in Figure 1. It is the main migration path for white crane in China and the total area is 1440 km^2. In recent years, the area of the Momoge Wetland has shrunk significantly. The local government has developed a series of wetland restoration projects. Baihe Lake is one of the largest lakes in the Momoge Wetland. The area of the lake is 15 km^2 and the eastern region of the lake is next to the Nenjiang River. The lake is also the main fishing area for local fishermen and its upstream is surrounded by local farmland. Every year a large amount of irrigation water recedes to Baihe Lake as a replenishment. At the same time, the lake also plays a role in purifying the water quality. The lake's outlet is next to the Taoer River. Therefore, Baihe Lake plays an extremely important role in local services such as economic activities, water quality purification, hydrological connectivity, and ecological protection.

2.2. Governing Equations

Baihe Lake is a shallow lake, where the horizontal scale is much larger than the vertical scale. According to field measurements, the largest water depth of Baihe Lake is 2.82 m. Therefore, the two-dimensional hydrodynamic model can be used to study its water flow. The depth-averaged hydrodynamic equations are derived from the incompressible Navier–Stokes equations, which are called the shallow water equations [25].

$$\frac{\partial h}{\partial t} + \frac{\partial (hu_j)}{\partial x_j} = 0 \tag{1}$$

$$\frac{\partial(hu_i)}{\partial t} + \frac{\partial(hu_iu_j)}{\partial x_j} = -\frac{g}{2}\frac{\partial h^2}{\partial x_i} + \nu\frac{\partial^2(hu_i)}{\partial x_j \partial x_j} + F_i \tag{2}$$

where h is the water depth; t is the time; the subscripts i and j are the space directions based on the Einstein summation convention; x_j and u_j are the distance and instantaneous velocity components in the j direction; g is the gravitational acceleration and equals 9.81 m^2/s; ν is the kinematic viscosity; and F_i is the external force term.

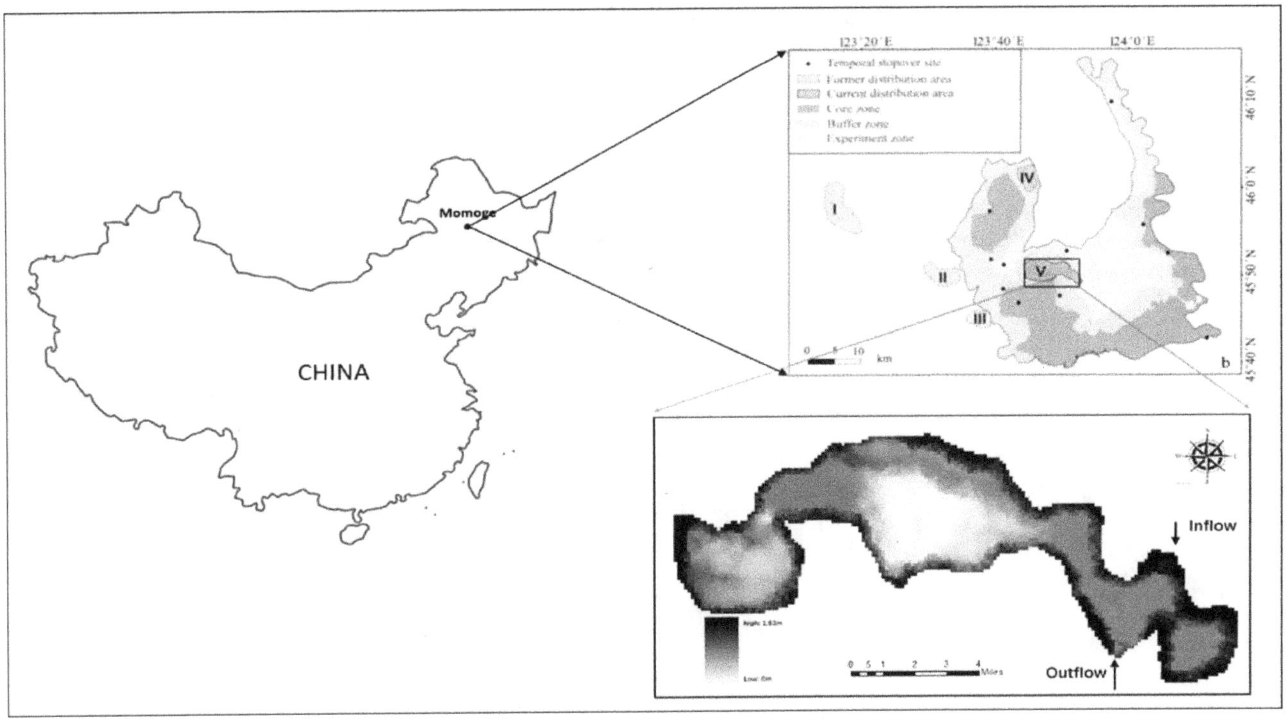

Figure 1. Geographical location and topographic conditions of Baihe Lake (Datum: 45°56′ N, 122°45′ E).

2.3. External Force Term

Shallow water can be significantly influenced by external forces. Generally, the external forces caused by gravity, wind speed, and riverbed friction should be considered [27]. In Equation (2), ignoring the Coriolis force, F_i is the force term and can be expressed as:

$$F_i = -gh\frac{\partial z_b}{\partial x_i} + \frac{\tau_{wi}}{\rho} - \frac{\tau_{bi}}{\rho} + S_{vi} \tag{3}$$

ρ is the water density, which is equal to 1000 kg/m^3. Z_b is the bed elevation; τ_{bi} is the bed friction and can be expressed as:

$$\tau_{bi} = \rho c_b u_i \sqrt{u_j u_j} \tag{4}$$

where $c_b = gn^2/h^{1/3}$; n is Manning's coefficient; and τ_{wi} is the wind shear stress that can be expressed as:

$$\tau_{wi} = \rho_a c_w u_{wi} \sqrt{u_{wj} u_{wj}} \tag{5}$$

where ρ_a is the air density; c_w is the resistance coefficient; and u_{wi} and u_{wj} are the wind velocities in the i and j directions.

In addition, if there is aquatic vegetation in lakes, the drag effect of vegetation on the water body should be considered as an external force [8,28]. A large amount of vegetation is distributed in Baihe Lake, where reeds and bulrush are dominant [29]. This type of aquatic vegetation shown in Figure 2 with high toughness is generally higher than the water free surface and can be simplified as unsubmerged rigid vegetation [29,30]. A rigid vegetation model based on two-dimensional shallow water equations is presented in Reference [8]. The model treated the unsubmerged rigid vegetation as vertical cylinders and the drag force can be expressed as:

$$S_{vi} = -\frac{1}{2}\lambda C_d h u_{vi}\sqrt{u_{vj}u_{vj}} \tag{6}$$

where u_{vi} is the average velocity on the vegetation elements in the i direction; C_d is the drag force coefficient and is usually in the range of 1 and 1.5 [31]; and λ is the projected area (normal to the flow) of vegetation per unit volume of water and is calculated by:

$$\lambda = \frac{4\alpha_v c}{\pi D_v} \tag{7}$$

where α_v represents the shape factor; c is the density of the vegetation zones and represents the projected area of vegetation per unit bed area; D_v is the vegetation stems diameter; and u_{vi} is equal to the average velocity u_i.

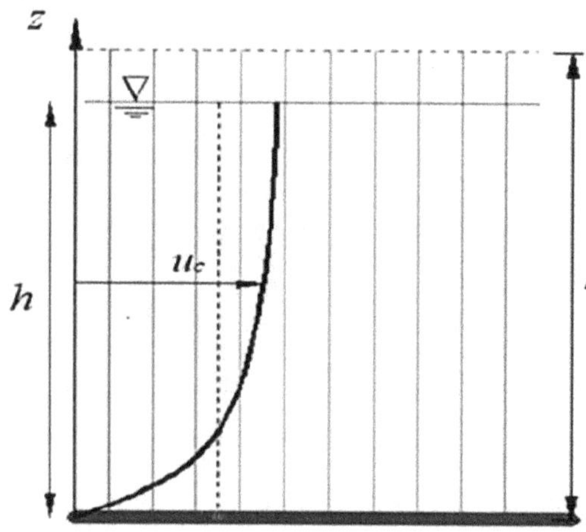

Figure 2. Flow over unsubmerged rigid vegetation.

2.4. Lattice Boltzmann Method (LBM)

The LBM is a modern numerical technique for computational fluid dynamics. The LBM for non-linear two-dimensional shallow water equations has been widely used [10]. It is a discrete computational method based on the lattice gas automata—a simplified, fictitious molecular model. The three main components in the LBM are lattice pattern, kinetic equation, and equilibrium distribution. The lattice Boltzmann equation can be expressed as:

$$f(x + e_\alpha \Delta t, t + \Delta t) = f_\alpha(x,t) - \frac{1}{\tau_t}[f_\alpha(x,t) - f_\alpha^{eq}(x,t)] + \Delta t F_\alpha \tag{8}$$

where f_α is the particle distribution function; Δx is the lattice size; Δt is time step; the external force F_α is calculated by:

$$F_\alpha = 3\omega_\alpha \frac{1}{e^2} e_{\alpha i} F_i \tag{9}$$

where F_i is force term computed by Equation (3); $e = \Delta x/\Delta t$; ω_α is the weight factor: $\omega_\alpha = 4/9$ for $\alpha = 0$; $\omega_\alpha = 1/9$ for $\alpha = 1, 3, 5, 7$; and $\omega_\alpha = 1/36$ for $\alpha = 2, 4, 6, 8$.

Lattice pattern in the LBM has two functions: indicating grid points and resolving particle motions. The former represents a similar role in the traditional numerical simulation methods. The latter shows a microscopic model for molecular dynamics. $e_{\alpha i}$ is the particle velocity in the i direction. The nine-velocity square lattice is shown in Figure 3. Each particle moves one lattice unit at its velocity along the eight links represented by numbers 1–8, while 0 represents a particle at rest with zero speed. The velocity vector of the particles is defined by:

$$e_\alpha = \begin{cases} (0,0) & \alpha = 0 \\ e\left[\cos\frac{(\alpha-1)\pi}{4}, \sin\frac{(\alpha-1)\pi}{4}\right] & \alpha = 1,\ 3,\ 5,\ 7 \\ \sqrt{2}e\left[\cos\frac{(\alpha-1)\pi}{4},\ \sin\frac{(\alpha-1)\pi}{4}\right] & \alpha = 2,\ 4,\ 6,\ 8 \end{cases} \quad (10)$$

A local equilibrium distribution function decides what flow equations are solved by means of the lattice Boltzmann equation. For 2D shallow water Equations (1) and (2), the local equilibrium distribution function f_α^{eq} is defined as:

$$f_\alpha^{eq} = \begin{cases} h - \frac{5gh^2}{6e^2} - \frac{2h}{3e^2}u_iu_i, & \alpha = 0 \\ \frac{gh^2}{6e^2} + \frac{h}{3e^2}e_{\alpha i}u_i + \frac{h}{2e^4}e_{\alpha i}e_{\alpha j}u_iu_j - \frac{h}{6e^2}u_iu_i, & \alpha = 1,\ 3,\ 5,\ 7 \\ \frac{gh^2}{24e^2} + \frac{h}{12e^2}e_{\alpha i}u_i + \frac{h}{8e^4}e_{\alpha i}e_{\alpha j}u_iu_j - \frac{h}{24e^2}u_iu_i, & \alpha = 2,\ 4,\ 6,\ 8 \end{cases} \quad (11)$$

Then the remaining task is to determine the physical quantities. The macroscopic variables, water depth h and flow velocity u_i can be expressed as:

$$h = \sum_\alpha f_\alpha, u_i = \frac{1}{h}\sum_\alpha e_{\alpha i}f_\alpha \quad (12)$$

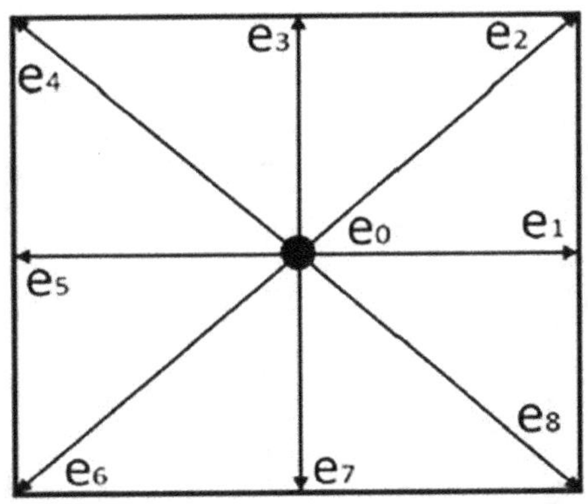

Figure 3. D2Q9 lattice pattern (D2 represent two-dimensional and Q9 represent nine velocity directions in each particle).

2.5. Rainfall

Rainfall has an effect on the surface water and runoff, causes soil erosion and floods, and contaminants transport. A numerical model should be applied to investigate how rainfall influences water flow [32]. In the LBM, the local equilibrium distribution function g_α^{eq} for shallow water equations with source term was developed. It can be expressed as follows:

$$g_\alpha^{eq} = \begin{cases} h - R\Delta t - \frac{5gh^2}{6e^2} - \frac{2h}{3e^2u_iu_i} & \alpha = 0 \\ \frac{gh^2}{6e^2} + \frac{h}{3e^2}e_{\alpha i}u_i + \frac{h}{2e^2}e_{\alpha i}e_{\alpha j}u_iu_j - \frac{h}{6e^2}u_iu_j & \alpha = 1,\ 3,\ 5,\ 7 \\ \frac{gh^2}{24e^2} + \frac{h}{12e^2}e_{\alpha i}u_i + \frac{h}{8e^2}e_{\alpha i}e_{\alpha j}u_iu_j - \frac{h}{24e^2}u_iu_j & \alpha = 2,\ 4,\ 6,\ 8 \end{cases} \quad (13)$$

where R is the rainfall intensity and the influence of dynamic pressure caused by precipitation is neglected.

2.6. Boundary Conditions

A general treatment at the inlet is to set a constant velocity and a water depth, whereas a specific water depth is imposed at the outlet. In addition, a zero-gradient condition is used to obtain the velocity components u and v at the outlet. The standard bounce-back scheme, in which an incoming particle towards the boundary is bounced back into the fluid, is widely used. At the upper boundary:

$$f_6 = f_2, f_7 = f_3, f_8 = f_4 \tag{14}$$

and at the lower boundary:

$$f_2 = f_6, f_3 = f_7, f_4 = f_8 \tag{15}$$

The grids in the corner are involved with changes in boundary conditions, as shown in Figure 4. At the boundary corner point, there will be multiple directions without particle input due to the proximity of the model boundary. It cannot be calculated using the bounce-back scheme or macro variable boundary conditions. It is necessary to calculate the missing particle distribution according to the depth and velocity of the adjacent grids. The computational formula can be expressed as:

$$\begin{cases} f_1 = \frac{2h_{x+1,y+1}u_{x,y}}{3e} + f_5, \\ f_2 = f_6 + \frac{h_{x+1,y+1}u_{x,y}}{6e} + \frac{h_{x+1,y+1}v_{x,y}}{2e}, \\ f_3 = f_7, \\ f_4 = \frac{1}{2}(h_{x+1,y+1} - f_1 - f_2 - f_3 - f_5 - f_6 - f_7 - f_9), \\ f_8 = f_4. \end{cases} \tag{16}$$

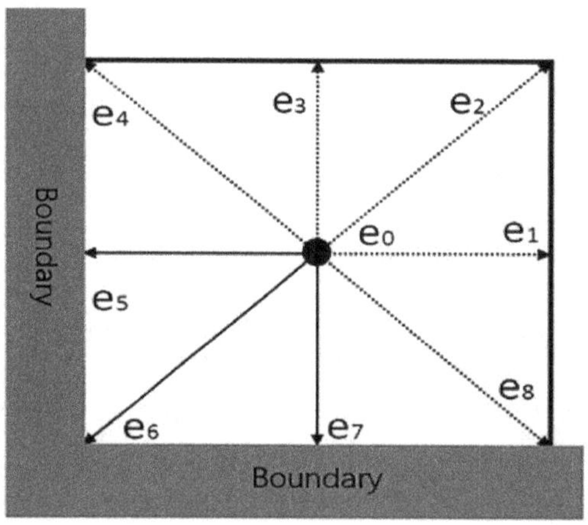

Figure 4. Corner grid.

3. Results

3.1. Initial Conditions

Based on data from 2017, the drainage of farmland converged towards the lake inlet and discharged into Baihe Lake at an average rate of 26.5 m^3/s from 11 May to 25 June, and at an average rate of 14.4 m^3/s from 10 October to 21 November. The average wind speed was equal to 1.78 m/s in the northeast (132°) from 11 May to 25 June, and average wind speed was equal to 2.36 m/s in the northeast (110°) from 10 October to 21 November.

Figure 5 shows the vegetation distribution obtained by a geographic information system and a field investigation. There are a large number of reeds in Baihe Lake that are higher than the water surface and have high rigidity. From inspection of Figure 5, it can be seen that the vegetation is mainly distributed in the northwest and southeast regions of the lake. Based on a field survey, it was found that the vegetation density varies significantly throughout the year. The density is highest from July to

August and lowest from January to March. In order to confirm the effect of drag force caused by the vegetation, two situations involving high- and low-vegetation density were tested in the present model. In the computation, the vegetation element shape factor and drag coefficient C_d are equal to 1.0 [28].

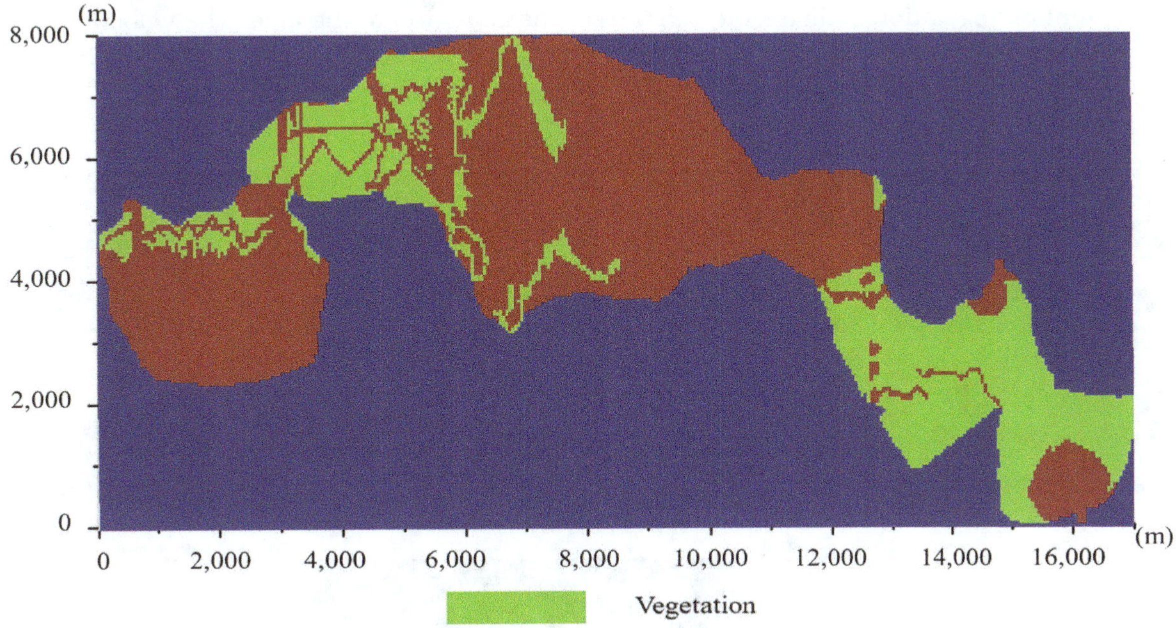

Figure 5. Vegetation distribution in Baihe Lake.

3.2. Numerical Tests

A 2D hydrodynamic model was established using the LBM. Baihe Lake is covered by 340 × 161 grids with each lattice being 50 m × 50 m. The inflow velocity from the northeast inlet is given by 0.5 m/s, which is controlled by sluice gate. The initial water depth in June is shown in Figure 6. Overall, the water depth in the western and eastern regions was relatively shallow compared to the middle region. The average water depth was 1.4 m, and the deepest area was 2.14 m.

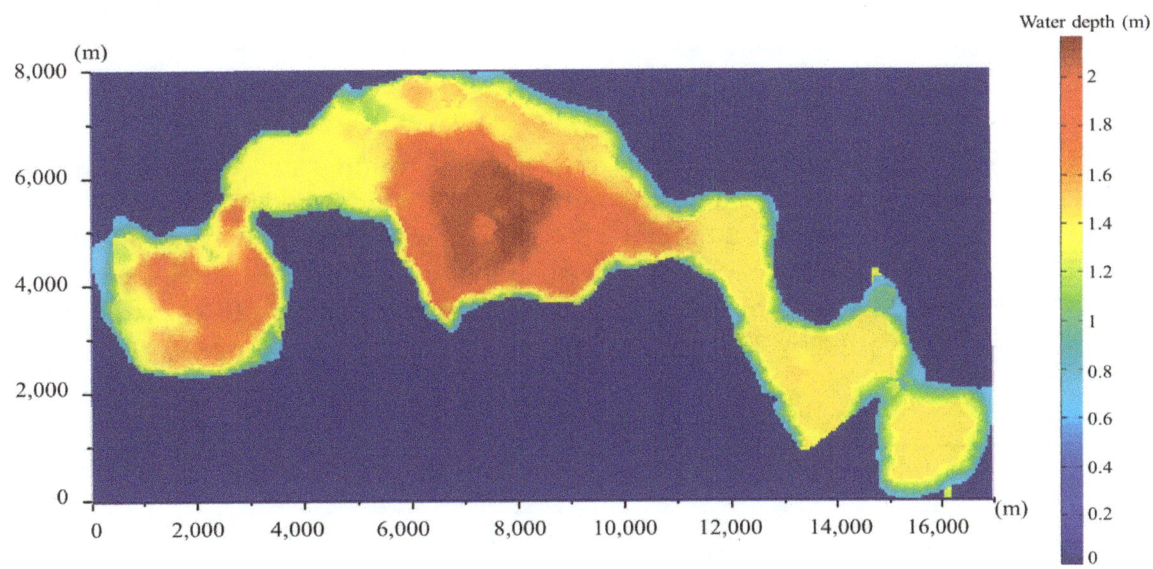

Figure 6. Initial water depth in Baihe Lake.

In order to discuss the hydrodynamic characteristics of the lake, the present study simulated the flow field with time from one day to five days. Figure 7 presents the flow field in Baihe Lake after five days for vegetation density $c = 0.2$. The western region of the lake is far away from the inlet, resulting in a very small flow velocity. The average velocity is about 0.0016 m/s, with an average water depth of 1.32 m. The middle region of the lake is the main area for local fishery. It has sparse vegetation and a larger water depth. The average velocity is about 0.36 cm/s, with an average water depth of 1.89 m. A large amount of vegetation is distributed in the southeast region of the lake. The velocity of the flow is small except for the areas near the outlet or the inlet where the velocities are about 26 cm/s and 0.063 m/s, respectively. The average velocity in the east area is about 0.32 cm/s.

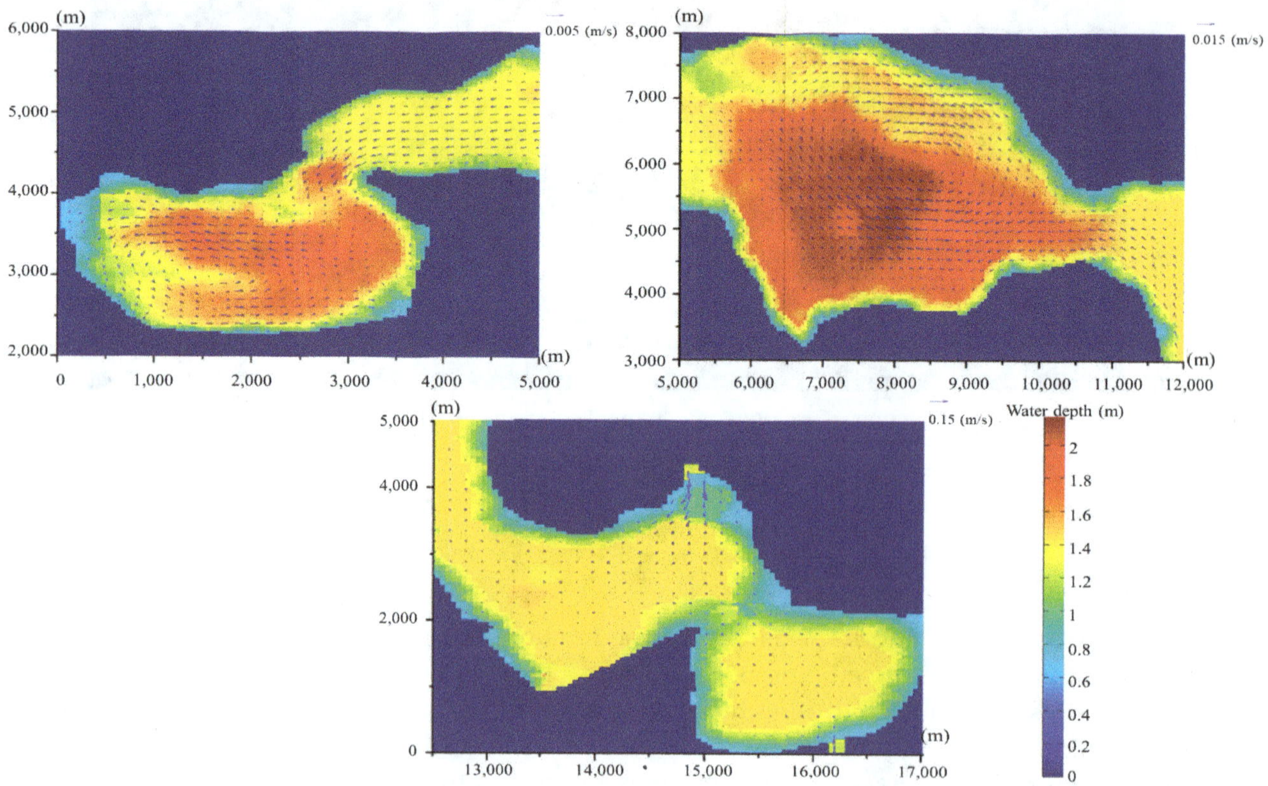

Figure 7. Water depth and velocity vectors after 5 days.

3.3. Sensitivity Analysis

3.3.1. Wind Speed

The wind speed in the Momoge Wetland varies all year round. As a result, it is necessary to investigate the flow field under different wind conditions. The wind speed in the northeast direction was varied by 10%, 20%, 30%, −10%, −20%, and −30%. The flow field became steady after the running time reached five days. The simulation results revealed that the greater the wind speed, the greater the velocity. Wind speed fluctuated by 30%, the outflow velocity varied by 18.77%, and the average velocity of the whole lake varied by 22.16% (see Table 1).

Table 1. Effect of wind speed on flow velocity.

Variation Range	Wind Speed (m/s)	Outflow Velocity (m/s)	Outflow Velocity Variation	Average Velocity (m/s)	Average Velocity Variation
−30%	1.25	0.0381	−5.93%	0.00149	−7.18%
−20%	1.42	0.0386	−4.69%	0.00152	−5.01%
−10%	1.60	0.0395	−2.47%	0.00155	−2.98%
0%	1.78	0.0405	0.00%	0.00160	0.00%
10%	1.96	0.0411	1.48%	0.00164	2.55%
20%	2.14	0.0433	6.91%	0.00172	7.81%
30%	2.31	0.0457	12.84%	0.00184	14.98%

3.3.2. Inflow Discharge

The inflow discharge is significantly affected by drainage from farmlands around the Momoge Wetland and varies across the seasons. The variation percentage of the inflow discharge was varied by 10%, 20%, 30%, −10%, −20%, and −30%. The velocity at the outlet and the flow field in Baihe Lake were the focus. The simulation results revealed a significant positive correlation between inflow and outflow velocity. Due to the strong drag effect of the reeds, a 30% fluctuation in inflow velocity could give rise to only up to 24.11% variation in outflow velocity. Average velocity in the whole lake varied from −21.99% to 27.65% (see Table 2).

Table 2. Effect of inflow velocity on flow velocity.

Variation Range	Inflow Discharge (m^3/s)	Outflow Velocity (m/s)	Outflow Velocity Variation	Average Velocity (m/s)	Average Velocity Variation
−30%	18.55	0.0338	−16.56%	0.00125	−21.99%
−20%	21.20	0.0358	−11.53%	0.00137	−14.58%
−10%	23.85	0.0370	−8.71%	0.00144	−10.24%
0%	26.50	0.0405	0.00%	0.00160	0.00%
10%	29.15	0.0436	7.66%	0.00178	11.42%
20%	31.80	0.0467	15.23%	0.00188	17.63%
30%	34.45	0.0503	24.11%	0.00204	27.65%

3.3.3. Vegetation Density

There is a large amount of aquatic vegetation distributed in Baihe Lake. Aquatic vegetation density becomes higher in summer but lower in autumn and winter. The simulation results revealed that there was a negative correlation between vegetation density and outflow velocity. The velocity with higher vegetation density was slower than that with lower density. In addition, a 30% density fluctuation would result in around −5% variation in outflow velocity, which indicates that the vegetation drag force can affect the water flow. Meanwhile, because of the uneven distribution of vegetation, more vegetation is distributed at the inlet and outlet. The average velocity in the whole lake varied from −3.72% to 2.97%, and was less than the variation at the outlet (see Table 3).

Table 3. Effect of vegetation density on flow velocity.

Variation Range	Vegetation Density	Outflow Velocity (m/s)	Outflow Velocity Variation	Average Velocity (m/s)	Average Velocity Variation
−30%	0.14	0.0424	4.71%	0.00165	2.97%
−20%	0.16	0.0417	3.03%	0.00164	2.57%
−10%	0.18	0.0411	1.49%	0.00162	1.09%
0%	0.20	0.0405	0.00%	0.00160	0.00%
10%	0.22	0.0399	−1.57%	0.00158	−1.23%
20%	0.24	0.0392	−3.11%	0.00155	−2.98%
30%	0.26	0.0385	−4.88%	0.00154	−3.72%

3.3.4. Rain Density

Due to the temperate continental climate, rainfall is less in autumn compared to summer. Therefore, the effect of rainfall on the flow field should be considered. The changes in rain density were 10%, 20%, 30%, −10%, −20%, and −30%. The flow field became steady when the simulation time reached five days. The simulation results revealed a positive correlation between rainfall density and outflow velocity. The outflow velocity varied from −14.23% to 21.11%, and the average velocity in the flow field varied from −14.17% to 20.44% (see Table 4).

Table 4. Effect of rainfall on flow velocity.

Variation Range	Rainfall (mm)	Outflow Velocity (m/s)	Outflow Velocity Variation	Average Velocity (m/s)	Average Velocity Variation
−30%	50.82	0.0347	−14.23%	0.00137	−14.17%
−20%	58.08	0.0370	−8.56%	0.00147	−8.34%
−10%	65.34	0.0385	−4.78%	0.00153	−4.66%
0%	72.60	0.0405	0.00%	0.00160	0.00%
10%	79.86	0.0433	6.97%	0.00171	6.85%
20%	87.12	0.0446	10.23%	0.00176	9.98%
30%	94.38	0.0490	21.11%	0.00193	20.44%

3.4. Scenario Simulation

Scenario simulations of different seasons were also run using data from July 2017 and September 2017.

In July 2017, the peak inflow discharge of the drainage from the surrounding farmland was 32 m^3/s. The monthly rainfall was 184 mm; the average wind speed was 2.13 m/s in the northeast direction (132°); and the vegetation density was 0.26. When the simulation was stable and the result was steady, the outflow rate was 0.54 m/s. The change in velocity and depth is shown in Figure 8a.

In September 2017, the inflow discharge was 12.6 m^3/s in the receding trough of the farmland drainage. The monthly rainfall was 38 mm; the average wind speed was 2.85 m/s in the northeast direction (110°); and the vegetation density was 0.12. The change in velocity and depth is shown in Figure 8b.

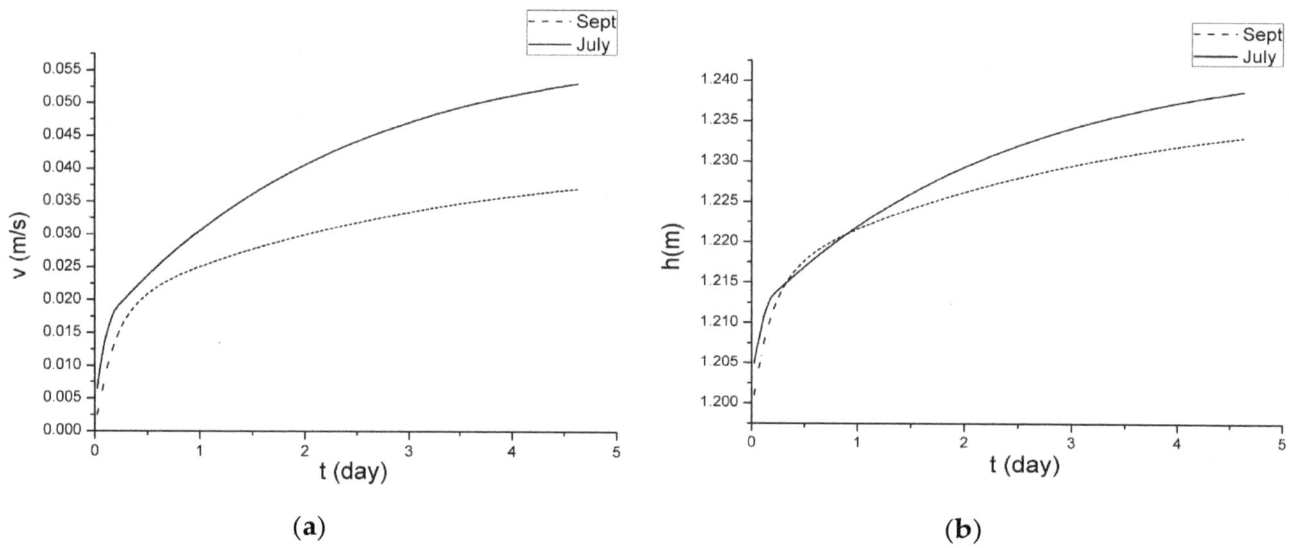

Figure 8. Comparison of the simulated results with different initial conditions: (**a**) Case 1 (water velocity); (**b**) Case 2 (water depth).

4. Discussion

In this paper, the LBM method is applied to Baihe Lake to reveal the temporal and spatial characteristics of water body. Overall, the water body of Baihe Lake suffered poor exchange, where the lowest flow rate based on simulation reached just 0.016 m/s. Simulations of different impact factors (e.g., wind speed, rainfall, vegetation density) were conducted here. The simulation results revealed that changes in inflow velocity, primarily as the result of water drainage from surrounding farmland, have the strongest effect. A 30% variation in the impact factors resulted in a 24.11% variation in the outflow rate at most. In addition, when the impact of wind speed was relatively small, the variation of the outflow rate reached just 18.87%. Relatedly, given the high distribution of aquatic vegetation in some areas of Baihe Lake, and that vegetation density varies with the season, the impact of vegetation on the flow field should be considered separately. The present model includes the change in vegetation density and investigates its effect on the flow fields. The simulation result revealed that vegetation density was negatively correlated with outflow rate. However, the influence of aquatic vegetation on flow distribution was relatively small, being only 9.59% when the vegetation density fluctuated 30%. On the whole, the outflow of Baihe Lake strongly depends on the drainage from upstream farmland, followed by the rainfall, and then by local wind speed. Aquatic vegetation can also have an impact. The inflow rate is the only factor that is easy to control. As a result, ecological protection and hydraulic engineering construction can be established upstream to control the water level and flow rate.

The monitored rainfall, wind speed, flowrate, and vegetation density data in different months was input into the present model in order to simulate and obtain the flow characteristics of Baihe Lake in July and September. Although the influence of wind speed and vegetation was relatively small in the single factor analysis, they can still have an impact on the flow as these two factors have potential uncertainty due to the change in climate. The monitored data revealed large drainage from the surrounding farmland in July, together with high rainfall. Consequently, based on the simulation results, the impact of wind speed and vegetation density was relatively small. In contrast, a reduction in rainfall with a similar farmland drainage happens in September. Importantly, this increases the influence of wind speed and vegetation density. The results indicated that the mean flow rate is much bigger in July than September, which also verifies the significance of different impact factors. Relatedly, the simulation results also demonstrate that the water exchange of Baihe Lake during the flood season from June to August is much better than it is in the dry season from September to November. Therefore, when developing an ecological restoration program, both the farmland drainage and the season should be emphasized.

5. Conclusions

Based on the lattice Boltzmann method (LBM), the present study has established a two-dimensional shallow water model that includes vegetation drag force. This paper for the first time puts forward a method to study the variation of flow field by establishing a numerical model of the Baihe Lake, which plays a guiding role in the construction of the ecological engineering of the Baihe Lake. The hydrodynamic model in this paper can be widely used in shallow lakes. Meanwhile, the drag force of aquatic vegetation to Baihe Lake is specially considered in the model, leading to that the model is most suitable for lakes with aquatic vegetation.

The model was used to explore the hydrodynamic impact of different influencing factors in Baihe Lake. Through a sensitivity test, drainage from surrounding farmland showed a dominant influence on the flow field, followed by rainfall, wind speed, and vegetation density. Furthermore, these factors may produce a noticeable variation in local flow characteristics in different seasons. As a result, the impact of these factors should be considered when developing an ecological restoration program. We should pay attention to different schemes corresponding to different seasons and control hydraulic construction to influence farmland receding water and improve lake fluidity.

Author Contributions: Conceptualization, H.L.; Data curation, Z.Z.; Formal analysis, H.L.; Investigation, Z.Z. and Q.L.; Methodology, H.L. and J.L.; Project administration, J.L.; Resources, J.L. and Q.L.; Visualization, Z.Z.; Writing–original draft, Z.Z.; Writing–review & editing, H.L.

References

1. Zeng, L.; Chen, G.Q. Flow distribution and environmental dispersivity in a tidal wetland channel of rectangular cross-section. *Commun. Nonlinear Sci.* **2009**, *17*, 4192–4209. [CrossRef]
2. Calhoun, A.J.; Mushet, D.M.; Bell, K.P.; Boix, D.; Fitzsimons, J.A.; Isselin-Nondedeu, F. Temporary wetlands: Challenges and solutions to conserving a 'disappearing' ecosystem. *Biol. Conserv.* **2017**, *211*, 3–11. [CrossRef]

3. Hui, F.; Xu, B.; Huang, H.; Yu, Q.; Gong, P. Modelling spatial-temporal change of Poyang Lake using multitemporal Landsat imagery. *Int. J. Remote. Sens.* **2008**, *29*, 5767–5784. [CrossRef]
4. Yang, R.; Cui, B. A wetland network design for water allocation based on environmental flow requirements. *Clean Soil Air Water* **2012**, *40*, 1047–1056. [CrossRef]
5. Stephan, U.; Gutknecht, D. Hydraulic resistance of submerged flexible vegetation. *J. Hydrol.* **2002**, *269*, 27–43. [CrossRef]
6. Kouwen, N. Friction factors for coniferous tress along rivers. *J. Hydraul. Eng.* **2000**, *126*, 732–740. [CrossRef]
7. Huai, W.X.; Zeng, Y.H.; Xu, Z.G. Three-layer model for vertical velocity distribution in open channel flow with submerged rigid vegetation. *Adv. Water. Resour.* **2009**, *32*, 487–492. [CrossRef]
8. Yang, Z.; Bai, F.; Huai, W.; An, R.; Wang, H. Modelling open-channel flow with rigid vegetation based on two-dimensional shallow water equations using the lattice Boltzmann method. *Ecol. Eng.* **2017**, *106*, 75–81. [CrossRef]
9. Yang, S.; Bai, Y.; Xu, H. Experimental analysis of river evolution with riparian vegetation. *Water* **2018**, *10*, 1500. [CrossRef]
10. Zhou, J.G. A lattice Boltzmann model for the shallow water equations. *Int. J. Mod. Phys. C* **2002**, *13*, 1135–1150. [CrossRef]
11. Liu, Q.; Qin, Y.; Zhang, Y.; Li, Z. A coupled 1D–2D hydrodynamic model for flood simulation in flood detention basin. *Nat. Hazards* **2015**, *75*, 1303–1325. [CrossRef]
12. Patro, S.; Chatterjee, C.; Mohanty, S.; Singh, R.; Raghuwanshi, N.S. Flood inundation modeling using MIKE FLOOD and remote sensing data. *J. Indian Soc. Remote Sens.* **2009**, *37*, 107–118. [CrossRef]
13. Arifin, R.R.; James, S.C.; de Alwis Pitts, D.A.; Hamlet, A.F.; Sharma, A.; Fernando, H.J. Simulating the thermal behavior in Lake Ontario using EFDC. *J. Great Lakes Res.* **2016**, *42*, 511–523. [CrossRef]
14. Bai, H.; Chen, Y.; Wang, D.; Zou, R.; Zhang, H.; Ye, R.; Ma, W.; Sun, Y. Developing an EFDC and numerical source-apportionment model for nitrogen and phosphorus contribution analysis in a Lake Basin. *Water* **2018**, *10*, 1315. [CrossRef]
15. Chen, S.; Doolen, G.D. Lattice Boltzmann method for fluid flows. *Ann. Rev. Fluid Mech.* **1998**, *30*, 329–364. [CrossRef]
16. Succi, S.; Santangelo, P.; Benzi, R. High-resolution lattice-gas simulation of two-dimensional turbulence. *Phys. Rev. Lett.* **1988**, *60*, 2738–2740. [CrossRef] [PubMed]
17. Salmon, R. The lattice Boltzmann method as a basis for ocean circulation modeling. *J. Mar. Res.* **1999**, *57*, 503–535. [CrossRef]
18. Dellar, P.J. Non-hydrodynamic modes and a priori construction of shallow water lattice Boltzmann equations. *Phys. Rev. E* **2002**, *65*, 036309. [CrossRef]
19. Liu, H.; Ding, Y.; Wang, H.; Zhang, J. Lattice Boltzmann method for the age concentration equation in shallow water. *J. Comput. Phys.* **2015**, *299*, 613–629. [CrossRef]
20. Liu, H.; Zhou, J.G. Inlet and outlet boundary conditions for the Lattice-Boltzmann modelling of shallow water flows. *Prog. Comput. Fluid. Dyn.* **2012**, *12*, 11–18. [CrossRef]
21. Ottolenghi, L.; Prestininzi, P.; Montessori, A.; Adduce, C.; La Rocca, M. Lattice Boltzmann simulations of gravity currents. *Eur. J. Mech. B-Fluid* **2018**, *67*, 125–136. [CrossRef]
22. O'Brien, G.S.; Bean, C.J.; Mcdermott, F. A comparison of published experimental data with a coupled lattice Boltzmann-analytic advection–diffusion method for reactive transport in porous media. *J. Hydrol.* **2002**, *268*, 143–157. [CrossRef]
23. Zhou, J.G. A rectangular lattice Boltzmann method for groundwater flows. *Mod. Phys. Lett. B* **2007**, *21*, 531–542. [CrossRef]
24. Tubbs, K.R.; Tsai, T.C. Multilayer shallow water flow using lattice Boltzmann method with high performance computing. *Adv. Water. Resour.* **2009**, *32*, 1767–1776. [CrossRef]
25. Liu, H.; Li, M.; Shu, A. Large eddy simulation of turbulent shallow water flows using multi-relaxation-time lattice Boltzmann model. *Int. J. Numer. Meth. Fluids* **2012**, *70*, 1573–1589. [CrossRef]
26. Prestininzi, P.; Montessori, A.; Rocca, M.; Sciortino, G. Simulation of arrested salt wedges with a multi-layer Shallow Water Lattice Boltzmann model. *Adv. Water Resour.* **2016**, *96*, 282–289. [CrossRef]
27. Chávarri, E.; Crave, A.; Bonnet, M.P.; Mejía, A.; Da Silva, J.S.; Guyot, J.L. Hydrodynamic modelling of the Amazon River: Factors of uncertainty. *J. S. Am. Earth. Sci.* **2013**, *44*, 94–103. [CrossRef]

28. Stone, B.M.; Shen, H.T. Hydraulic resistance of flow in channels with cylindrical roughness. *J. Hydraul. Eng.* **2002**, *128*, 500–506. [CrossRef]
29. Konings, A.G.; Katul, G.G.; Thompson, S.E. A phenomenological model for the flow resistance over submerged vegetation. *Water. Resour. Res.* **2012**, *48*, 2478. [CrossRef]
30. Poggi, D.; Krug, C.; Katul, G.G. Hydraulic resistance of submerged rigid vegetation derived from first-order closure models. *Water. Resour. Res.* **2009**, *45*, 2381–2386. [CrossRef]
31. Guan, M.; Liang, Q. A two-dimensional hydro-morphological model for river hydraulics and morphology with vegetation. *Environ. Modell. Softw.* **2017**, *88*, 10–21. [CrossRef]
32. Ding, Y.; Liu, H.; Peng, Y.; Xing, L. Lattice Boltzmann method for rain-induced overland flow. *J. Hydrol.* **2018**, *562*, 789–795. [CrossRef]

6

Spatial Distribution Characteristics of Rainfall for Two-Jet Collisions in Air

Hao Yuan, Weilin Xu *, Rui Li, Yanzhang Feng and Yafeng Hao

State Key Laboratory of Hydraulics and Mountain River Engineering, Sichuan University, Chengdu 610065, China; 18996152721@163.com (H.Y.); li.rui.sai@gmail.com (R.L.); 13835141175@163.com (Y.F.); 13258387216@163.com (Y.H.)

* Correspondence: xuwl@scu.edu.cn

Abstract: Many researchers have studied the energy dissipation characteristics of two-jet collisions in air, but few have studied the related spatial rainfall distribution characteristics. In this paper, in combination with a model experiment and theoretical study, the spatial distributions of rainfall intensity of two-jet collisions, with different collision angles and flow ratios, are systematically studied. The experimental results indicated that a larger collision angle corresponds to a larger rainfall intensity distribution. The dimensionless maximum rainfall intensity sharply decreased with the flow ratio, while the maximum rainfall intensity slightly increased when the flow ratio was greater than 1.0. A theoretical equation to compute the location of maximum rainfall intensity is presented. The range of rainfall intensity distribution sharply increased with the flow ratio. When the flow ratio was greater than 1.0, the range of longitudinal distribution slightly increased, whereas the lateral distribution remained unchanged or slowly decreased. A formula to calculate the boundary lines of the x-axis is proposed.

Keywords: rainfall intensity distribution; two water jets; collision in air; Gaussian distribution; trajectory line

---

1. Introduction

At present, the high dam projects that have been built, are under construction, or are planned in China have the common features of large drops, a narrow river valley, and a large flood discharge flow. The flood discharge energy dissipation mode of the dam body has been mostly adopted by two-jet collisions and downstream water reservoirs, such as Ertan (Figure 1), Xiaowan, Xiluodu, Goupitan, and the Jinping grade I project. Two-jet collisions in air change the motion track of each of the two jets, significantly disperse the water flow, effectively alleviate the dynamic pressure impact of the water flow on the bottom plate of the water cushion reservoir, and have a remarkable energy dissipation effect [1–7]. However, compared with the traditional energy dissipation method, two-jet collisions in air appear to be a serious problem with regard to flood discharge atomization in the existing engineering examples, which has caused great damage to the normal operation, traffic safety, surrounding environment, and even the stability of the downstream bank slopes. It is necessary to study the spatial distribution characteristics of rainfall intensity in two-jet collisions in air, in order to improve the prediction accuracy of the discharge atomization rain strength and influence range, as well as make a relevant protection classification and protection design scheme in the design planning [8–10].

Figure 1. Simultaneous discharge of the surface and the deep hole of the Ertan hydropower station in 2010.

Many former researchers who have studied two-jet collisions in air have focused on the energy dissipation effect of collisions. Xiong [11] introduced the main factors that affect energy dissipation during an in-air collision between the surface and the deep hole. Guo [12] theoretically analyzed the effect of the collision angle on energy dissipation. Using the momentum integral equation of fluid mechanics, Liu [13] derived the related formula of energy dissipation for the collision of two water jets in air in detail, and introduced the concept and calculation method for the energy dissipation rate. Diao [14] studied the relationship between the collision angle, the flow ratio, and the collision energy dissipation effect through model testing, and proposed that the main function of jet collisions is to disperse the water flow. Sun [15] analyzed hydraulic characteristics, such as the energy loss of up- and down-collisions, three-dimensional (3-D) diffusion and leakage collision of the surface, and deep orifice slugs. The 3-D collision velocity, collision efficiency, and collision energy loss were derived using the momentum equation and variation law of the air diffusion width of the water tongue, and the optimal hydraulic conditions for collision were obtained. Sun [16] applied the turbulence jet theory and flow momentum equation, in order to introduce the calculation formula of the collision velocity vector and collision energy dissipation efficiency of the combined water tongue when it collides with the left and right sides of the air.

At present, the research on flood discharge atomization is mainly about ski-jump atomization. Liang [17] proposed a calculation model of atomized water flow, and obtained various calculation formulas and methods in the field of atomized water flow influence. Liu [18,19] studies the diffusion law, velocity distribution, and energy loss of a water jet. According to the prototype observation data of the flood discharge atomization and some experimental data of the model, based on the comprehensive analysis of various factors that affect the atomization range, Li [20] discusses the rough estimation method of the range of the energy dissipation atomization precipitation area. Liu [21–24] introduces several key problems with regard to the shape and the numerical simulation of atomized flows, such as the jet calculation of the water tongue, collision between the water tongue and the water surface, and calculation of the fog source quantity. By collecting, inducing, and summarizing the partial flood discharge atomization of engineering prototype observation data, Sun [25] found the longitudinal boundary flood discharge atomization average discharge flow, and a good relationship between the water flow velocity and the water entry angle of the water tongue. Based on the dimensional analysis, a method was established to estimate the rainfall intensity, flood discharge atomization, and longitudinal boundary experience relationship. Liu [26] proposed an atomization forecast model based on artificial neural networks. The Monte Carlo method was applied to add the effect of environmental, wind, and topographic factors to the atomization mathematical model of overpass discharge by Lian [27]. Sun [28] revised the resistance coefficient of particles, using the research results of raindrop movement, and obtained the resistance coefficient of the splashing water droplet movement. Then the effect of the diameter of a splashing water droplet on the splashing length is preliminarily discussed, which deepens the understanding of the splashing movement. Using the generalized model test in 2013 by Wang [29]—under different hydraulic conditions, to select a tongue that fell into the water downstream of an atomized water source area of rainfall intensity—the flood discharge atomization fog source area of the plane distribution features was measured and analyzed, to determine the reasons for the formation of different areas around the atomization source and the plane distribution of rainfall intensity. Lian [30] proposes a method to predict the spray atomization of spillage by combining a physical model and theoretical analysis. Zhang [31] experimentally studied the motion track, distribution range, water point shape, and velocity of single-strand jets under different air dosages. Although their methods are accurate in predicting the atomization of a ski-jump flow, they are not good at predicting the atomization of two-jet collisions, which add a collision zone and change the trajectory of two jets. In rocket propulsion, extensive experiments—mostly based on physical model investigations—have been conducted on two-jet collisions, including the liquid membrane [32–35], droplet size, and velocity distribution [36–41].

Most domestic and foreign scholars have focused on the atomization of flood discharge by single-strand pick flows. However, compared with the single-strand pick flow, two-jet collisions in air increase the area of collision atomization, where both range and intensity of the atomization increase. For example, Ertan applied this method to flood discharge and cause landslide by rainfall. Therefore, in this paper, through theoretical analysis and a physical model experiment, we studied the distribution characteristics of rainfall intensity after the collision of two jets, the trajectory lines of the water tongue, and the range of the x-axis after collision for different flow ratios and collision angles.

2. Model Design and Experiment

The experiment was conducted in the State Key Laboratory of Hydrodynamics and Mountain River Engineering in Sichuan University. Figure 2 shows the schematic diagram of the model test. The water flux was supplied by a large rectangular tank (width of 4.0 m, length of 4.0 m, and height of 5.0 m). When laying the rainfall receiving platform, we considered the symmetry of the surface and deep hole, and only one side was arranged. Water was fed into the iron box through the pump behind the iron box, and the water level of the iron box was maintained through the valve in order to achieve a stable deep and surface flow.

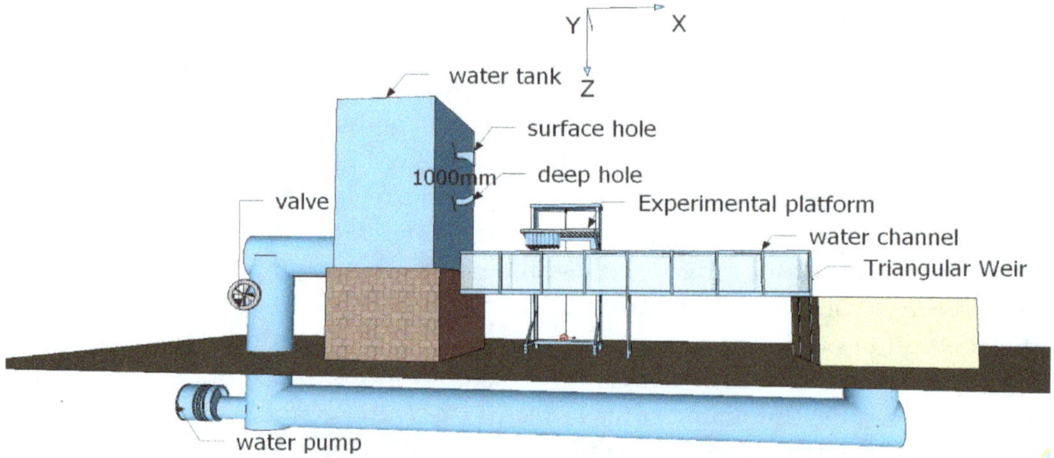

(A)

Figure 2. *Cont.*

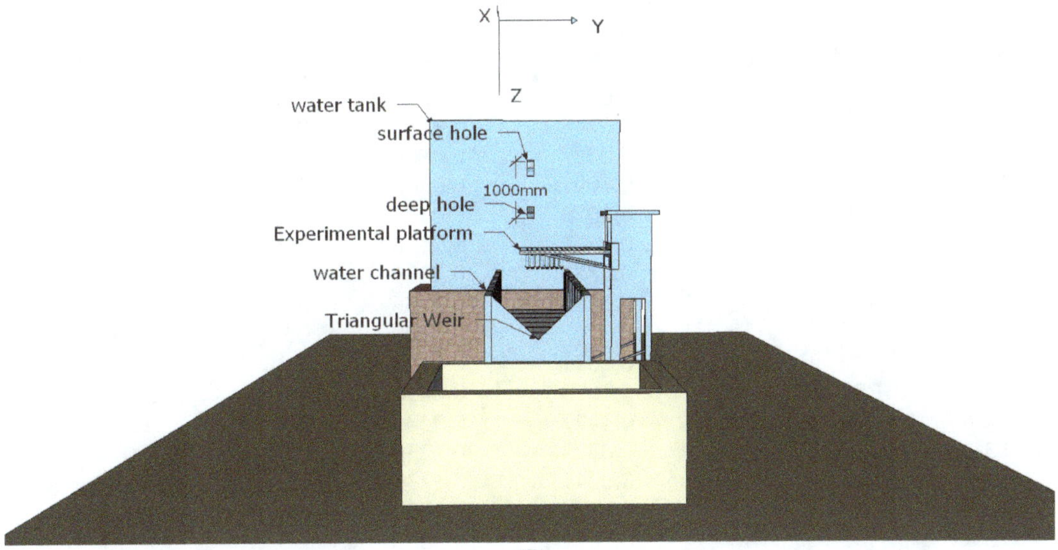

(B)

Figure 2. (**A**) Experiment layout: side view; (**B**) experiment layout: front view.

In order to be consistent with the prototype which adopts two-jet collisions in air, this experiment used a constant surface hole angle θ_1 (a pitch-down angle of $-30°$) and adjusted the deep hole pick angle θ_2 ($0°$~$45°$) to change the collision angle β. The width of the surface deep hole was identical, the height of the deep hole could be adjusted, and the range was 0~5 cm. The maximum single-width flux of the model table hole was 0.1236 m^2/s, the maximum single-width flux of the deep hole was 0.1431 m^2/s, and the flux was measured by a triangular thin-walled weir ($\pm1\%$). Four horizontal planes, under the collision points of 36 cm, 56 cm, 76 cm, and 96 cm, were set to measure the rainfall intensity. The platform was supported by a slide rail system, and the accuracy in the vertical platform position was less than 0.5 cm. Table 1 lists 18 working conditions in the experiment, including almost 30,000 measurement points. To obtain the four different sections, we first used image processing to find the collision point, and then adjusted the height of the measuring platform. Detailed visual observations of the collision points were conducted using a Canon EOS80D DSLR camera (Cannon, Chengdu, China).

Rainwater was collected in the horizontal plane (xy plane in Figure 3) at the different elevations below the collision point and 0.15 m above the faceplate. The rainfall collection platform included the movable base, support, and rainfall collection panel, which was made of an organic glass panel and a PVC pipe. In each plane, up to 121 PVC pipes (each with a diameter of 2.0 cm) were used simultaneously to collect rain within a period varying from 60 s to 1000 s, depending on the rain intensity in that region, which was found to be sufficient to ensure stable results. Zhang and Zhu [31] indicated that the measurement error could be ignored with different pipe diameters. In their test, the bottles (25 mL with a bottle-neck diameter of 1.52 cm) were tested to give the same (3% difference) result of rain intensity distribution as those of much larger plastic bottles (250 mL with a bottle-neck diameter of 3.26 cm). To avoid the impact of water splashing on the back of the faceplate, the faceplate was hollow, except for the holes and supporting frames. The lower end of the short pipe, whose range was 15~3000 mL, used a measuring cylinder to collect rainfall. Rainfall intensity varies from point to point, so different cylinders were used. To obtain the whole region's rainfall intensity, the platform was first placed in one region to collect rainwater, then the entire platform was moved 1.2 m in the longitudinal direction to the next region. To ensure measurement accuracy, two measurements were made in the same working conditions, and the average was taken.

Table 1. Working conditions in experiments.

Run	θ_1 (°)	θ_2 (°)	q_m (m²/s)	q_c (m²/s)	f	$Q \times 2/(q_m + q_c)$
T1-1	−30	0	0.0373	0.1236	0.3	103%
T1-2	−30	0	0.0745	0.1236	0.6	95%
T1-3	−30	0	0.0727	0.0800	0.9	102%
T1-4	−30	0	0.1091	0.0800	1.3	104%
T1-5	−30	0	0.1073	0.0436	2.5	98%
T1-6	−30	0	0.1427	0.0436	3.2	104%
T2-1	−30	30	0.0373	0.1236	0.3	95%
T2-2	−30	30	0.0745	0.1236	0.6	110%
T2-3	−30	30	0.0727	0.0800	0.9	102%
T2-4	−30	30	0.1091	0.0800	1.3	105%
T2-5	−30	30	0.1073	0.0436	2.5	103%
T2-6	−30	30	0.1427	0.0436	3.2	92%
T3-1	−30	45	0.0373	0.1236	0.3	111%
T3-2	−30	45	0.0745	0.1236	0.6	94%
T3-3	−30	45	0.0727	0.0800	0.9	103%
T3-4	−30	45	0.1091	0.0800	1.3	106%
T3-5	−30	45	0.1073	0.0436	2.5	108%
T3-6	−30	45	0.1427	0.0436	3.2	111%

q_m is the flow discharge per unit width of deep hole. q_c is the flow discharge per unit width of surface hole. f is the ratio of deep hole flow to surface hole flow, that is $f = q_m/q_c$. Q is the integral of measurement value.

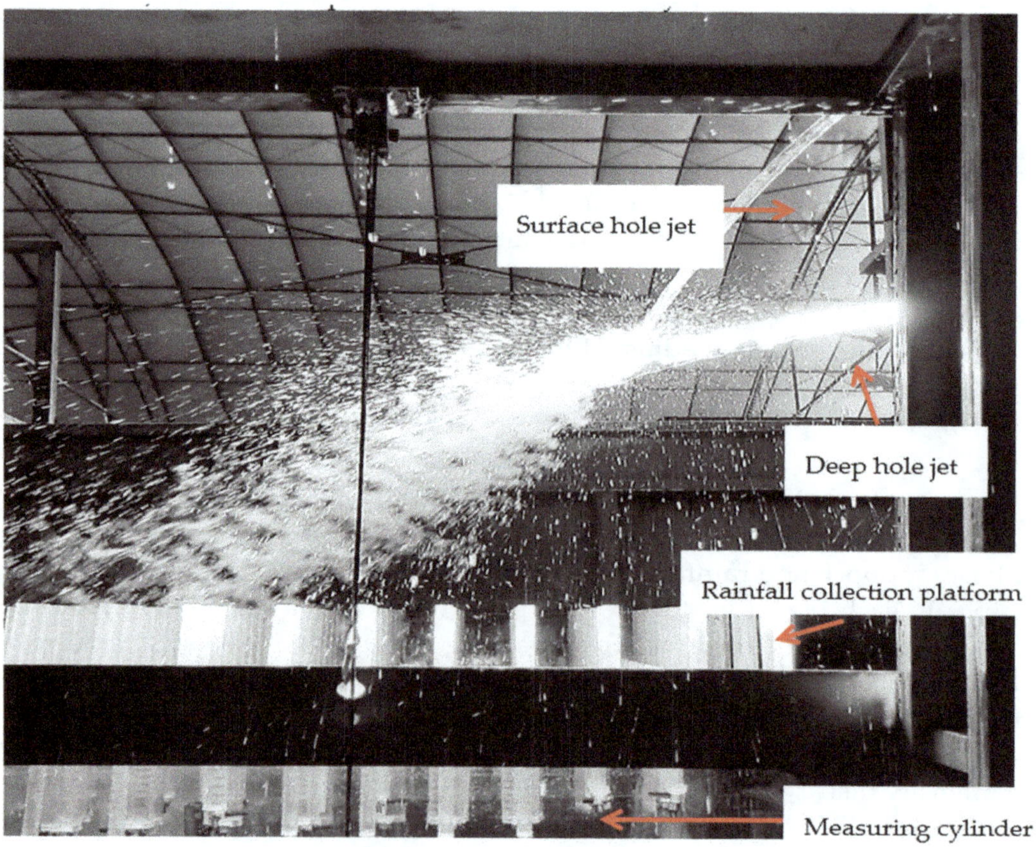

Figure 3. Rainfall collection platform in 2017.

In each test, the total rain flux in the horizontal plane was integrated and compared to the total water flux of the surface and the deep holes. The total water/rain flux Q was calculated from $Q = \int_{-\infty}^{+\infty}\int_{-\infty}^{+\infty} I dx dy$, where I was the measured rain intensity. Comparisons of the integrated values with the water flux at the surface and the deep holes were listed in Table 1. For all tests, the conservation ratio of rain flux on average was 101.4 ± 7.0%. In runs T2-6 and T3-6, the loss of rain flux was slight larger (±9.3%), which may be related to the fact that the rain had a wider range in the test. Hence, the above comparison showed the reliability of the measurements.

3. Experimental Results and Discussion

3.1. Spatial Distribution Characteristics of Rainfall Intensity

Figures 4–6 show the distribution of rainfall intensity for several flow ratios, collision angles, and heights of the rainfall intensity collection platform from the collision point. The rainfall intensity boundary defined 5% of the maximum rainfall intensity under each working condition [31].

In Figure 4, the flow ratios vary, and the height and collision angles from the collision point to the measurement platform are constant. Note that on the section with the same vertical distance from the collision point, the rainfall intensity range showed an increasing trend on the y-axis when the flow ratio increased. On the x-axis, the rainfall intensity range increased first, and subsequently decreased slowly. For f = 0.3, the rainfall intensity range was concentrated in the range 0 cm $\leq x \leq$ 22 cm; for f = 0.9, the rainfall intensity range was concentrated in the range of 10 cm $\leq x \leq$ 100 cm; while when f = 3.2, the rainfall intensity range was concentrated in the range of 40 cm $\leq x \leq$ 115 cm. Under all of the working conditions, the maximum rainfall intensity was on the x-axis.

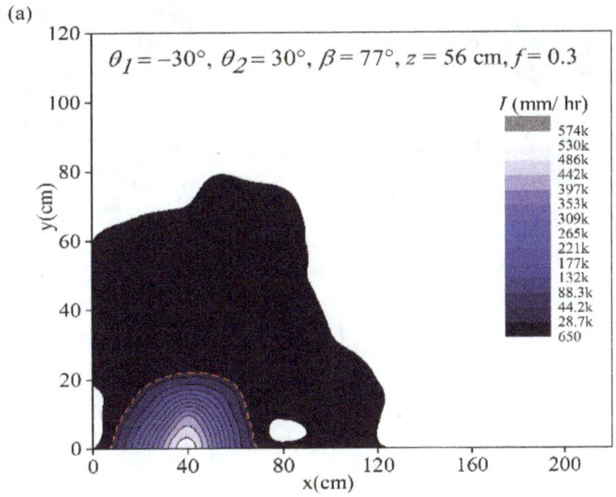

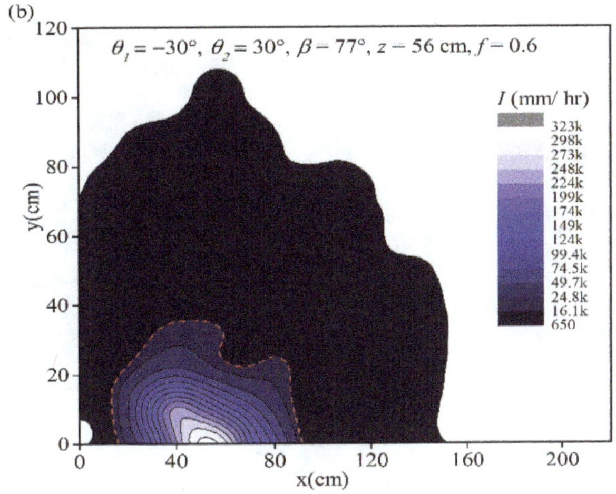

Figure 4. *Cont.*

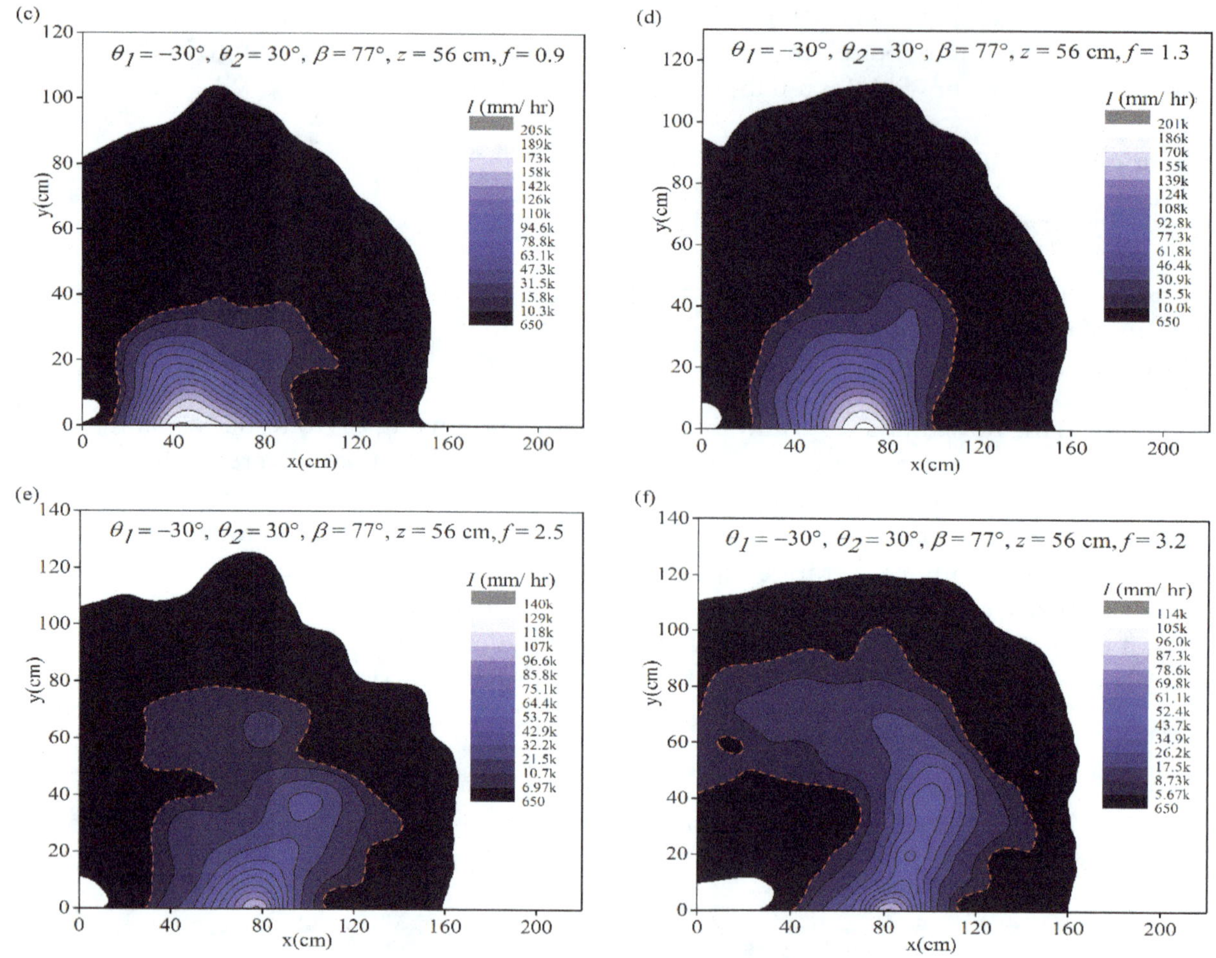

Figure 4. Rainfall intensity distribution at different flow ratios on the *x-o-y* measuring platform. (**a**) $\theta_1 = -30°$, $\theta_2 = 30°$, $\beta = 77°$, $z = 56$ cm, and $f = 0.3$; (**b**) $\theta_1 = -30°$, $\theta_2 = 30°$, $\beta = 77°$, $z = 56$ cm, and $f = 0.6$; (**c**) $\theta_1 = -30°$, $\theta_2 = 30°$, $\beta = 77°$, $z = 56$ cm, and $f = 0.9$; (**d**) $\theta_1 = -30°$, $\theta_2 = 30°$, $\beta = 77°$, $z = 56$ cm, and $f = 1.3$; (**e**) $\theta_1 = -30°$, $\theta_2 = 30°$, $\beta = 77°$, $z = 56$ cm, and $f = 2.5$; (**f**) $\theta_1 = -30°$, $\theta_2 = 30°$, $\beta = 77°$, $z = 56$ cm, and $f = 3.2$.

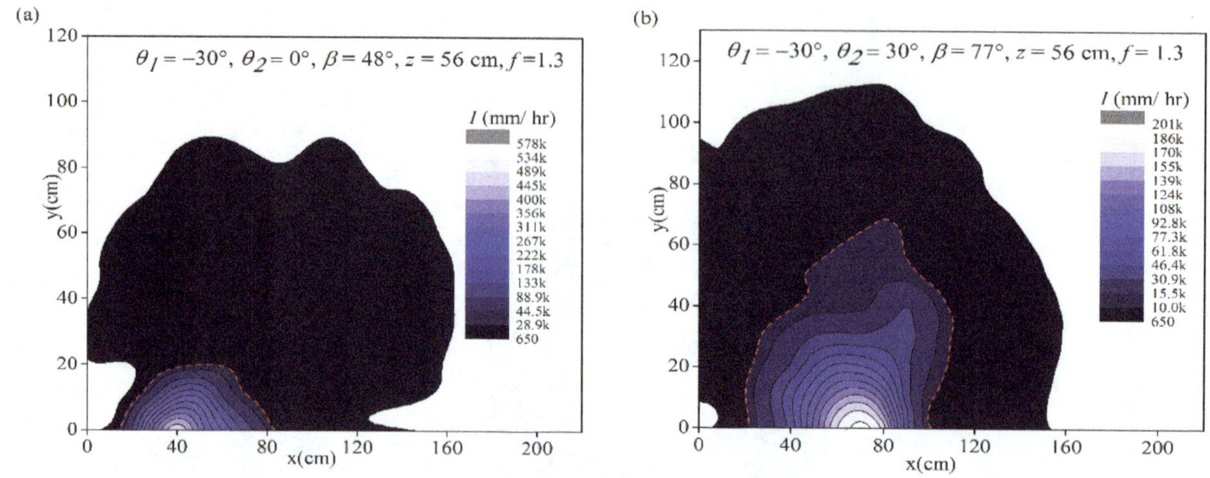

Figure 5. *Cont.*

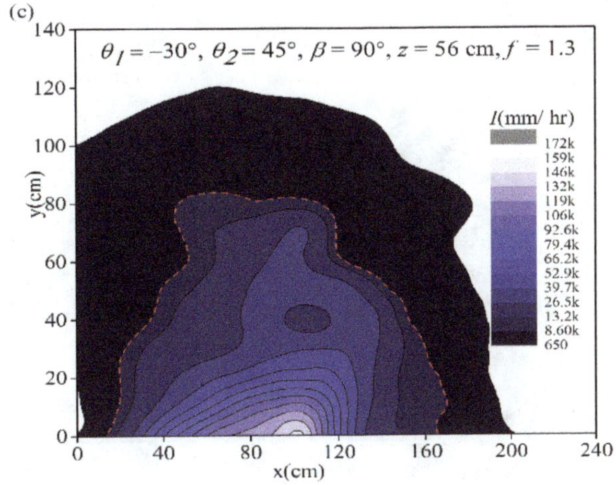

Figure 5. Rainfall intensity distribution at different collision angles on the $x\bar{o}\bar{y}$ measuring platform. (**a**) $\theta_1 = -30°$, $\theta_2 = 0°$, $\beta = 48°$, $z = 56$ cm, and $f = 1.3$; (**b**) $\theta_1 = -30°$, $\theta_2 = 30°$, $\beta = 77°$, $z = 56$ cm, and $f = 1.3$; (**c**) $\theta_1 = -30°$, $\theta_2 = 45°$, $\beta = 90°$, $z = 56$ cm, and $f = 1.3$.

In Figure 5, the flow ratios and height from the collision point to the measurement platform are constant, and the collision angles vary. Note that with the increase in the collision angle, the rainfall intensity range on the x- and y-axes increased—for example, when $\beta = 48°$, $L_x = 70$ cm, and $L_y = 20$ cm; when $\beta = 77°$, $L_x = 108$ cm, and $L_y = 68$ m; when $\beta = 90°$, $L_x = 163$ cm, and $L_y = 80$ cm. However, the maximum rainfall intensity decreased with the increased collision angle—for example, when $\beta = 48°$, $I_{max} = 5.78 \times 10^5$ mm/h; and for $\beta = 90°$, $I_{max} = 5.78 \times 10^5$ mm/h.

In Figure 6, the flow ratio and collision angle were constant, and the collision point varied from the height of the measurement platform. Note that (1) the range of rainfall intensity on the x- and y-axes gradually increases along the z direction, and (2) the maximum rainfall intensity value continues decreasing. At $z = 56$ cm, the maximum value is 5.74×10^5 mm/h; at $z = 76$ cm, the maximum value is 5.32×10^5 mm/h; and at $z = 96$ cm, the maximum value is 4.22×10^5 mm/h. The maximum value decreases from 5.74×10^5 mm/h to 4.22×10^5 mm/h, which indicates that the maximum rainfall was continuously spreading during the falling process.

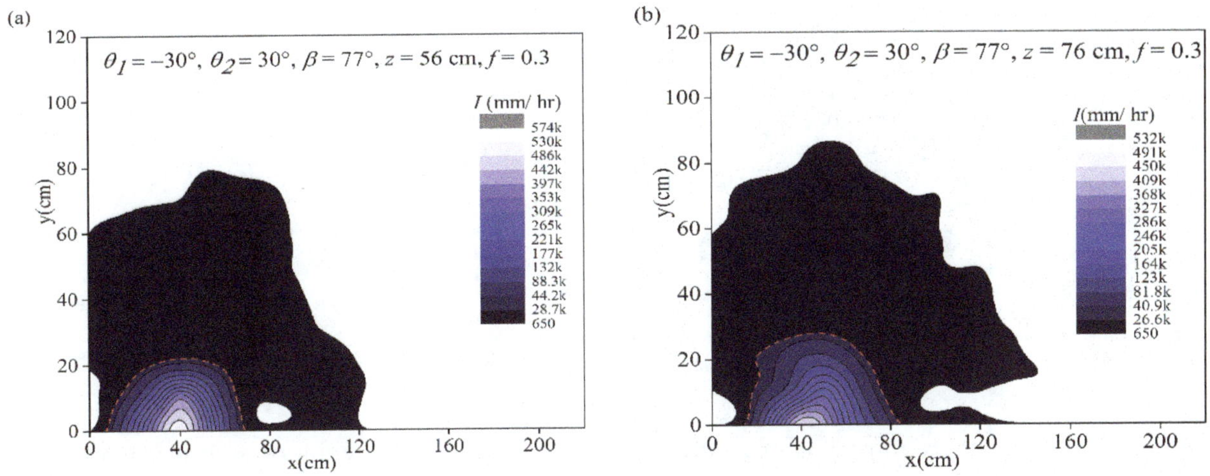

Figure 6. *Cont.*

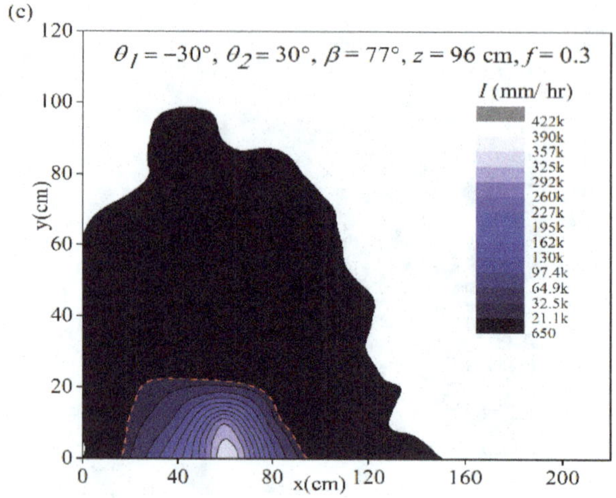

Figure 6. Rainfall intensity distribution at different section heights on the *x-o-y* measuring platform. (**a**) $\theta_1 = -30°$, $\theta_2 = 30°$, $\beta = 77°$, $z = 56$ cm, and $f = 0.3$; (**b**) $\theta_1 = -30°$, $\theta_2 = 30°$, $\beta = 77°$, $z = 76$ cm, and $f = 0.3$; (**c**) $\theta_1 = -30°$, $\theta_2 = 30°$, $\beta = 77°$, $z = 96$ cm, and $f = 0.3$.

The maximum rainfall of each section (q_{max}) is shown in Figure 7. Note that (1) the maximum rainfall intensity decreases with an increase in distance of z, (2) q_{max} sharply decreases for $f < 1.4$ and slightly increases for $f > 1.4$, and (3) the results indicated that q_{max} decreases with an increase of β. In Figure 7a,c, for $f = 1.3$ and $z = 96$ cm, $q_{max}/Q_t \times 10^3 = 1.6$ at $\beta = 48°$, while $q_{max}/Q_t \times 10^3$ decreased to 0.45 at $\beta = 90°$, where Q_t is the sum flow discharge of surface hole and deep hole outlets.

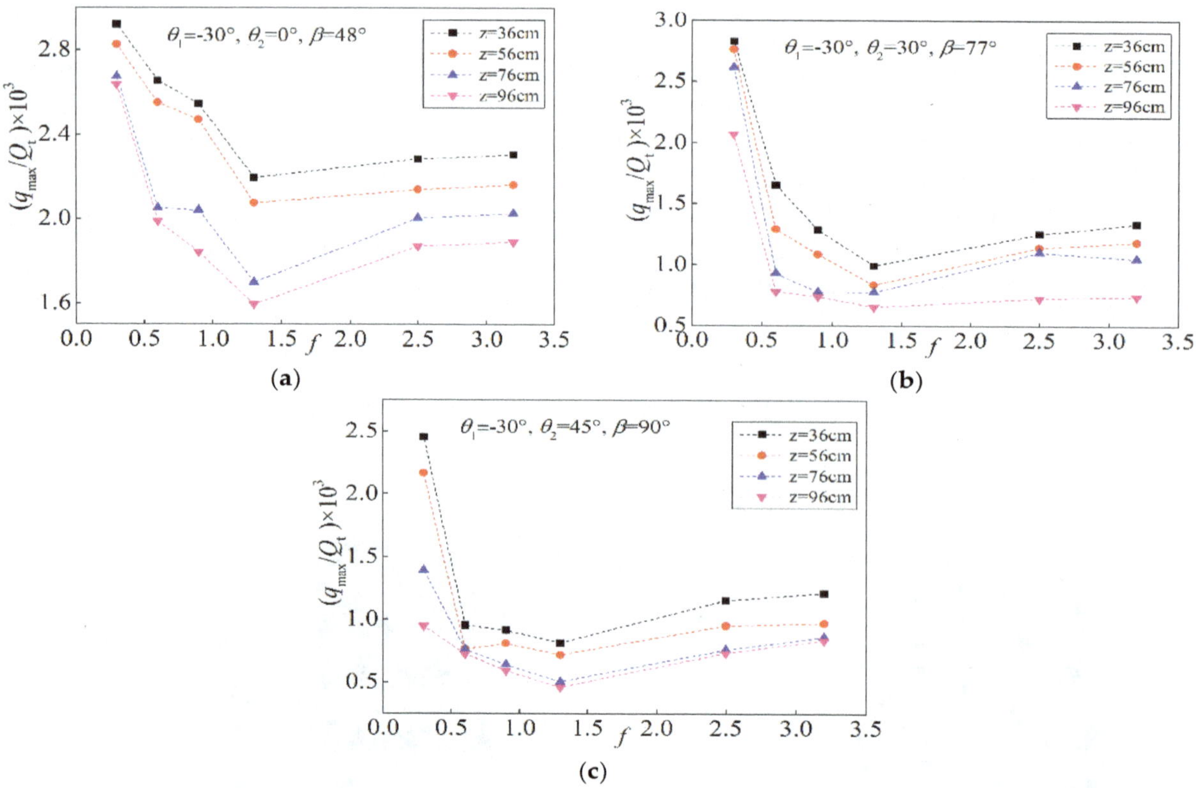

Figure 7. The maximum rainfall varies with the collision angles and flow ratios at different sections. (**a**) $\theta_1 = -30°$, $\theta_2 = 0°$, and $\beta = 48°$; (**b**) $\theta_1 = -30°$, $\theta_2 = 30°$, and $\beta = 77°$; (**c**) $\theta_1 = -30°$, $\theta_2 = 45°$, and $\beta = 90°$.

3.2. Rainfall Intensity Distribution on the x-Axis

The rainfall intensity distributions on the x-axis are shown in Figure 8. Note that the rainfall intensity data along the x-axis was best correlated with the Gaussian distribution. In addition, the maximum rainfall intensity value gradually decreases with an increase along the z-axis. For example, in Figure 8a, the height of the section increases from 36 cm to 96 cm, and the maximum rainfall intensity value decreases from 2.48×10^5 mm/h to 1.43×10^5 mm/h. With an increase in section height, the rainfall intensity distribution along the x-axis flattens, and the range on the x-axis is bigger.

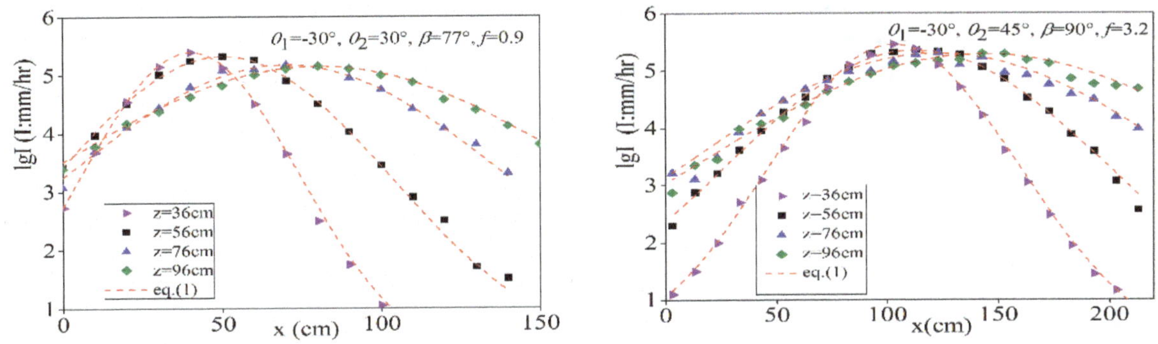

(a) (b)

Figure 8. Rainfall intensity distributions along the x-axis and comparison with Equations (1) and (2). (**a**) $\theta_1 = -30°$, $\theta_2 = 30°$, $\beta = 77°$, and $f = 0.9$; (**b**) $\theta_1 = -30°$, $\theta_2 = 45°$, $\beta = 90°$, and $f = 3.2$.

For two-jet collisions in air on the x- and z-axis, the rainfall intensity distribution along the x-axis after the collision can be expressed by the following formulas (Figure 8):

$$\lg I = \lg I_{\max} \times e^{\left(-a \times \left(\frac{x - x_{\max}}{z}\right)^2\right)} \quad (1)$$

$$\alpha = 1.7e^{(-2.1 \times f)} + 2\cos\beta \quad (2)$$

where I is the rainfall intensity and $I_{\max}$ is the maximum rainfall intensity, α is a coefficient.

3.3. Trajectory Line after the Collision on the x-Axis

As shown in Figure 9, based on previous research on collision flow [9], the surface water tongue and deep water tongue completely collide in air; then, two water tongues combine and shoot in one direction. V_1 is the surface hole water tongue velocity, θ_1 is the angle of depression, and q_1 is the discharge per unit width; in addition, V_1 is the deep hole water tongue velocity, θ_2 is the pick angle, and q_2 is the discharge per unit width. Two water jets meet at the unit of M. At the confluence, the flow velocity of the upper edge of the water tongue at the confluence is V_{1M}; the angle between the upper edge of the water tongue and the horizontal direction is β_1; the flow velocity of the inner edge of the water tongue is V_{2M}; the angle between the inner edge of the water tongue and the horizontal direction is β_2; the flow velocity of the mixed water tongue after the phase confluence is V_M, and the angle between the water tongue and the horizontal direction is β_M.

The spatial position of collision point M can be obtained by calculating the trajectories of the inner edge of the surface-hole jet and outer edge of the deep-hole jet. The trajectories of the deep and surface water tongues were obtained from Equation (3).

$$z(x) = z_0 + \tan(\alpha)x - \frac{gx^2}{2V_0^2\cos^2(\alpha)} \quad (3)$$

where V_0 is the initial speed, g is gravity acceleration.

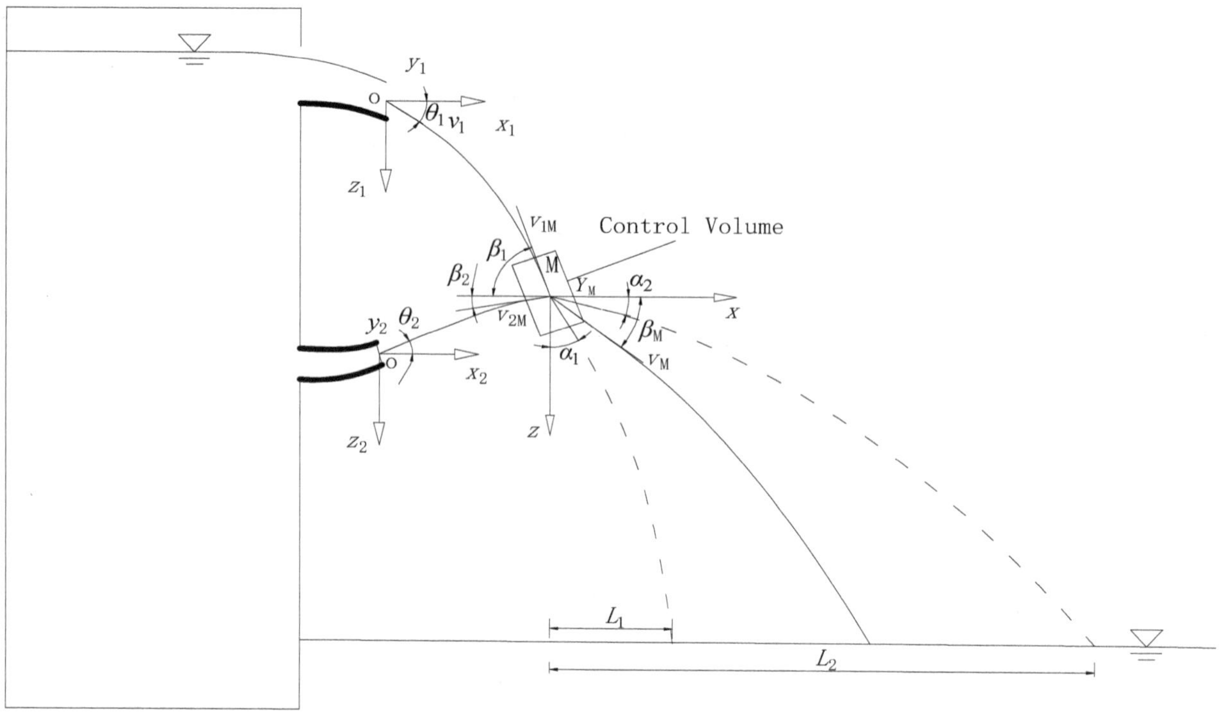

Figure 9. Schematic diagram of the collision between the surface-hole jet and the deep-hole jet in air.

The coordinate systems $x_1 0 z_1$ and $x_2 0 z_2$, in Figure 9 have the following geometric relationship:

$$x_1 = x_2 + \Delta x_0, z_1 = z_2 + \Delta z_0 \tag{4}$$

Using the projectile principle, the trajectory equation of the surface water jet was:

$$z_1 = \frac{g}{2{V_1}^2 \cos^2\theta_1} {x_1}^2 + x_1 tg\theta_1 = A_1 {x_1}^2 + x_1 tg\theta_1 \tag{5}$$

where V_1 is the surface hole jet initial speed, x_1 is the transverse distance, z_1 is the vertical distance, $A_1 = \frac{g}{2V_1^2 \cos^2\theta_1}$.

The energy velocities of V_{1M} and V_{2M} before the impact of the surface water jet and deep water jet became:

$$V_{1M} = \varphi_a \sqrt{2g(z_{1M} + \frac{{V_1}^2}{2g})} \tag{6}$$

$$V_{2M} = \varphi_a \sqrt{2g(z_{2M} + \frac{{V_2}^2}{2g})} \tag{7}$$

where φ_a is the velocity coefficient of air resistance ($\varphi_a \approx 0.95$), z_{1M} is the vertical distance from the surface hole to control volume, z_{2M} is the vertical distance from the deep hole to control volume. The control volume was taken around point M, as shown in Figure 9, and the momentum integral equation and continuous equation of fluid mechanics were used.

$$\iiint\limits_M \frac{\partial \rho}{\partial t} \vec{v} dV + \oiint\limits_M \rho \vec{v}(\vec{v} \cdot \vec{n}) dA = \sum \vec{F} \tag{8}$$

$$\iiint\limits_M \frac{\partial \rho}{\partial t} dV + \oiint\limits_M \rho(\vec{v} \cdot \vec{n}) dA = 0 \tag{9}$$

where ρ is the density of water, and $\vec{F}$ is the external force that acts on the control volume. In the absence of air resistance, β_M and V_M could be obtained with

$$\tan\beta_M = \frac{V_{1M}q_1\sin\beta_1 - V_{2M}q_2\sin\beta_2}{V_{1M}q_1\cos\beta_1 + V_{2M}q_2\cos\beta_2} \tag{10}$$

$$V_M = \frac{V_{1M}q_1\cos\beta_1 + V_{2M}q_2\cos\beta_2}{q_M\cos\beta_M} \tag{11}$$

with the assumption that $V_{1M} = V_{2M}$. Thus, Equations (10) and (11) become

$$\tan\beta_M = \frac{q_1\sin\beta_1 - q_2\sin\beta_2}{q_1\cos\beta_1 + q_2\cos\beta_2} \tag{12}$$

$$V_M = \frac{V_{1M}(q_1\cos\beta_1 + q_2\cos\beta_2)}{q_M\cos\beta_M} \tag{13}$$

The comparisons of the measured locations of $I_{\max}$ with Equations (12) and (13) are shown in Figure 10, which indicates good agreement.

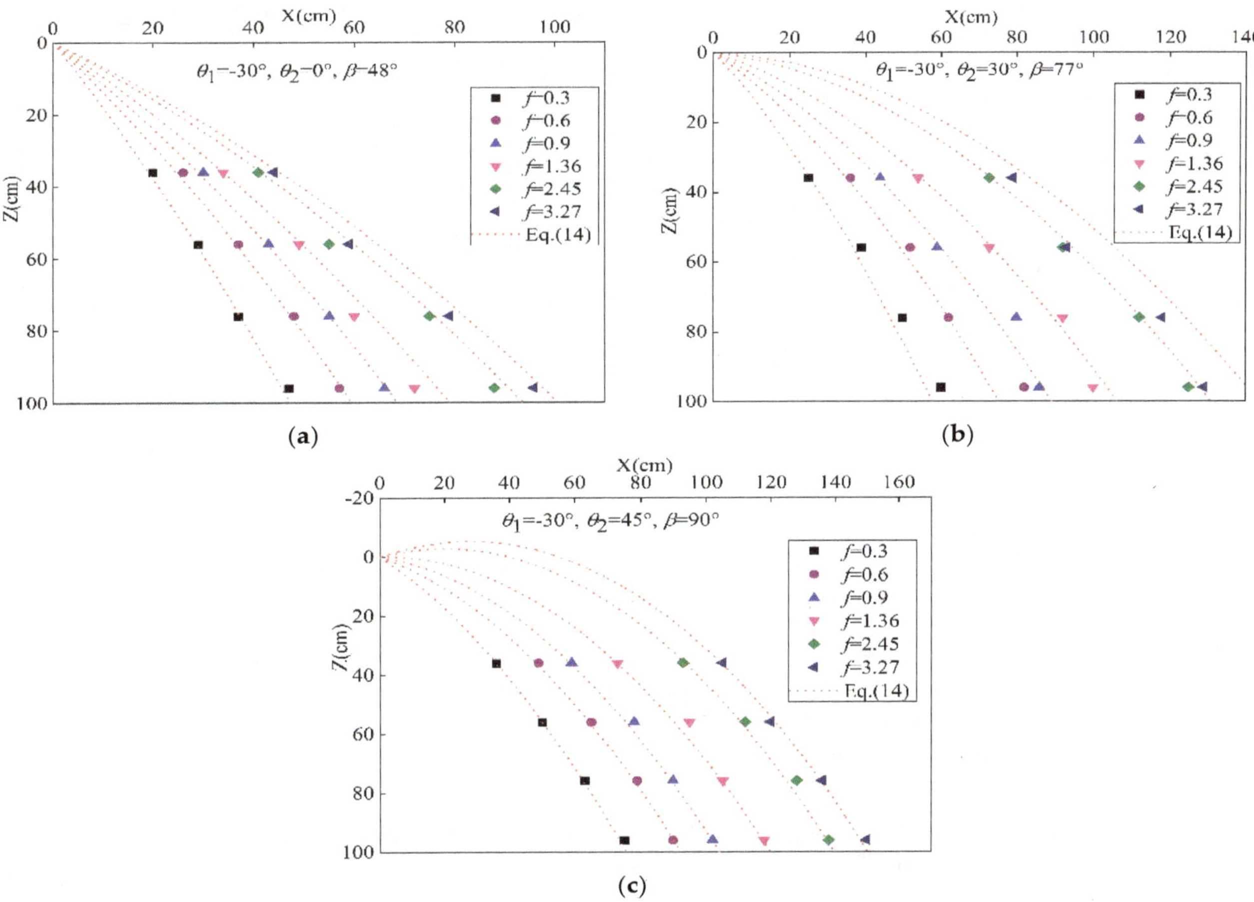

Figure 10. Comparison of experimental and calculated values of the maximum rainfall intensity points. (**a**) $\theta_1 = -30°$, $\theta_2 = 0°$, and $\beta = 48°$; (**b**) $\theta_1 = -30°$, $\theta_2 = 30°$, and $\beta = 77°$; (**c**) $\theta_1 = -30°$, $\theta_2 = 45°$, and $\beta = 90°$.

Based on the above results, we assume that the velocity magnitude was identical in all directions after the collision, but that the directions vary. Therefore, the velocities in all directions were V_M. Downstream of the collision, the trajectory boundaries ($5\% \times I_{\max}$) are shown in Figure 11. Hence, using Equation (3), α_1 and α_2 can be obtained—α_1 is the angle between the z-axis and the inner edge line, and α_2 is the angle between the x-axis and the outer edge line.

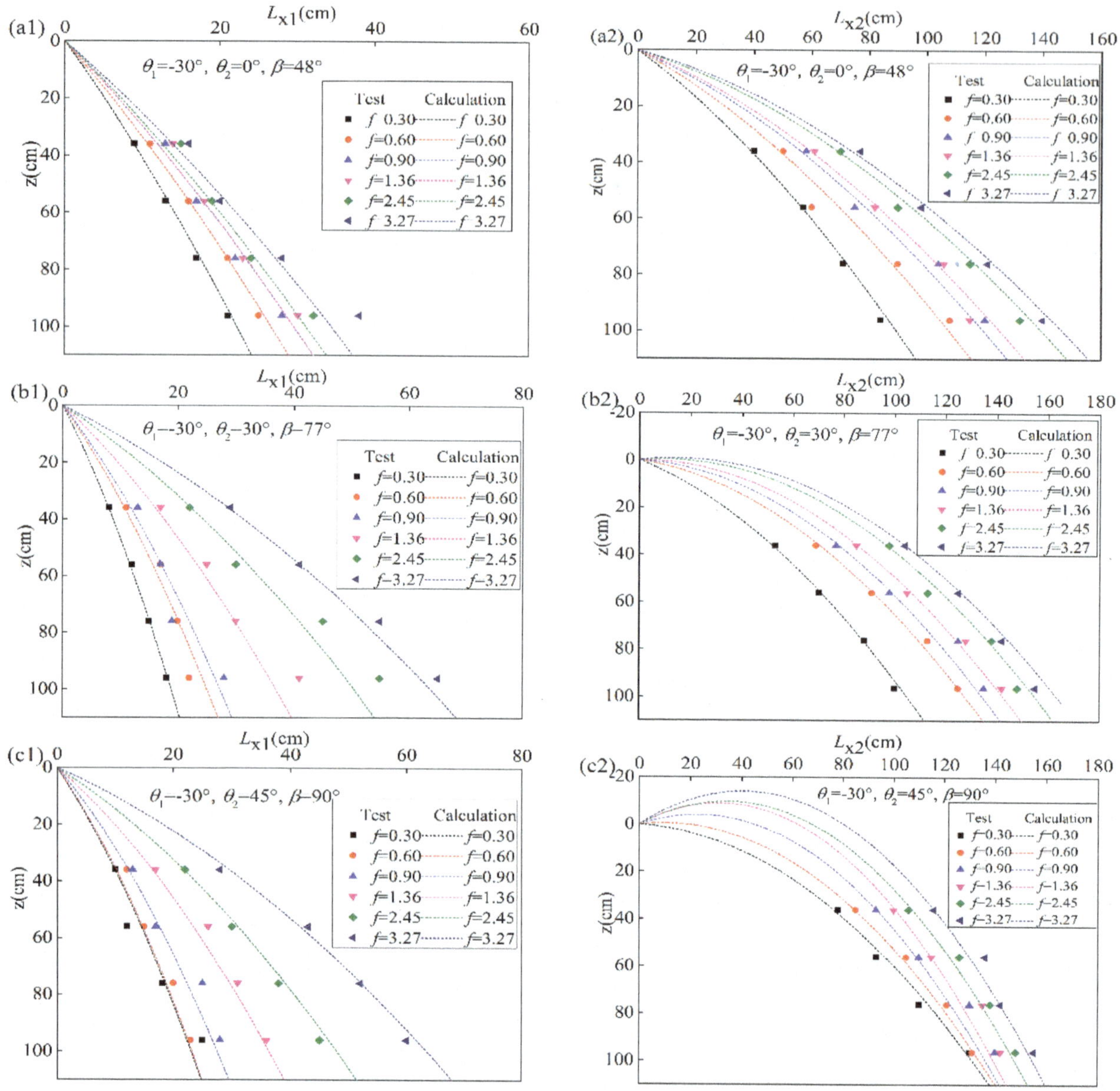

Figure 11. Comparison of the experimental and calculated values of L_{X1} and L_{X2}. L_{X1} is the distance of inner edge water tongue in x axis, L_{X2} is the distance of outer edge water tongue in x axis. (**a1**) $\theta_1 = -30°$, $\theta_2 = 0°$, and $\beta = 48°$; (**a2**) $\theta_1 = -30°$, $\theta_2 = 0°$, and $\beta = 48°$; (**b1**) $\theta_1 = -30°$, $\theta_2 = 30°$, and $\beta = 77°$; (**b2**) $\theta_1 = -30°$ and $\theta_2 = 30°$; (**c1**) $\theta_1 = -30°$, $\theta_2 = 45°$, and $\beta = 90°$; (**c2**) $\theta_1 = -30°$, $\theta_2 = 45°$, and $\beta = 90°$.

The calculated values of α_1 and α_2 are shown in Figure 12. Note that (1) with the identical collision angle β, α_1 increased, and with the increase in flow ratio, α_2 decreased; and (2) with the identical flow ratio, α_1 increased with the increase in collision angle β, and α_2 decreased with the increase in collision angle β. It can also be seen from Figure 12 that α_1 approximately shows exponential growth with the flow ratio, and α_2 shows logarithmic growth with the flow ratio. Data analysis indicates that the values of α_1 and α_2 are obtained as

$$\alpha_1 = 16e^{(f \sin^2 \beta)} \tag{14}$$

$$\alpha_2 = \beta \times (-0.182) \times \ln(f) - \beta \times \sin(\beta - 80°) \tag{15}$$

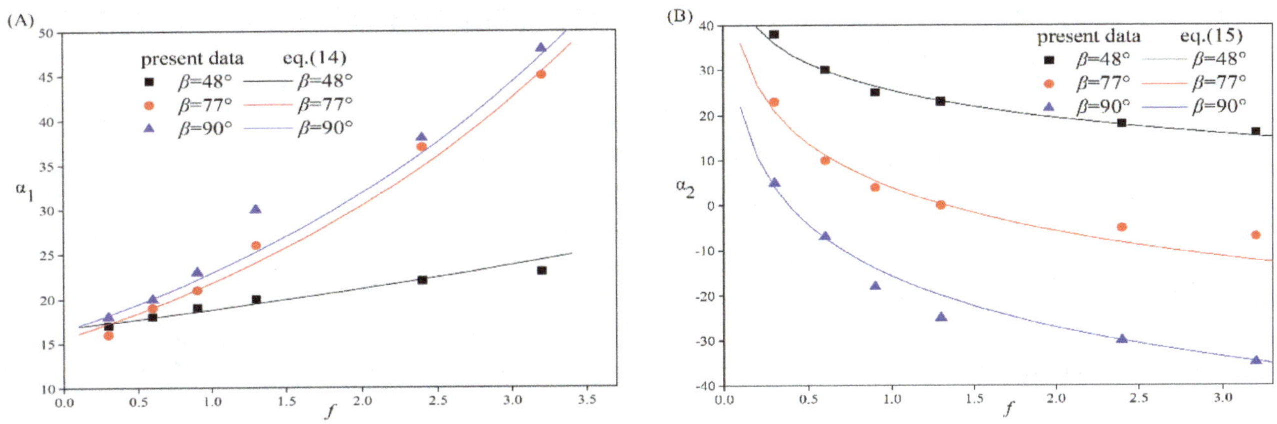

Figure 12. Comparison of the values obtained from Equations (14) and (15) with the present data. (**A**) Inner angle of α_1, and (**B**) outer angle of α_2.

The computed and measured values correlated to $R^2 = 0.97$ (Figure 12A) and 0.96 (Figure 12B). However, up until now, there was a paucity of information on the rainfall characteristics of the two-jet collisions. This present study provides a new method for understanding the rainfall characteristics of the two-jet collision, which will be useful to extend to other models or even prototypes for further research.

4. Conclusions

In this paper, the flow characteristics of two-jet collisions with different collision angles and flow ratios were investigated systematically by a series of model tests. The present investigation illustrated the development of the spatial distribution of rainfall intensity. Observations indicated that the maximum rainfall intensity sharply decreased with an increase in flow ratio, while the maximum rainfall intensity slightly increased, as the flow ratio was greater than 1.0.

On the x-axis, the distribution of rainfall intensity followed a Gaussian distribution. The range of rainfall intensity tended to reach the maximum for the flow ratio = 1.0. A theoretical equation to compute the locations of maximum rainfall intensity was presented, which is in good agreement with the model tests. The formulas to calculate the boundary lines of the x-axis were proposed. The present analysis was based upon experiments performed in a model with $48° < \beta < 90°$, $0.3 < f < 3.2$, and 36 cm $< z <$ 96 cm. The rainfall intensity distributions on the y-axis will be further studied.

Author Contributions: Conceptualization, H.Y. and W.X.; Methodology, W.X.; Software, Validation, H.Y.; W.X. and R.L.; Formal Analysis, H.Y.; Experiment, H.Y., Y.F., Y.H.; Writing-Original Draft Preparation, H.Y.; Writing-Review & Editing, H.Y.; Supervision, W.X.; Project Administration, W.X.; Funding Acquisition, W.X.

References

1. Li, N.-W. *Study on the Non-Collision Way of Flood Discharge and Energy Dissipation in High Arch Dams with Contracting Spillways and Middle Level Outlets*; Sichuan University: Chengdu, China, 2008.
2. Li, N.-W.; Xu, W.-L.; Zhou, M.-L.; Tian, Z. Experimental study on energy dissipation of flood discharge in high arch dams without impact of jets in air. *J. Hydraul. Eng.* **2008**, *39*, 927–933.
3. Shuangke, S. Summary of research on flood discharge and energy dissipation of high dams in China. *J. China Inst. Water Resour. Hydropower Res.* **2009**, *7*, 89–95.
4. Bai, R.; Zhang, F.; Liu, S.; Wang, W. Air concentration and bubble characteristics downstream of a chute aerator. *Int. J. Multiph. Flow* **2016**, *87*, 156–166. [CrossRef]
5. Bai, R.; Liu, S.; Tian, Z.; Wang, W.; Zhang, F. Experimental investigations on air-water flow properties of offse-aerator. *J. Hydraul. Eng.* **2017**, *144*, 04017059. [CrossRef]
6. Li, S.; Zhang, J.; Xu, W.; Chen, J.; Peng, Y. Evolution of Pressure and Cavitation on Side Walls Affected by Lateral Divergence Angle and Opening of Radial Gate. *J. Hydraul. Eng.* **2016**, *142*, 05016003. [CrossRef]

7. Bai, Z.; Peng, Y.; Zhang, J. Three-dimensional turbulence simulation of flow in a V-shaped stepped spillway. *J. Hydraul. Eng.* **2017**, *143*, 06017011. [CrossRef]
8. Huang, G.; Wu, S.; Chen, H. Model test studies of atomized flow for high dams. *Hydro-Sci. Eng.* **2008**, *2008*, 91–94.
9. Xue, L.-F. Influence of atomization on environment during flood discharge from dam and reducing measures. *Sichuan Water Power* **2005**, *24*, 104–107.
10. Bai, R.; Liu, S.; Tian, Z.; Wang, W.; Zhang, F. Experimental investigation of the dissipation rate in a chute aerator flow. *Exp. Therm. Fluid Sci.* **2019**, *101*, 201–208. [CrossRef]
11. Xiong, X.-L.; Ge, G. The collision energy dissipation between surface hole's water tongue and middle hole's water tongue of Ertan hydropower station. *J. Hydropower Eng. Res.* **1991**, 1–10. (In Chinese)
12. Chi-gong, G.Y.W. An optimal research on the energy dissipation by collision of two free jets in ertan hydroelectric plant. *J. Chengdu Univ. Sci. Technol.* **1992**, *18*, 17–24.
13. Liu, P.; Dong, J.; Liu, Y. Computation of energy dissipation for impact of two jets in air. *J. Hydraul. Eng.* **1995**, *7*, 38–44.
14. Diao, M.J.; Yang, Y.Q. Experimental study on two jets impact in air for energy dissipation. *J. Sichuan Univ. (Eng. Sci. Ed.)* **2002**, *34*, 13–15.
15. Sun, J.; Li, Y.-Z. Hydraulic characteristics of water jets by air-collision in lateral direction and estimation on limiting riverbed scour. *Chin. J. Appl. Mech.* **2004**, *21*, 134–137.
16. Sun, J.; Li, Y.-Z.; Yu, C.-Z. Limiting scour depth by the interaction of water jets from surface outlets with jets from mid-level outlets in high arch dams. *J. Tsinghua Univ. (Sci. Technol.)* **2002**, *42*, 564–568.
17. Liang, Z.-C. A Computation Model for Atomization Flow. *J. Hydrodyn.* **1992**, *7*, 247–255.
18. Liu, X.-L.; Zhang, W. The investigation of dynamic characteristics of jet-flow in open air. *J. Hydroelectr. Eng.* **1988**, *2*, 46–54. (In Chinese)
19. Liu, X.-L.; Liu, J. Experiment study on the diffusion and aeration of three-dimensional jet. *J. Hydraul. Eng.* **1989**, *11*, 9–17. (In Chinese)
20. Li, W.-X.; Wang, W.; Xu, W.-L.; Diao, M.-J. A rough method to calculate the range of atomized flow caused by jet impact. *J. Sichuan Union Univ. (Eng. Sci. Ed.)* **1999**, *3*, 17–23.
21. Liu, S.-H.; Liang, Z.-C. On atomizing problems by jumping jet in hydropower station at narrow and deep valley. *J. Wuhan Univ. Hydraul. Electr. Eng.* **1997**, *30*, 6–9.
22. Liu, S.-H.; Qu, B. Investigation on splash length of aerated jet. *J. Wuhan Univ. Hydraul. Electr. Eng.* **2003**, *36*, 5–8.
23. Liu, S.-H. Study of the atomized flow in hydraulic engineering. *J. Hydrodyn.* **1999**, *2*, 77–83.
24. Liu, S.-H.; Yin, S.-R.; Luo, Q.-S. Numerical simulation of atomized flow diffusion in deep and narrow goeges. *Conf. Glob. Chin. Scholar Hydrodyn.* **2006**, *18*, 515–518.
25. Sun, S.-K.; Liu, Z.-P. Longitudinal range of atomized flow forming by discharge of spillways and outlet works in hydropower stations. *J. Hydraul. Eng.* **2003**, *34*, 53–58.
26. Liu, H.-T.; Sun, S.-K.; Liu, Z.-P.; Wang, X.-S. Atomization prediction based on artificial neural networks for flood releasing of high dams. *J. Hydraul. Eng.* **2005**, *36*, 1241–1245.
27. Lian, J.-J.; Liu, F.; Huang, C.-Y. Numerical study on effects of environmental wind and terrain on spray caused by jet from flip bucket. *J. Hydraul. Eng.* **2005**, *36*, 1147–1152.
28. Sun, X.-F.; Liu, S.-H. Investigation on the motion of splash droplets in atomized flow. *J. Hydrodyn.* **2008**, *23*, 61–66.
29. Wang, S.-Y.; Chen, D.; Hou, D.-M. Experimental Research on the Rainfall intensity in the Source Area of Flood Discharge Atomization. *J. Yangtze River Sci. Res. Inst.* **2013**, *30*, 70–74.
30. Lian, J.-J.; Li, C.-Y.; Liu, F.; Wu, S.-Q. A prediction method of flood discharge atomization for high dams. *J. Hydraul. Res.* **2014**, *52*, 274–282. [CrossRef]
31. Zhang, W.; Zhu, D. Far-field properties of aerated water jets in air. *Int. J. Multiph. Flow* **2015**, *76*, 158–167. [CrossRef]
32. Taylor, G. Formation of thin flat sheets of water. *Proc. R. Soc.* **1960**, *259*, 1–17. [CrossRef]
33. Dombrowski, N.; Hooper, P.C. A study of the sprays formed by impinging jets in laminar and turbulent flow. *J. Fluid Mech.* **1963**, *18*, 392–440. [CrossRef]
34. Anderson, W.E.; Ryan, H.M.; Pal, S.; Santoro, R.J. Fundamental studies of impinging liquid jets. In Proceedings of the 30th AIAA, Aerospace Sciences Meeting and Exhibit, Reno, NV, USA, 6–9 January 1992; Volume 1, pp. 6–9.

35. Ibrahim, E.A.; Przekwas, A.J. Impinging jets atomization. *Phys. Fluids Fluid Dyn.* **1991**, *3*, 2981–2987. [CrossRef]
36. Choo, Y.J.; Kang, B.S. Parametric study on impinging-jet liquid sheet thickness distribution using an interferometric method. *Exp. Fluids* **2001**, *31*, 56–62. [CrossRef]
37. Choo, Y.J.; Kang, B.S. The velocity distribution of the liquid sheet formed by two low-speed impinging jets. *Phys. Fluids* **2002**, *14*, 622–627. [CrossRef]
38. Couto, H.S.; Bastos-Netto, D. Modeling droplet size distribution from impinging jets. *J. Propul. Power* **1989**, *7*, 654–656. [CrossRef]
39. Choo, Y.J.; Kang, B.S. A study on the velocity characteristics of the liquid elements produced by two impinging jets. *Exp. Fluids* **2003**, *34*, 655–661. [CrossRef]
40. Lee, C.H. An experimental study on the distribution of the drop size and velocity in asymmetric impinging jet sprays. *J. Mech. Sci. Technol.* **2008**, *22*, 608. [CrossRef]
41. Fore, L.B.; Dukler, A.E. The distribution of drop size and velocity in gas-liquid annular flow. *Int. J. Multiph. Flow* **1995**, *21*, 137–149. [CrossRef]

7

Assessing the Analytical Solution of One-Dimensional Gravity Wave Model Equations using Dam-Break Experimental Measurements

Wenjun Liu, Bo Wang *, Yunliang Chen, Chao Wu and Xin Liu

State Key Laboratory of Hydraulics and Mountain River Engineering, Sichuan University, Chengdu 610065, China; Liuwenjun_SCU@163.com (W.L.); liangyunchen@163.com (Y.C.); wuchao_w@163.com (C.W.); Liuxin_SCU@163.com (X.L.)

* Correspondence: wangbo@scu.edu.cn.

Abstract: The one-dimensional gravity wave model (GWM) is the result of ignoring the convection term in the Saint-Venant Equations (SVEs), and has the characteristics of fast numerical calculation and low stability requirements. To study its performances and limitations in 1D dam-break flood, this paper verifies the model using a dam-break experiment. The experiment was carried out in a large-scale flume with depth ratios (initial downstream water depth divided by upstream water depth) divided into 0 and 0.1~0.4. The data were collected by image processing technology, and the hydraulic parameters, such as water depth, flow discharge, and wave velocity, were selected for comparison. The experimental results show that the 1D GWM performs an area with constant hydraulic parameters, which is quite different from the experimental results in the dry downstream case. For a depth ratio of 0.1, the second weak discontinuity point, which is connected to the steady zone in the 1D GWM, moves upstream, which is contrary to the experimental situation. For depth ratios of 0.2~0.4, the moving velocity of the second weak discontinuity point is faster than the experimental value, while the velocity of the shock wave is slower. However, as the water depth ratio increases, the hydraulic parameters calculated by 1D GWM in the steady zone gradually approach the experimental value.

Keywords: gravity wave model; dam-break flood; image processing technology; experimental study

---

1. Introduction

The dam-breaking flood problem has been of wide concern in academic and engineering circles due to its importance and complexity; once the dam fails, it will cause huge numbers of casualties and property losses [1–3]. Ritter derived a theoretical solution of the dam-breaking flood problem in a rectangular prismatic channel that has a flat bottom and no resistance with a dry bed downstream [4]. Based on Ritter's solution, Stoker derived the theoretical solution of instantaneous dam failure in the rectangular prismatic channel for stationary water downstream [5]. Considering that actual rivers often have a certain slope, many scholars have tried to add this factor into the analytical model. Hunt derived an approximate solution for the dam failure of a rectangular prismatic channel in the case of a declining bottom [6]. In addition, actual river channels are mostly irregular, and their cross-sections cannot be simplified into rectangles. Wang et al. derived an analytical solution for the collapse of a prismatic channel with an arbitrary cross-sectional shape under a tilted channel [7]. Due to the occurrence of undular bore waves in nature, some physical models have been proposed based on a nonhydrostatic pressure assumption in dam break flows [8–10]. Since the current analytical model is still quite simplified compared to actual situations, it is necessary to conduct much experimental research. Stansby et al. systematically studied the initial shape of 1D dam failure. The analysis shows that at the initial stage of dam break, a mushroom-like jet occurs downstream [11]. Aleixo et al. used particle tracking velocimetry (PTV) to measure the cross-sectional velocity of the 1D dam-breaking flow and obtained a more accurate distribution of the cross-sectional flow velocity [12]. LaRocque et al. used an ultrasonic velocity profiler (UVP) to study the vertical velocity distribution in a rectangular flume and compared the different turbulence models to simulate the dam flow, and the large eddy simulation (LES) performed better than other models [13]. Many dam break experiments have also been carried out in the case of a moving bed to further approach the actual situation [14–16]. Recently, image processing technology has been used to experimentally study dam-break flood movements [17–21]. Because the equipment is not immersed, the measurement does not interfere with the flow field, and the whole measuring process is easy to program; furthermore, the cost is relatively low compared with that of high-precision equipment such as particle image velocimetry (PIV) or PTV, and its application prospects are quite extensive [22].

Certain problems, such as a long calculation time and unstable or divergent calculations, often exist in solving shallow water equations (SWEs). To solve these problems, on the one hand, many scholars use some excellent numerical schemes or commercial software to simulate them [23–25], and the lattice boltzmann model (LBM) performs very well for dealing with SWEs under many different boundary conditions [26–34]; smoothed particle hydrodynamics (SPH) can also give sufficient simulation accuracy in urban flood and 2D dam break flood problems [35–37], but on the other hand, when some items in the equation are negligible compared to others in some actual physical phenomena, many researchers have also tried to simplify the equations [38–43]. Among the simplified models, zero-inertia (or diffusive wave) models seem to need more computing time for flood simulation, and their reliability is significantly affected by complex topographic information; they give poor predictions of events around buildings, especially in urban districts [38,40,41,43]. Local inertial approximation (or gravity wave model) has good agreement with full-dynamic models when $Fr < 0.5$, leads to milder water depth gradients in steady flow, and has slower flood propagation speeds in unsteady flow. However, it requires less computing time, and is a faster alternative for gradually varied subcritical flows where domain-average friction typically exceeds $n = 0.03$ [39,40,42]. Most previous methods for studying these simplified models include numerical analysis; Martins et al. derived a simplified analytical solution for the 1D gravity wave model (GWM) and compared it with the analytical solution of the Saint-Venant Equation (SVE) in dam break flow [44]. After using the MacCormack and Roe scheme to simulate the simplified equations [45], the author found that the 1D GWM was similar to the analytical solution of the SVE when there was a subcritical flow downstream and the shock wave velocity was relatively small, whereas the difference was large when there was no water downstream. The numerical simulation does not need special initial conditions to remain steady and has good agreement with the derived simplified analytical solution.

However, the work of Martins did not analyze the relationship between the 1D GWM and the actual dam-breaking flow in detail, and the only two cases (dry and wet bed downstream) selected for comparison with SVE seemed to be insufficient. The ability of the 1D GWM to capture the 1D dam break flow characteristics and whether it can be used as a simplified alternative model outside the SVE need to be verified by experiment. This study carried out a large number of dam-break experiments in a large-scale water flume and used a non-immersion image processing technology to collect experimental data. Depth ratios (initial downstream water depth divided by upstream water depth) of 0 and 0.1~0.4 were considered for an upstream head of 0.6 m. Hydraulic parameters, such as water depth, flow discharge, average flow velocity, Froude number, and wave velocity, were measured or calculated. Through detailed comparison of the experimental data and the analytical solution of the 1D GWM, the ability of the 1D GWM to describe the characteristics of 1D dam-break flood movement was analyzed, and a valuable reference for using this model to calculate 1D dam-break flows in the future was created.

2. Materials and Methods

2.1. Flume and Other Instruments

The flume was made of tempered glass supported by a steel frame. The upstream was 8.37 m long, and the downstream 9.63 m long. The width was 1 m, and the height was 1.09 m. The flume had a rectangular cross section and a flat bottom. The gate material was a 15-mm-thick fiberglass board, which was lifted instantly by the electromagnetic brake asynchronous motor. This method was used to simulate the instantaneous break of the dam. The lifting time was 0.27 s when the upstream water depth was 0.6 m, which satisfies the following transient failure condition [46]: $t < (2h_u/g)^{1/2} = 0.3499$ s (h_u is the upstream water depth, and g is the gravity acceleration, 9.8 m/s^2). Eight CCD (charge coupled device) cameras were arranged in parallel at the side of the flume at a distance of 1.4 m to capture the evolution of the water flow. To ensure the quality of the video and make it more convenient to use the image processing technology, the background was covered with a smooth white car sticker, and the water was dyed in the experiment. To avoid noise from the reflection of the flume glass during image processing, a light-colored curtain was arranged on both sides of the flume to eliminate the shadow caused by natural light. In the experiment, 8 CCD cameras were simultaneously turned on by a wireless remote control. A wide-angle lens with a resolution of 1440 × 1920 was used for shooting, and the shooting speed was 48 frames per second. The device schematic is shown in Figure 1.

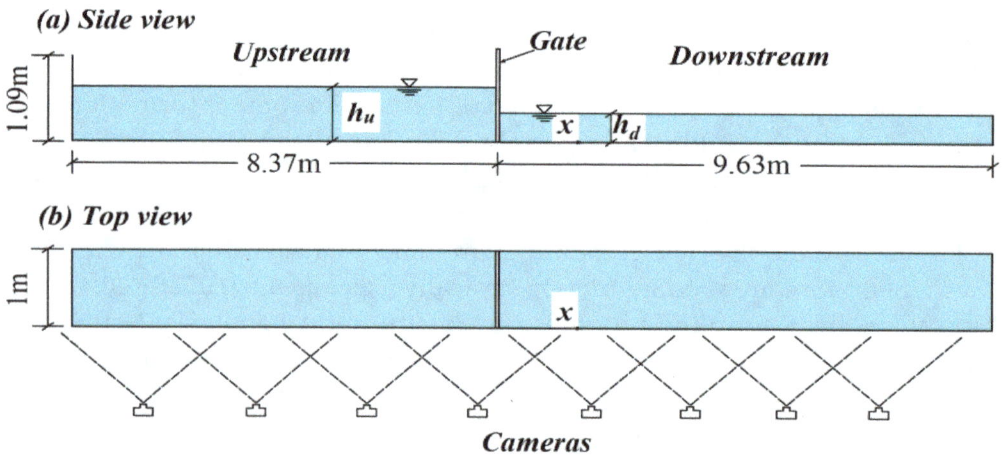

Figure 1. Schematic of the experimental device: (**a**) side view; (**b**) top view

2.2. Data Acquisition Technology and Experimental Repeatability Verification

The process of experimental data acquisition is divided into two main steps: the first step is to obtain the external and internal parameters of each camera by using the calibration plate and the calibration toolbox in MATLAB (MATLAB2016a, Mathworks company, Natick, MA, USA) and then to use the data and the corresponding calibration function to re-project and obtain the undistorted image. The second step is to analyze the image based on the undistorted image. In addition, the calibrated picture is then cropped and merged using the ruler above the flume. After binarizing the image, the water depth at each location can be obtained by boundary recognition functions. The calibration plate is shown in Figure 2a. The size of the square lattice in the plate is 100 mm × 100 mm. The number of grids is 11 on one side and 10 on the other. A margin of 50 mm is left around the calibration plate for the calibration toolbox to recognize the corner points so that the internal and external parameters of each camera can be accurately obtained to better correct the distortion. A set of 25 pictures in different orientations and angles is used to calibrate each camera, and the pictures before and after calibration are shown in Figure 2b,c. The actual length represented by one pixel is approximately 1.8 mm, and the depth of the part blocked by the flume steel column is obtained by interpolation. The picture after clipping and binarization is shown in Figure 3.

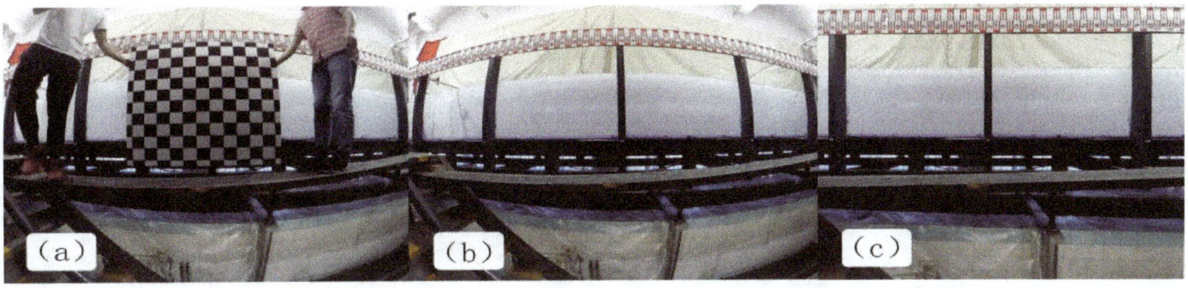

Figure 2. Picture calibration: (**a**) calibration board; (**b**) original image; (**c**) calibrated image.

Figure 3. Picture binarization. (**a**) original image; (**b**) binarized image.

The depth ratio is defined as $\alpha = h_d/h_u$, where h_u and h_d are the upstream and downstream initial water depths, respectively. The experimental conditions are shown in Table 1:

Table 1. Experimental conditions.

h_u (m)	h_d (m)				
	$\alpha = 0$	$\alpha = 0.1$	$\alpha = 0.2$	$\alpha = 0.3$	$\alpha = 0.4$
0.6	0	0.06	0.12	0.18	0.24

To ensure the stability and accuracy of the experimental data collection, the repeatability of the experiment is verified here by repeating the conditions of $h_u = 0.6$ m and $h_d = 0$ m and comparing the depth data at positions of $x = 0.3$ m (near the dam) and $x = 4.5$ m (far from the dam). The results are shown in Figure 4, where *Run* 1 and *Run* 2 represent the first and second experiments for the conditions of $h_u = 0.6$ m and $h_d = 0$ m, respectively. It can be seen that the overlap of the data between the two groups is quite good, and R^2 is close to 1, which proves that the experiment has good repeatability.

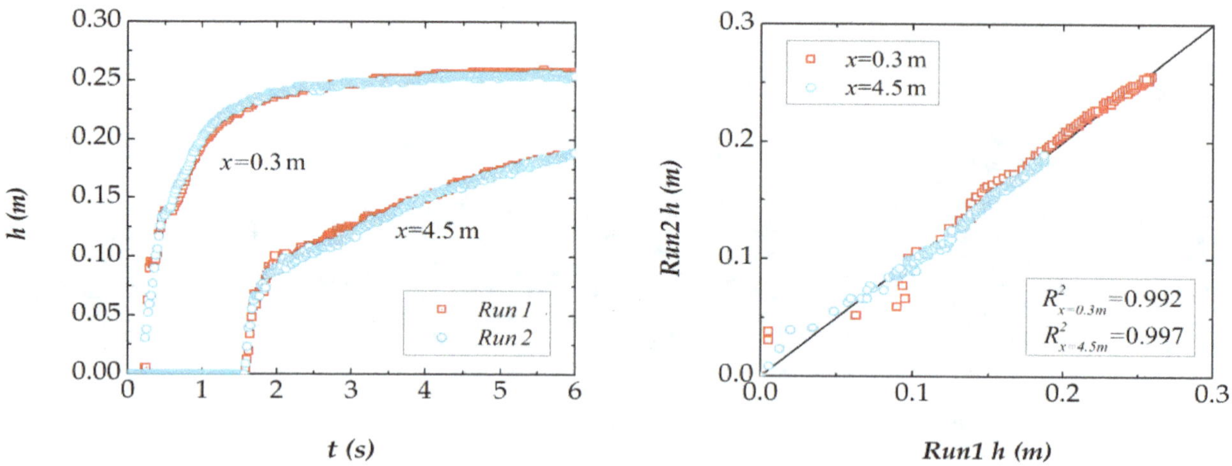

Figure 4. Experimental repeatability verification.

3. Results

3.1. Dry Bed Downstream Condition

3.1.1. Water Depth, Flow Discharge, Average Velocity and *Fr*

Here, the hydraulic parameters involved are dimensionless in the following formulas:

$$X = \frac{x}{t\sqrt{gh_u}} \tag{1}$$

$$T = \frac{t}{\sqrt{h_u/g}} \tag{2}$$

$$H = \frac{h}{h_u} \tag{3}$$

$$U = \frac{u}{\sqrt{gh_u}} \tag{4}$$

where h, u are water depth and flow velocity; X, T, H, and U are the dimensionless position, time, water depth, and flow velocity, respectively,.

The method for calculating flow discharge in the experiment is similar to those of Bento et al. [47] and Cestero et al. [48]; that is, calculate the water body area of the façade by the water depth data at each position for a certain time, because it is a one-dimensional case, then multiply the width of the flume to obtain the volume of the cross section at a certain moment. The following formula is used to calculate the flow discharge:

$$Q_{(t)} = \frac{V_{(t+\Delta t)} - V_{(t)}}{\Delta t} \tag{5}$$

where $Q_{(t)}$ is the flow discharge for time t, $V_{(t)}$ is the water volume between a cross section and the end of the upstream reservoir at time t, $V_{(t+\Delta t)}$ is the water volume between a cross section and the end of the upstream reservoir at time $t + \Delta t$, and Δt is the time difference of 1/48 s.

For convenience of later description, the analytical solution of 1D GWM is simply explanation here. Martins divided the structure of dam break into four areas, as shown in Figure 5, where P_{1u} and P_{12} represent the first and second weak discontinuity point, respectively; P_{2d} represents the strong discontinuity point; ξ represents the downstream wave-front evolution velocity; h_1 and h_2 represent the water depth in the rarefaction wave zone and the steady zone, respectively; C_u and C_2 represent the velocity of the first and second weak discontinuity point, respectively.

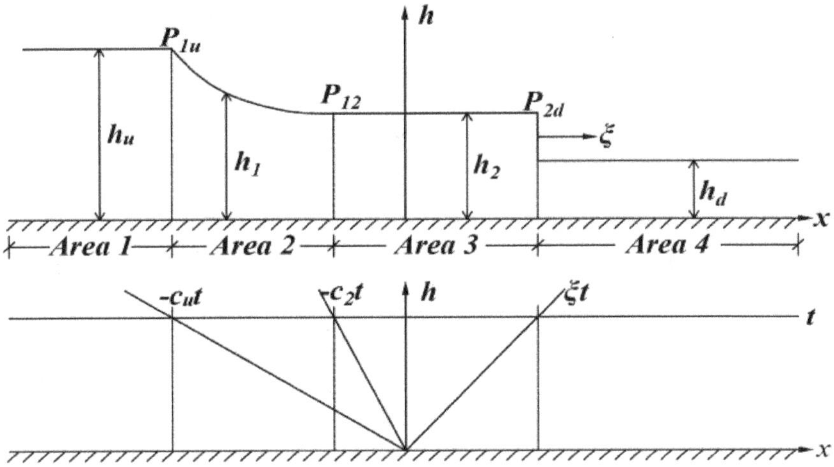

Figure 5. Dam break flood evolution model.

Using the method of characteristics (MOC) and performing some formula operations, the analytical formulas for the 1D GWM in four areas are shown in Table 2, where C_1 and C_d are celerity-velocity in Area2 and Area4, respectively; U_1 and U_2 are average flow velocity in Area2 and Area3, respectively.

Table 2. Analytical formulas for 1D GWM in four areas.

Variable	Area 1	Area 2	Area 3	Area 4
h	h_u	$h_1 = x^2/gt^2$	$h_2 = c_2^2/g$	h_d
c	$c_u = \sqrt{gh_u}$	$c_1 = -x/t$	$c_2^3 = c_u^3 + 3/2(c_d^2 - c_2^2)\sqrt{(c_d^2 + c_2^2)/2}$	$c_d = \sqrt{gh_d}$
u	0	$u_1 = 2(c_1^3 t^3 + x^3)/3x^2 t$	$u_2 = 2/3(c_u^3/c_2^2 - c_2)$	0
ξ	-	-	$\xi = \sqrt{(c_d^2 + c_2^2)/2}$	-
P	$-c_u t$	$-c_2 t$	ξt	-

Under the dry bed downstream condition, the hydraulic parameters calculated by the 1D GWM are very different from those of the experiment, as shown in Figure 6, where Exp. and *Fr* represent the experimental results and Froude number. According to the analytical solution of Martins, the water depth of area 3 still exists when there is no water in the downstream, which is quite different from the actual situation. The 1D GWM cannot accurately describe the evolution of 1D dam-break flow when there is no water downstream.

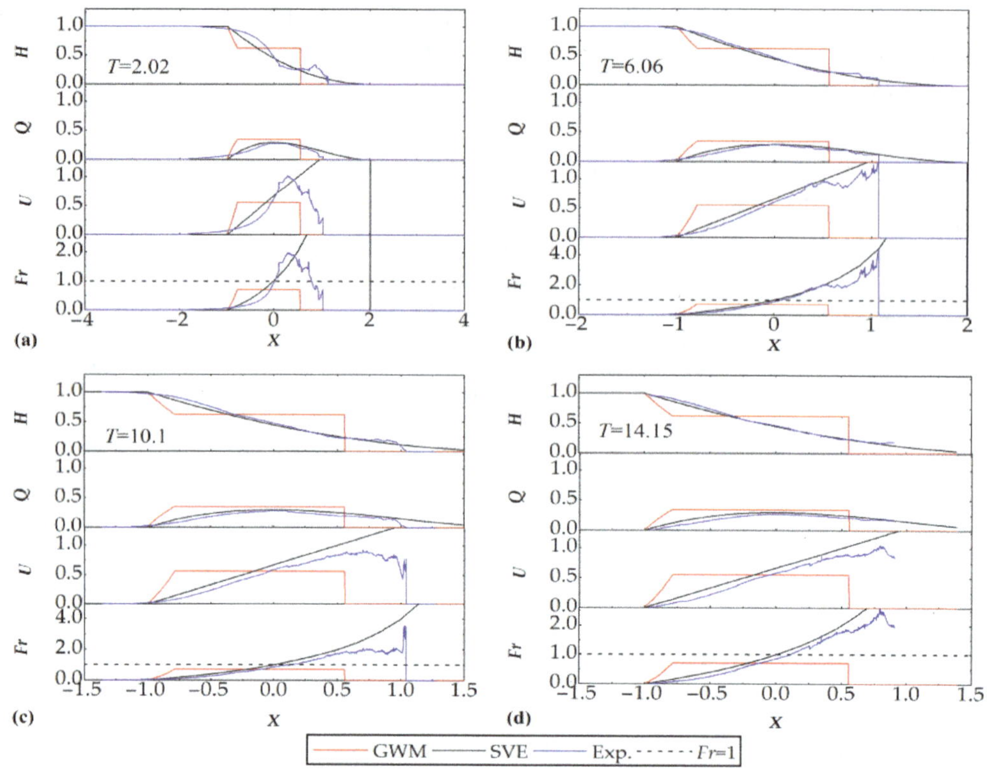

Figure 6. Water depth, flow discharge, average velocity and Fr for dry bed downstream: (**a**) $T = 2.02$; (**b**) $T = 6.06$; (**c**) $T = 10.1$; (**d**) $T = 14.15$.

3.1.2. Upstream Negative Wave and Downstream Wave-Front Velocities for Dry Bed Downstream

The calculation results of the upstream negative wave and the downstream wave-front velocities are shown in Figure 7, where U_n represents the dimensionless upstream negative wave evolution velocity and ξ_n represents the dimensionless downstream wave-front evolution velocity. Here, the relative root mean square error ($RRMSE$) is used to compare the error between the calculated value and the experimental value. The formula is as follows:

$$RRMSE = \sqrt{\frac{1}{n}\sum_{i=1}^{n}\left(\frac{y_i - \mu_i}{\mu_i}\right)^2} \times 100\% \tag{6}$$

where $RRMSE$ represents the relative root mean square error, n is the total number of all measured data points, y_i is the model calculated value, and μ_i is the measured value.

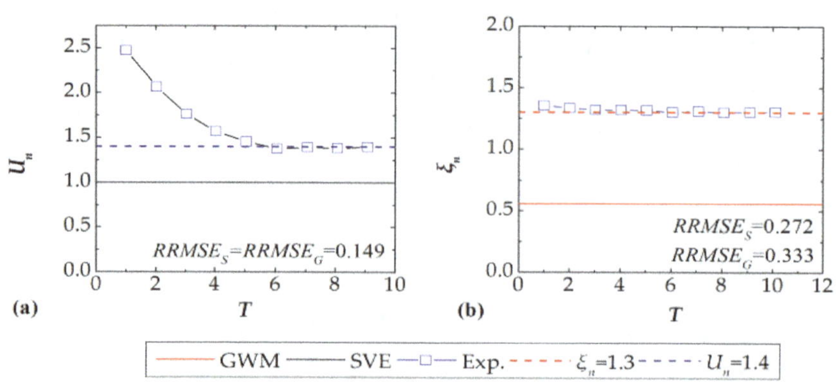

Figure 7. Velocities of (**a**) the upstream negative wave and (**b**) the downstream wave-front.

The result is shown in Figure 7, where $RRMSE_S$ and $RRMSE_G$ represent the relative root mean square error for SVE and 1D GWM, respectively. In the dry bed downstream case, the upstream negative wave velocities calculated by the 1D GWM and SVE are the same. The negative wave velocity gradually decreases and then stabilizes at approximately 1.4 in the experiment, Lauber and Hager explained that it is caused by excessive curvature of the streamline [46]. While the evolution velocity of the downstream wave-front basically remains stable, the dimensionless velocity is approximately 1.3 due to the resistance, which is smaller than the value of 2.0 from the SVE and much larger than the value of 0.55 calculated by the 1D GWM. The upstream negative wave and downstream wave-front speeds obtained by Leal et al. are all average values [49], and the characteristics of their propagation speed are not obtained over time. However, from the results of the stable speed, we can see that they are basically consistent with the results of Leal et al. [49].

3.2. *Wet Bed Downstream Condition*

3.2.1. Water Depth

Different degrees of water jump will occur downstream for the initial dam break (T = 2.02). It can be seen from Figure 8a that neither the SVE nor the 1D GWM describes this phenomenon well. Stansby et al. also observed that a mushroom-like jet occurs downstream of the initial dam break in the case of a wet bed downstream, and found that the movement form is quite complicated, but after the wave is broken, the motion of the water flow has good agreement with the analytical solution of the shallow water equations [11]. It can be seen from Figure 8 that the conclusion is still valid when the water depth ratio is between 0.1 and 0.4.

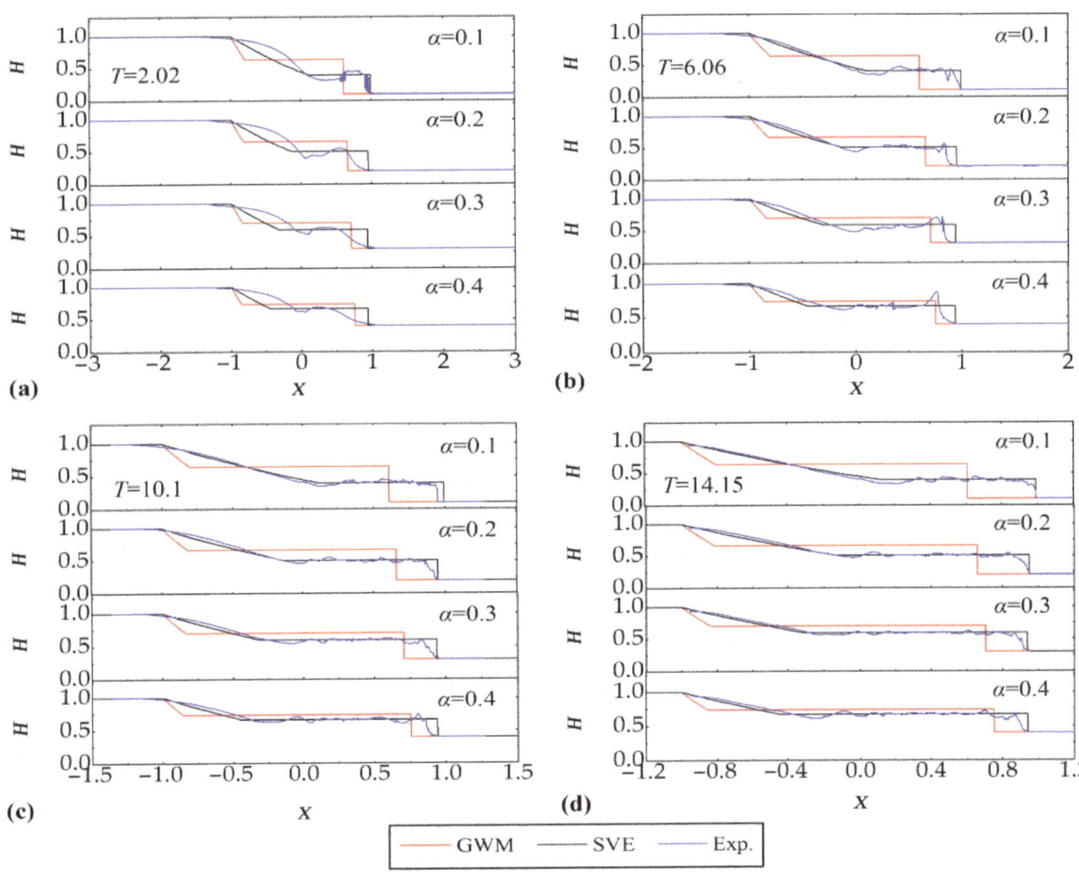

Figure 8. Water depth for wet bed downstream: **(a)** T = 2.02; **(b)** T = 6.06; **(c)** T = 10.1; **(d)** T = 14.15.

When $\alpha = 0.1$, the 1D GWM shows that the second weak discontinuity point moves upstream, but the experimental results are just the opposite. The evolution of water flow is much closer to that of the SVE. As Martins [44] stated, the 1D GWM shows only one subcritical flow state. When $\alpha = 0.2{\sim}0.4$, the second weak discontinuity point moves upstream in the experiment, but the range of the rarefaction wave zone calculated by the 1D GWM is shorter than that of the experiment. The speed at which the second weak discontinuity point moves upstream is much faster than the experimental value; the speed is 498%, 180%, and 94.7% faster than the experimental value for $\alpha = 0.2$, 0.3, and 0.4, respectively. In addition, the water depth of the steady zone calculated by the 1D GWM is higher than the experimental value, but it can be seen that as the water depth ratio increases, the water depth gradually approaches the experimental value.

3.2.2. Flow Discharge

In the case of wet bed downstream, the downstream water surface is distorted in the initial stage of dam break. The leaping water body will spin and aerate, and there will also be holes in the water body when it evolves downstream. Thus, the shape of the water surface line is complicated and not easy to capture. The flow discharge obtained by the above method cannot accurately describe the actual change, so the calculation of flow discharge in the case of wet bed downstream starts from $T = 6.06$. The calculation result is shown in Figure 9. It can be seen that due to the shortening of the rarefaction wave zone calculated by the 1D GWM, there is a significant deviation from the experimental value throughout the evolution period. The flow discharge in the steady zone is larger than the experimental value, but it can be seen that with the water depth ratio increased, the flow discharge value calculated by the 1D GWM in the steady zone also gradually approaches the experimental value.

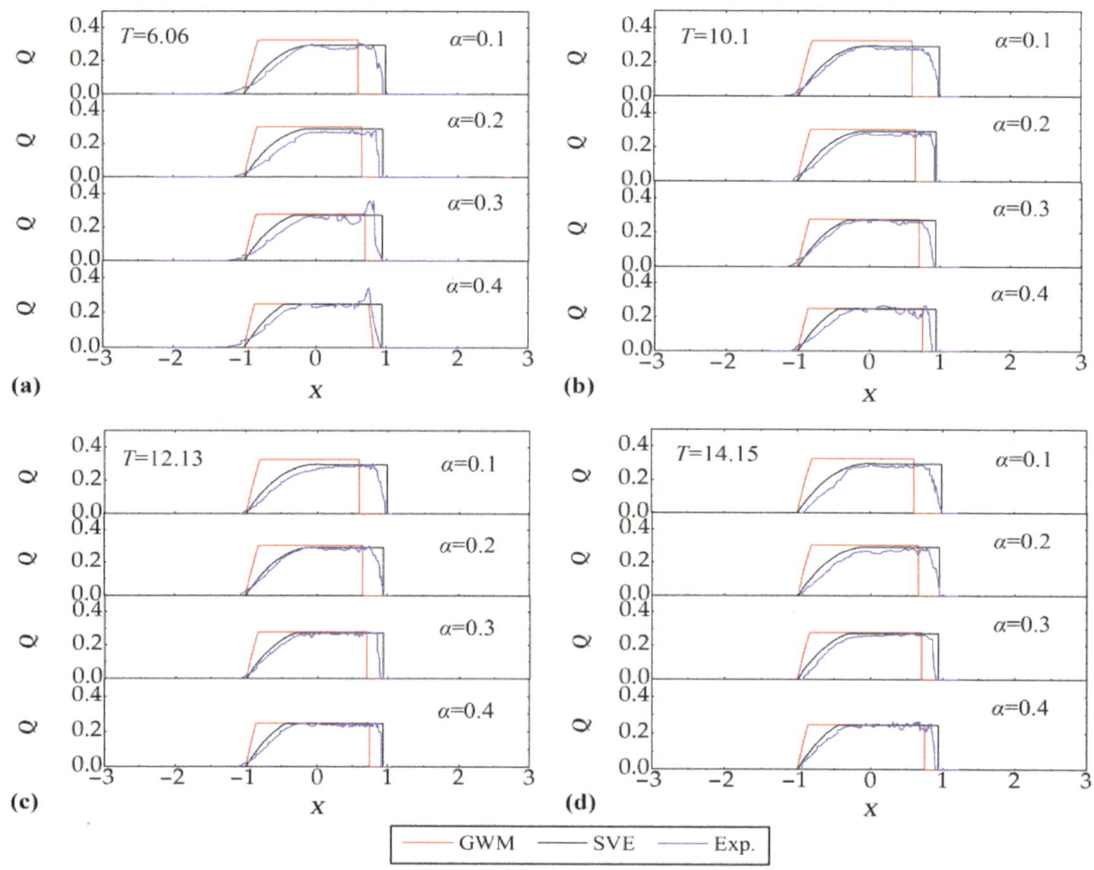

Figure 9. Flow discharge for wet bed downstream: (**a**) $T = 6.06$; (**b**) $T = 10.1$; (**c**) $T = 12.13$; (**d**) $T = 14.15$.

3.2.3. Average Velocity and *Fr*

The calculation results of the average velocity are shown in Figure 10. Similar to the flow discharge comparison, the average velocity calculated by the 1D GWM in the rarefaction wave zone is very different from the experimental value. The average velocity is smaller than the experimental value in the steady zone, especially when the water depth ratio is small, but it can be seen that as the water depth ratio increases, the average velocity gradually approaches the experimental value in the steady zone.

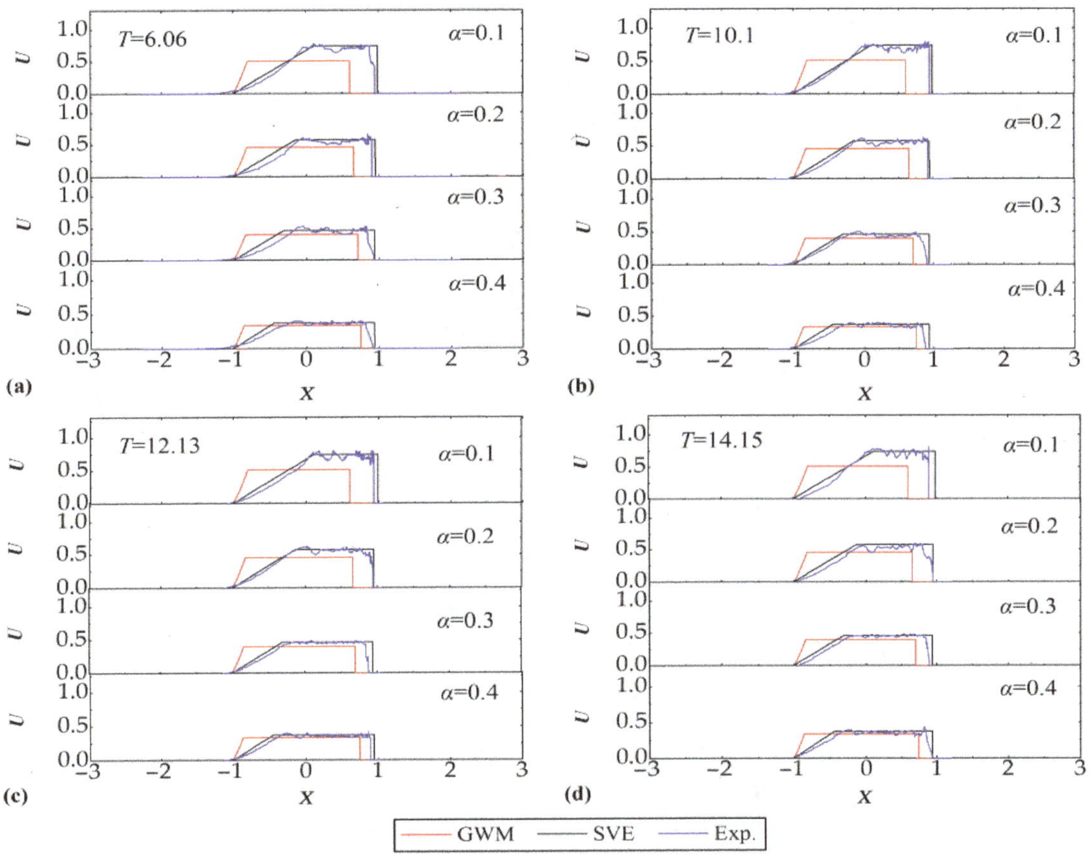

Figure 10. Average velocity for wet bed downstream: (**a**) $T = 6.06$; (**b**) $T = 10.1$; (**c**) $T = 12.13$; (**d**) $T = 14.15$.

The calculation results for *Fr* are shown in Figure 11. As described above, when $\alpha = 0.1$, the *Fr* value calculated by the 1D GWM is less than 1 in the entire flow region, and the *Fr* of the actual flow is greater than 1 in the steady zone. At this time, the second weak discontinuity point will not propagate upstream, and the description of the 1D GWM is flawed. When $\alpha = 0.2$~0.4, *Fr* is less than 1 in the whole flow region, and the flow is in a subcritical flow state. The result of the 1D GWM calculation still differs greatly in the rarefaction wave zone, and the *Fr* calculated in the steady zone is smaller than the experimental value. However, it also has the characteristic that its value gradually approaches the experimental value as the water depth ratio increases.

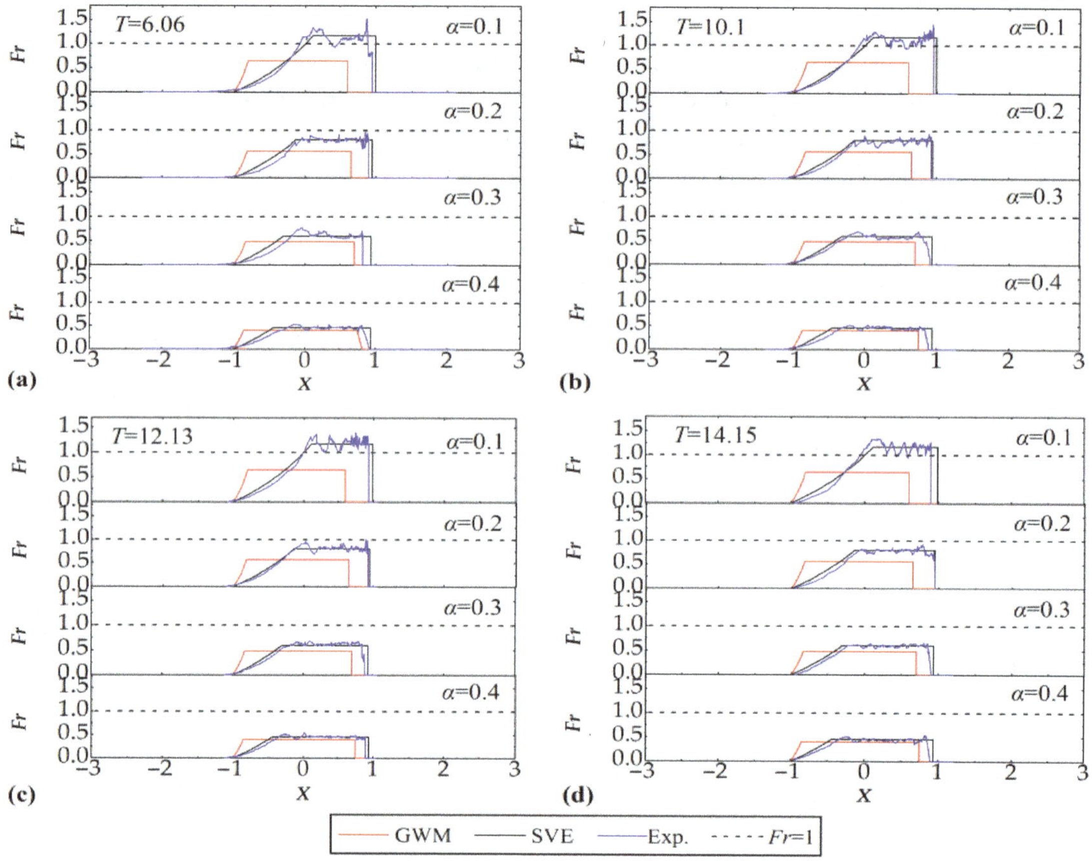

Figure 11. *Fr* for wet bed downstream: (**a**) $T = 6.06$; (**b**) $T = 10.1$; (**c**) $T = 12.13$; (**d**) $T = 14.15$.

3.2.4. Relative Error of Hydraulic Parameters in Steady Zone

In the previous sections, it was found that although the 1D GWM has large defects in the rarefaction wave zone, the results are much closer to the experimental value in the steady zone when the water depth ratio is relatively large. Here, the error is calculated by the following formula:

$$\delta = |y - \mu|/\mu \tag{7}$$

where y is the value calculated by the 1D GWM or SVE and μ is the experimental measurement value.

The calculation result is shown in Figure 12, where δ_H, δ_U, δ_{Fr}, and δ_Q represent the errors in the water depth, average velocity, *Fr* and flow discharge in the steady zone, respectively. It can be seen from Figure 12 that with increasing water depth ratio, the error between these hydraulic parameters and the experimental values tend to gradually decrease. When $\alpha = 0.4$, the errors for water depth, average velocity, *Fr* and flow discharge are 7.91%, 6.32%, 8.36%, and 3.94%, respectively.

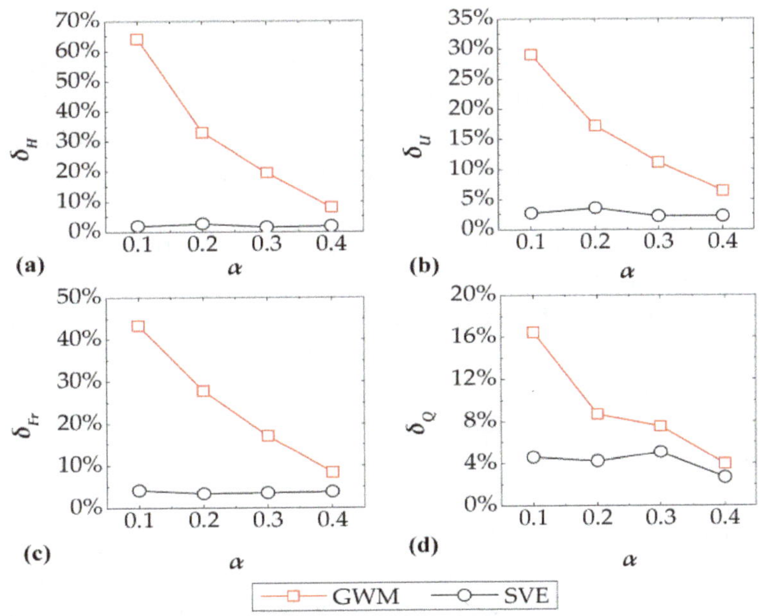

Figure 12. Error in the steady zone for different water depth ratios: (**a**) water depth; (**b**) average velocity; (**c**) Fr; (**d**) flow discharge.

3.2.5. Upstream Negative Wave and Downstream Wave-Front Velocities for Wet Bed Downstream

The calculation results of the velocity for upstream negative wave and downstream wave-front velocities for α = 0.1~0.4 are shown in Figure 13. It can be seen that the negative wave velocities calculated by the SVE and 1D GWM are the same for the wet bed downstream case, but the negative wave velocity is larger than those of the SVE or 1D GWM at the initial time point in the experiment. After the flow has evolved for a period of time, the velocity gradually decreases, and the error gradually decreases as the water depth ratio increases. The steady velocity is relatively stable in the experiment, and it is consistent with the SVE, the shock wave velocity is small for the 1D GWM. However, it should be noted that the error of shock wave velocity between the 1D GWM and the experimental value tends to gradually decrease as the water depth ratio increases.

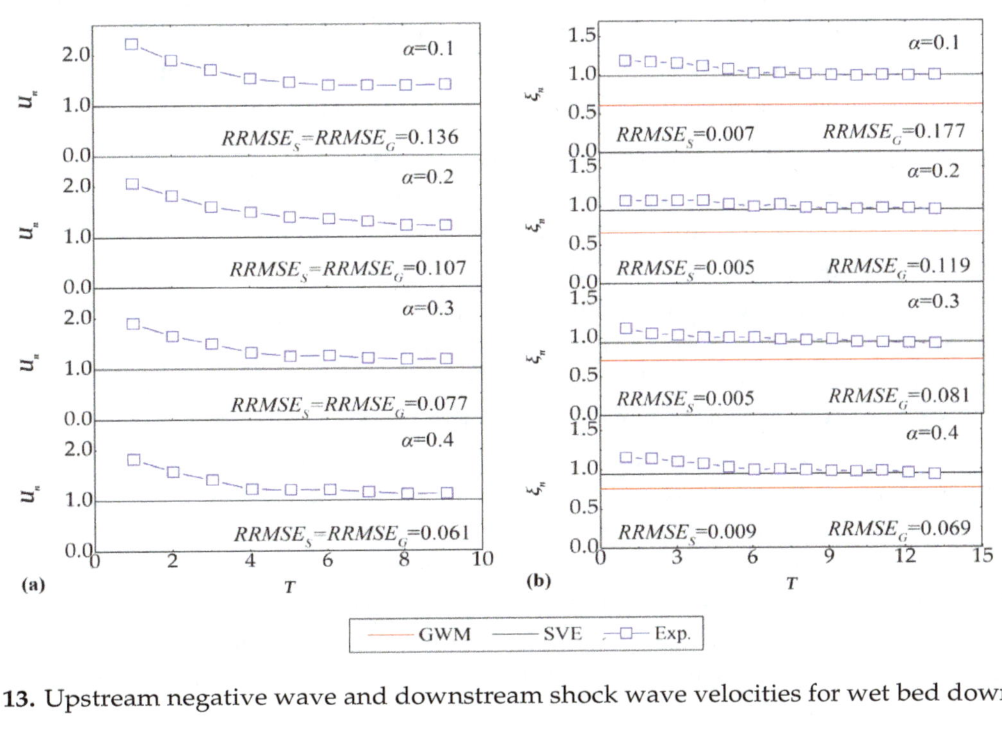

Figure 13. Upstream negative wave and downstream shock wave velocities for wet bed downstream.

4. Discussion

For the study of GWM, past researchers have mostly used different numerical calculation methods to solve the simplified governing equations directly [39,40,42]. Their work proves that the GWM requires less computing time and does not need to pay much effort to maintaining computational stability. It has certain prediction accuracy when the surface friction is not small ($n > 0.03$) in subcritical flows. In view of the above advantages of GWM, Martins tried to apply it to the 1D dam break flow and derived the analytical solution of 1D GWM under simplified conditions. However, the performances and limitations of analytical solution need to be further confirmed by experimental research [50,51]. This paper analyzes the analytical solution of 1D GWM by means of experimental research. Unlike the numerical model, which can arbitrarily control the boundary conditions and initial state to analyze the characteristics of the governing equations, both analytical and experimental research is performed under ideal boundary conditions here. 1D dam-breaking flow is a kind of unsteady flow in open channel, which has special evolution characteristics and need targeted analysis. In the experiment, we have observed that 1D GMW has a great difference with experimental conditions in the case of dry bed downstream, the calculated rarefaction wave zone is significantly shorter than the experimental one when the depth ratio is 0.1~0.4, and the wave-front evolution speed is slower. When the depth ratio is 0.4, the value calculated in the steady zone is close to the experimental value. It can be seen that the application of the 1D GWM to 1D dam break flow should be careful and significantly consider the influence of depth ratio.

Of course, unlike the previous numerical model studies, the experiment cannot capture the characteristics of dam break flood for very far from the dam (De Almeida and Bates 5000 m, Bates et al. 5000 m, Neal et al. 10,000 m). In this article, we compared the experimental result with the analytical solution of 1D GWM in a basic way. Future work could consider changing the boundary conditions for 1D GWM, such as the analytical solution of 1D GWM, considering the river bed slope or the shape of the cross-section (triangular, parabolic, etc.), and the corresponding experimental research could be carried out simultaneously. Because the frictional resistance has a great impact on GWM [39,40], considering different friction coefficients for the river bed material in the experiment is also a research direction.

5. Conclusions

This paper compares the data obtained by image processing technology with one-dimensional GWM in dam-break flood. Because the current test results are obtained under very simplified conditions, and for more complicated cases, further experiments or numerical simulation studies are needed in future, but at present we can draw the following conclusions:

(1) The calculated result of the 1D GWM is very different from the experimental value for dry bed downstream. There is still a region with the same hydraulic parameters value downstream that does not appear in the experiment. The 1D GWM gives a poor description for the dam-break flow movement in the dry bed downstream case.

(2) When $\alpha = 0.1$, the second weak discontinuity point calculated by the 1D GWM is continuously developed upstream, which is contrary to the experimental results. When $\alpha = 0.2{\sim}0.4$, the rarefaction wave zone of the 1D GWM is significantly shorter than that of the experiment, and the shock wave evolution distance is smaller than the experimental value.

(3) When $\alpha = 0.1{\sim}0.4$, the relative errors in the water depth, average velocity, Fr and flow discharge in the steady zone gradually decrease as the water depth ratio increases. When $\alpha = 0.4$, the errors are 7.91%, 6.32%, 8.36%, and 3.94%, respectively. The upstream negative wave speeds for the 1D GWM and SVE are the same, but both are smaller than the experimental value. The downstream shock speed for the 1D GWM is smaller than the experimental value. However, as the water depth ratio increases, the errors in the upstream negative wave and downstream shock wave velocities are also reduced. When $\alpha = 0.4$, the $RRMSE$ values are 0.061 and 0.069, respectively.

Author Contributions: Conceptualization, B.W. and C.W.; Methodology, W.L., B.W. and X.L.; Software, W.L. and X.L.; Validation, B.W. and Y.C.; Formal Analysis, B.W.; Investigation, W.L. and B.W.; Resources, Y.C.; Data Curation, W.L.; Writing-Original Draft Preparation, W.L. and B.W.; Writing-Review & Editing, W.L. and B.W.; Visualization, W.L. and X.L.; Supervision, B.W.; Project Administration, B.W.; Funding Acquisition, B.W.

References

1. Haltas, I.; Tayfur, G.; Elci, S. Two-dimensional numerical modeling of flood wave propagation in an urban area due to Ürkmez dam-break, İzmir, Turkey. *Nat. Hazards* **2016**, *81*, 2103–2119. [CrossRef]
2. Kim, B.; Sanders, B.F. Dam-Break Flood Model Uncertainty Assessment: Case Study of Extreme Flooding with Multiple Dam Failures in Gangneung, South Korea. *J. Hydraul. Eng.* **2016**, *142*, 05016002. [CrossRef]
3. Bosa, S. A Numerical Model of the Wave that Overtopped the Vajont Dam in 1963. *Water Resour. Manag.* **2013**, *27*, 1763–1779. [CrossRef]
4. Ritter, A. The propagation of water waves. *Ver Deutsh Ing. Z.* **1892**, *36*, 947–954.
5. Stoker, J.J. *Water Waves*; Interscience Publishers, Inc.: New York, NY, USA, 1957.
6. Hunt, B. Asymptotic solution for dam-break on sloping channel. *J. Hydraul. Eng.* **1983**, *109*, 1698–1706. [CrossRef]
7. Wang, B.; Chen, Y.; Wu, C.; Peng, Y.; Ma, X.; Song, J.J. Analytical solution of dam-break flood wave propagation in a dry sloped channel with an irregular-shaped cross-section. *J. Hydro-Environ. Res.* **2016**, *14*, 93–104. [CrossRef]
8. Frazao, S.S.; Zech, Y. Undular bores and secondary waves -Experiments and hybrid finite-volume modelling. *J. Hydraul. Res.* **2003**, *40*, 33–43. [CrossRef]
9. Kim, D.H.; Lynett, P.J. Dispersive and Nonhydrostatic Pressure Effects at the Front of Surge. *J. Hydraul. Eng.* **2011**, *137*, 754–765. [CrossRef]
10. Cantero-Chinchilla, F.N.; Castro-Orgaz, O.; Dey, S.; Ayuso, J.L. Nonhydrostatic dam break flows. I: Physical equations and numerical schemes. *J. Hydraul. Eng.* **2016**, *142*, 04016068. [CrossRef]
11. Stansby, P.K.; Tcd, B.; Chegini, A. The initial stages of dam-break flow. *J. Fluid Mech.* **1998**, *374*, 407–424. [CrossRef]
12. Aleixo, R.; Soares-Frazão, S.; Zech, Y. Velocity-field measurements in a dam-break flow using a PTV Voronoï imaging technique. *Exp. Fluids* **2011**, *50*, 1633–1649. [CrossRef]
13. Larocque, L.A.; Imran, J.; Chaudhry, M.H. Experimental and numerical investigations of two-dimensional dam-break flows. *J. Hydraul. Eng.* **2013**, *139*, 569–579. [CrossRef]
14. Qian, H.; Cao, Z.; Liu, H.; Pender, G. New experimental dataset for partial dam-break floods over mobile beds. *J. Hydraul. Res.* **2017**, *56*, 1–12. [CrossRef]
15. Soares-Frazão, S.; Canelas, R.; Cao, Z.; Cea, L.; Chaudhry, H.M.; Die Moran, A.; El Kadi, K.; Ferreira, R.; Cadórniga, I.F.; Gonzalez-Ramirez, N.; et al. Dam-break flows over mobile beds: Experiments and benchmark tests for numerical models. *J. Hydraul. Res.* **2012**, *50*, 364–375. [CrossRef]
16. Goutiere, L.; Soares-Frazão, S.; Zech, Y. Dam-break flow on mobile bed in abruptly widening channel: Experimental data. *J. Hydraul. Res.* **2011**, *49*, 367–371. [CrossRef]
17. Kocaman, S.; Ozmen-Cagatay, H. The effect of lateral channel contraction on dam-break flows: Laboratory experiment. *J. Hydrol.* **2012**, *432–433*, 145–153. [CrossRef]
18. Kocaman, S.; Ozmen-Cagatay, H. Investigation of dam-break induced shock waves impact on a vertical wall. *J. Hydrol.* **2015**, *525*, 1–12. [CrossRef]
19. Ozmen-Cagatay, H.; Kocaman, S.; Guzel, H. Investigation of dam-break flood waves in a dry channel with a hump. *J. Hydro-Environ. Res.* **2014**, *8*, 304–315. [CrossRef]
20. Soares-Frazão, S. Experiments of dam-break wave over a triangular bottom sill. *J. Hydraul. Res.* **2007**, *45*, 19–26. [CrossRef]
21. Aureli, F.; Maranzoni, A.; Mignosa, P.; Ziveri, C. An image processing technique for measuring free surface of dam-break flows. *Exp. Fluids* **2011**, *50*, 665–675. [CrossRef]
22. Eaket, J.; Hicks, F.E. Use of stereoscopy for dam break flow measurement. *J. Hydraul. Eng.* **2005**, *131*, 24–29. [CrossRef]
23. Wu, G.; Yang, Z.; Zhang, K.; Dong, P.; Lin, Y.T. A non-equilibrium sediment transport model for dam break flow over moveable bed based on non-uniform rectangular mesh. *Water* **2018**, *10*, 616. [CrossRef]

24. Lu, C. Simulations of shallow water equations with finite difference lax-wendroff weighted essentially non-oscillatory Schemes. *J. Sci. Comput.* **2011**, *47*, 281–302. [CrossRef]
25. Macchione, F.; Costabile, P.; Costanzo, C.; Lorenzo, G.D.; Razdar, B. Dam breach modelling: Influence on downstream water levels and a proposal of a physically based module for flood propagation software. *J. Hydroinform.* **2016**, *18*. [CrossRef]
26. Peng, Y.; Zhou, J.G.; Burrows, R. Modelling the free surface flow in rectangular shallow basins by lattice boltzmann method. *J. Hydraul. Eng.* **2011**, *137*, 1680–1685. [CrossRef]
27. Peng, Y.; Zhou, J.G.; Zhang, J.M.; Liu, H.F. Lattice boltzmann modelling of shallow water flows over discontinuous beds. *Int. J. Numer. Method Fluids* **2014**, *75*, 608–619. [CrossRef]
28. Peng, Y.; Zhang, J.M.; Meng, J.P. Second order force scheme for lattice boltzmann model of shallow water flows. *J. Hydraul. Res.* **2017**, *55*, 592–597. [CrossRef]
29. Peng, Y.; Meng, J.P.; Zhang, J.M. Multispeed lattice boltzmann model with stream-collision scheme for transcritical shallow water flows. *Math. Probl. Eng.* **2017**, *2017*. [CrossRef]
30. Meng, J.P.; Gu, X.J.; Emerson, D.R.; Peng, Y.; Zhang, J.M. Discrete boltzmann model of shallow water equations with polynomial equilibria. *Int. J. Mod. Phys.* **2018**. [CrossRef]
31. Peng, Y.; Zhou, J.G.; Burrows, R. Modelling solute transport in shallow water with the lattice Boltzmann method. *Comput. Fluids* **2011**, *50*, 181–188. [CrossRef]
32. Peng, Y.; Zhou, J.G.; Zhang, J.M.; Burrows, R. Modeling moving boundary in shallow water by LBM. *Int. J. Mod. Phys. C* **2013**, *24*, 1–17. [CrossRef]
33. Peng, Y.; Zhou, J.G.; Zhang, J.M. Mixed numerical method for bed evolution. *Proc. Inst. Civil Eng. Water Manag.* **2015**, *168*, 3–15. [CrossRef]
34. Peng, Y.; Zhang, J.M.; Zhou, J.G. Lattice Boltzmann Model Using Two-Relaxation-Time for Shallow Water Equations. *J. Hydraul. Eng.* **2016**, *142*, 06015017. [CrossRef]
35. Ata, R.; Soulaïmani, A. A stabilized SPH method for inviscid shallow water flows. *Int. J. Numer. Methods Fluids* **2005**, *47*, 139–159. [CrossRef]
36. Amicarelli, A.; Albano, R.; Mirauda, D.; Agate, G.; Sole, A.; Guandalini, R. A smoothed particle hydrodynamics model for 3d solid body transport in free surface flows. *Comput. Fluids* **2015**, *116*, 205–228. [CrossRef]
37. Albano, R.; Sole, A.; Mirauda, D.; Adamowski, J. Modelling large floating bodies in urban area flash-floods via a smoothed particle hydrodynamics model. *J. Hydrol.* **2016**, *541*, 344–358. [CrossRef]
38. Moussa, R.; Bocquillon, C. On the use of the diffusive wave for modelling extreme flood events with overbank flow in the floodplain. *J. Hydrol.* **2009**, *374*, 116–135. [CrossRef]
39. Bates, P.D.; Horritt, M.S.; Fewtrell, T.J. A simple inertial formulation of the shallow water equations for efficient two-dimensional flood inundation modelling. *J. Hydrol.* **2010**, *387*, 33–45. [CrossRef]
40. Neal, J.; Villanueva, I.; Wright, N.; Fewtrell, T.; Bates, P. How much physical complexity is needed to model flood inundation? *Hydrol. Process.* **2012**, *26*, 2264–2282. [CrossRef]
41. Costabile, P.; Costanzo, C.; Macchione, F. Performances and limitations of the diffusive approximation of the 2-d shallow water equations for flood simulation in urban and rural areas. *Appl. Numer. Math.* **2016**. [CrossRef]
42. Almeida, G.A.M.D.; Bates, P. Applicability of the local inertial approximation of the shallow water equations to flood modeling. *Water Resour. Res.* **2013**, *49*, 4833–4844. [CrossRef]
43. Aricò, C.; Nasello, C. Comparative analyses between the zero-inertia and fully dynamic models of the shallow water equations for unsteady overland flow propagation. *Water* **2018**, *10*, 44. [CrossRef]
44. Martins, R.; Leandro, J.; Djordjević, S. Analytical solution of the classical dam-break problem for the gravity wave–model equations. *J. Hydraul. Eng.* **2016**, *142*, 06016003. [CrossRef]
45. Martins, R.; Leandro, J.; Djordjević, S. Analytical and numerical solutions of the local inertial equations. *Int. J. Non-Linear Mech.* **2016**, *81*, 222–229. [CrossRef]
46. Lauber, G.; Hager, W.H. Experiments to dambreak wave: Horizontal channel. *J. Hydraul. Res.* **1998**, *36*, 291–307. [CrossRef]
47. Bento, A.M.; Amaral, S.; Viseu, T.; Cardoso, R.; Rui, M.L.F. Direct estimate of the breach hydrograph of an overtopped earth dam. *J. Hydraul. Eng.* **2017**, *143*. [CrossRef]

48. Cestero, J.A.F.; Imran, J.; Chaudhry, M.H. Experimental investigation of the effects of soil properties on levee breach by overtopping. *J. Hydraul. Eng.* **2014**, *141*, 04014085. [CrossRef]
49. Leal, J.G.A.B.; Ferreira, R.L.; Cardoso, A.H. Dam-break waves on movable bed. In Proceedings of the Fluvial Hydraulics: River Flow 2002, Louvain-la-Neuve, Louvain-la-Neuve, Belgium, 4–6 September 2002; Bousmar, D., Zech, Y., Eds.; Balkema: Rotterdam, The Netherlands; pp. 553–563.
50. Bukreev, V.I.; Degtyarev, V.V.; Chebotnikov, A.V. Experimental verification of methods for calculating partial dam-break waves. *J. Appl. Mech. Tech. Phys.* **2008**, *49*, 754–761. [CrossRef]
51. Bukreev, V.I.; Gusev, A.V. Gravity waves due to discontinuity decay over an open-channel bottom drop. *J. Appl. Mech. Tech. Phys.* **2004**, *44*, 506–515. [CrossRef]

8

Application of Python Scripting Techniques for Control and Automation of HEC-RAS Simulations

Tomasz Dysarz

Department of Hydraulic and Sanitary Engineering, Poznan University of Life Sciences, ul. Wojska Polskiego 28, 60-637 Poznan, Poland; dysarz@up.poznan.pl or tdysarz@gmail.com.

Abstract: The purpose of the paper was to present selected techniques for the control of river flow and sediment transport computations with the programming language Python. The base software for modeling of river processes was the well-known and widely used HEC-RAS. The concepts were tested on two models created for a single reach of the Warta river located in the central part of Poland. The ideas described were illustrated with three examples. The first was a basic simulation of a steady flow run from the Python script. The second example presented automatic calibration of model roughness coefficients with Nelder-Mead simplex from the SciPy module. In the third example, the sediment transport was controlled by Python script. Sediment samples were accessed and changed in the sediment data file stored in XML format. The results of the sediment simulation were read from HDF5 files. The presented techniques showed good effectiveness of this approach. The paper compared the developed techniques with other, earlier approaches to control of HEC-RAS computations. Possible further developments were also discussed.

Keywords: river flow modeling; sediment transport simulation; automation of flow modeling; HEC-RAS controller; python scripting

---

1. Introduction

River flow modeling is a very popular approach in science and education, as well as in small, medium-size, and large technical projects in the areas of water management [1,2], river regulation [3,4], sediment transport [5], flood protection [6], and many others [7,8]. There are several professionally prepared and widely used commercial and non-commercial models for river flow, e.g., HEC-RAS [9], MIKE 11 [10], Delft Sobek [11], BASEMENT by ETH [12], and SRH1D [13]. The first of the mentioned models, HEC-RAS, is very specific. The acronym HEC means Hydrologic Engineering Center, and it is the name of the software developer. The second element, RAS, means River Analysis System. It defines the software application area which is focused on modeling of flow and transport processes in rivers, floodplains, and reservoirs. All of the modeled elements are solved with highly accurate numerical methods [9]. Additionally, HEC-RAS is freeware with a professionally developed graphical user interface. It makes application of the software easier, and it is considered to be the reason that HEC-RAS is so popular. Today a strong focus on modeling of uncertainty related to river processes is observed. It is a particularly crucial problem in the areas of model calibration [14,15], sediment transport [16,17], and flood hazard mapping [18,19]. The effective management of such computational tasks is not possible without automation of simulation and integration with other tools, e.g., GIS software. The examples analyzed in the paper present crucial automation techniques for the mentioned applications.

The main purpose of this paper is to present selected techniques for controlling HEC-RAS computations with the programming language Python. Control of river flow simulations with external scripts opens several new opportunities in the fields of flood hazard assessment, water management, and river hydraulics. It also enables application of more sophisticated methods in these fields, such as sensitivity analysis, automatic model calibration, and probabilistic flood mapping. The control of the HEC-RAS computations is possible due to the fact that this package is installed with the Component Object Module (COM). This COM library of methods is called HECRASController. The COM library enables outer access to the computational elements of HEC-RAS. The discussion presented here is focused on three specific examples. The ideas presented are not limited to the basic running of flow model simulation. Some concepts are aimed at integration of HEC-RAS with specific tools available in Python modules, e.g., SciPy, while others extend the capabilities of HECRASController and open access to more sophisticated data formats, e.g., XML (eXtensible Markup Language) and HDF5 (Hierarchical Data Format 5).

The HECRASController is a collection of Visual Basic subroutines and functions within the HEC-RAS package. It may be effectively used with VBA (Visual Basic for Applications) in Excel [20,21]. The main capabilities of this collection include running HEC-RAS and particular HEC-RAS editors, running computations plans, reading results of flow simulation, etc. One very interesting feature is the possibility of manipulation of roughness coefficients. It is a very important problem studied by many researchers, e.g., Yang et al. [22] and Lacasta et al. [23]. This capability is implemented in AHYDRA software [24]. The AHYDRA supports calibration of roughness coefficients and boundary conditions for steady flow conditions. The software cooperates with the initially configured HEC-RAS model, enabling automatic change of settings.

The capabilities of HECRASController are still limited. Even the main author promoting the use of this programming tool, Goodell [20,21], advises manipulation of the HEC-RAS data files in ASCII format. Another problem is the use of Visual Basic/VBA. Although it is a very handy programming language, its use is rather limited and still narrowing, e.g., VBA was replaced in ArcGIS by Python and its extension ArcPy. Hence, other languages are also implemented to handle HECRASController, e.g., MATLAB [25].

Python [26] is one of the most popular programming languages used today worldwide. It is easily applied in practical cases, as well as in highly scientific problems [27,28]. Its popularity follows from two factors: (1) relative simplicity and (2) a broad range of application areas. Considering the specific problem considered here, Python has several advantages and unique features. They include very easy access to text files in standard form, e.g., [26]; several specific libraries included, e.g., [29–33]; opportunities to handle the components of the Windows operation system, e.g., [34,35]; access to sophisticated data formats, e.g., [36,37]; and linking with a number of Windows application, e.g., [38,39].

It is no surprise that the first approach to control of HEC-RAS with Python was developed. The PyRAS module prepared by Peña-Castellanos [40] is one such approach. Unfortunately, PyRAS only transforms objects and functions of HECRASController available in VBA to Python. Hence, the limitations of HECRASController still apply. Another disadvantage is that the module was prepared for older versions of HEC-RAS, namely 4.1 and 5.0.0. It has not been developed further to date. Despite these facts, the codes of PyRAS could be an inspiration for independent development. Another open-source application linking Python and HEC-RAS is the example included in the package PyFloods [41]. Unfortunately, it is the only example for the previous version of HEC-RAS with some important elements hard-coded.

No general approach to linking Python and HEC-RAS has been proposed as yet, and it seems to us that it should be done on the basis of available Python modules for management of COM libraries. Another requirement is to develop the potential of HECRASController with additional Python modules, e.g., optimization of model parameters, access to specific data in XML files, and reading and handling of results in HDF5 files. These problems are addressed in the present paper.

The next section, Methods, includes general description of all important elements in the presented manuscript. Theoretical aspects of open channel flow, as well as Python scripting issues related to the applied codes, are discussed. In the section Results and Discussion, examples illustrating the invented concepts are described in detail. At the beginning of this section, the HEC-RAS models used are presented. In the final section, the summary, concluding remarks, and further developments are discussed.

2. Methods

2.1. HEC-RAS

HEC-RAS is a well-known hydrodynamic model for rivers and water reservoirs. The concepts applied in the package are well described by Brunner [9]. The HEC-RAS is applied for simulation of flow and transport processes in river networks, including floodplains and reservoirs. The modeled flow conditions include steady and unsteady longitudinal flow. The first is based on the simple equation of energy balance. The form of this equation implemented in HEC-RAS is shown below:

$$z_1 + h_1 + \alpha_1 \frac{u_1^2}{2g} = z_2 + h_2 + \alpha_2 \frac{u_2^2}{2g} + h_e, \tag{1}$$

where z is bottom elevation, h is the depth, and u is mean velocity in the channel cross-section. α is called the St. Venant coefficient and plays the role of correction factor, including the effect of velocity profile non-uniformity. g is well known acceleration of gravity. The Equation (1) is written for gradually-varying flow, when the assumption of hydrostatic pressure distribution may be suitable. The subscripts 1 and 2 denote two different cross-sections in the same channel reach. The basic assumption is that the cross-section number 1 is located upstream of the cross-section number 2. The first three terms of both sides represent the potential energy of the stream, work of pressure forces, and kinematic energy of the stream. The last element of the right side, h_e, describes friction losses due to bed and banks influence on flowing water. It also includes effects of channel contraction and extension. If the depth is known in one cross-section, on this basis, the depth may be also determined in the second cross-section. To calculate the distribution of the depth along the channel, its value must be known in one cross-section a priori. In practical cases, this cross-section is the inlet or outlet boundary of the channel reach. Hence, the condition is frequently called "boundary condition", but it is rather hydraulic jargon than strict mathematical terms. The discharge Q is a very important parameter of the Equation (1). The kinematic energy terms, as well as friction losses, depend on the magnitude of the flow. The influence of floodplains is included in the calculation of the St. Venant coefficients and calculation of weighted distance between cross-sections. The last element is important for the determination of the total losses h_e. More detailed description of this equation and its implementation may be found in Reference [9].

For description of the second model, the unsteady flow, the numerical solution of the St. Venant system of equations for 1D flow is implemented. This system consists of two partial differential equations shown below.

$$\frac{\partial A}{\partial t} + \frac{\partial(\phi Q)}{\partial x_c} + \frac{\partial[(1-\phi)Q]}{\partial x_f} = 0 \tag{2}$$

$$\frac{\partial Q}{\partial t} + \frac{\partial}{\partial x_c}\left(\phi^2 \frac{Q^2}{A_c}\right) + \frac{\partial}{\partial x_f}\left[(1-\phi)^2 \frac{Q^2}{A_f}\right] + gA_c\left(\frac{\partial H}{\partial x_c} + S_{fc}\right) + gA_f\left(\frac{\partial H}{\partial x_f} + S_{ff}\right) = 0 \tag{3}$$

The Equation (2) describes mass balance in the open channel flow. The second equation represents the momentum balance. It is written for gradually-varying flow conditions in the channel. There are two independent variables: t—time and x—distance. The distance is measured not only along the channel, but also along the floodplains. Hence, there are denotations x_c and x_f for channel and floodplain flow paths, respectively. It is important that the pair of distance (x_c, x_f) is unique for each cross-section. Q is still the total discharge in the cross-section and g is acceleration of gravity. H is the water surface elevation, which is the sum of the bottom elevation z and the depth h shown in the previously discussed Equation (1). It is assumed that the transversal slope of the water surface is zero in the single cross-section. Hence, H is constant in the cross-section, but varies along the channel. Denotation A without subscript is the total cross-section area in a single cross-section. However, A_c and A_f are the area of channel and floodplain parts of the cross-section, respectively. The hydraulic slopes describing the friction losses along the channel and along the floodplains are denoted as S_{fc} and S_{ff}. The HEC-RAS specific element is coefficient ϕ. It describes the distribution of the discharge between the parts of the cross-section. Hence, ϕQ is the flow along the channel and $(1-\phi)Q$ is the flow along the floodplains. The coefficient ϕ depends on the hydraulic parameter called conveyance, which is the function of the water surface elevation H and other elements of the cross-section, such as shape and roughness. In the Equation (2), the first term represents the local storage of water, whilst two other terms describe the net inflow of water along the channel and floodplains. In Equation (3), the first three terms represent the inertia forces, the terms including gradient of H describe the gravity and pressure forces, and the terms with hydraulic slopes S_{fc} and S_{ff} represent the friction forces. The solution of the system (2)–(3) are two functions of time t and distance pair (x_c, x_f). These are discharge Q and water surface elevation H. To solve the system (2)–(3), the initial and boundary conditions have to be specified. The basic forms of the initial condition are known distributions of Q and H along the modeled reach. In each inlet and outlet cross-section, there should be one boundary condition imposed in gradually-varying flow. These conditions may have a basic form of discharge or water stage hydrographs. They can also be imposed as other more complex relationships of discharge and water stage, e.g., normal depth, rating curve, etc. The Preissmann scheme is used for the approximation. More detailed description of the St. Venant equations used in HEC-RAS and their implementation may be found in Reference [9].

The number of hydro-structures, such as bridges, culverts, etc., is also available for incorporation with the prepared water system. Both these mathematical models of open channel flow are well described in many books [42–45], as well as software manuals mentioned above [9–13]. In HEC-RAS, there is also a module enabling so-called quasi-unsteady flow simulation, i.e., simplified flow simulation in unsteady conditions on the basis of the energy equation. In the latest versions of the package, a 2D flow module is also available. The HEC-RAS modules for simulation of transport processes include different transport of solutes, heat, and sediments with deposition and erosion. Heat and solute transport is based on numerical solution of convection–dispersion equations with source terms. Models of this kind are described in some books on mathematical modeling of river processes, e.g., References [9,46,47]. The QUICKEST scheme with the ULTIMATE limiter is applied as an approximation method [48]. The sediment transport is modeled with the standard Exner equation, enabling simulation of different fractions of sediments. The Exner equation is shown below:

$$(1-p)B\frac{\partial z}{\partial t} + \frac{\partial Q_s}{\partial x} = 0 \tag{4}$$

where x is the distance along the channel and t is time. z represents bottom elevation, B is the width of this bottom part, where the sediments are deposited or removed from. p is porosity of the bed layer. Q_s is volumetric intensity of sediment transport. This element is described by subjectively chosen empirical formulae, e.g., Meyer-Peter and Müller (MPM). It depends on the hydraulic parameters and the sediment characteristics. The first term represents the local change of the sediment deposited. The second term is the net inflow of sediments. The solution of Equation (4) is the change of bottom elevation z in time t, and along the channel x. The initial and one boundary condition have to be imposed to solve the problem. The initial condition is simple initial bottom elevation profile along the modeled channel. The boundary condition is imposed in the inlet cross-section. It may have the form of bottom elevation changes, sediment inflow intensity varying in time, or equilibrium sediment load in the stream. In HEC-RAS, the applied solution method is the basic upwind scheme [9,45]. More details may be found in Reference [9].

The HEC-RAS package also includes several useful tools for data preparation and results processing. These tools include the module for GIS data processing called RAS Mapper [49].

2.2. *HECRASController*

HECRASController is a part of the HEC-RAS application programming interface (API). It is a collection of programming tools: Classes, functions and subroutines. Access to the HEC-RAS elements is possible because this package is compiled as the Component Object Module (COM). Any program able to read COM DLL (Dynamic-Link Library) may be used to control HEC-RAS computations. The most convenient approach is to use Visual Basic in any form, i.e., as a separate suite for developers or linked with another application, e.g., MS Excel.

The most detailed description of the above-mentioned collection of programming tools was presented by Goodell [20,21]. Appendix A of his book [20] is a very good source of information about syntax, and usage of HECRASController functions and subroutines. The examples presented there, illustrate the great potential of these concepts. VBA in MS Excel is very powerful in computing, as well as in data analysis [50]. Linking HECRASController with VBA in Excel opens new opportunities for handling simulation data and results, as presented by Goodell [20,21]. The controller makes it possible to open HEC-RAS projects, run simulations, read results, and store them in other specific formats. It is also possible to open HEC-RAS editing tools for manual configuration of input data. The controller also enables direct manipulation of project data, but its capabilities in this area are rather limited. There are three procedures for changing roughness coefficients, and one for setting the area of the element called storage area.

Although this concept is not new and seems to be very useful, there are still areas of HEC-RAS computations in which controller functions are not able to help. The basic HEC-RAS data, such as geometry and flow boundary conditions, are stored in ASCII files. They may be accessed by manually written code. Goodell [20] has advised such an approach. However, some HEC-RAS data are stored in formats other than pure text, e.g., sediment data in XML format. Access to such data requires application of additional modules. The results of more complex simulations are also beyond HECRASController access, e.g., 1D sediment and 2D flow simulation results are stored only in HDF files.

In addition, the opportunities afforded by the controller depend on the application areas of VBA. This powerful language is widely used. However, there are also very important areas where VBA has been withdrawn, e.g., scripting in ArcGIS. There are also areas in which VBA has never been present, e.g., scripting in QGIS. Relying on VBA for automation of HEC-RAS simulation could limit possible applications.

2.3. *Python Scripting and Applied Modules*

Python is one of the most popular programming languages today, and it is still under development. In this paper, version 2.7.12 was used. Its usefulness and popularity in many areas have been reported in References [51–53]. This scripting language is relatively simple for the beginner, but it is also very powerful if applied by an experienced coder. A description of this language may be found in many books or on internet websites, e.g., Downey et al. [54], Python Software Foundation [26]. Considering the problem considered here, Python has several specific advantages:

(1) The basic language is extended by additional modules for numerical simulation and general scientific analysis, e.g., NumPy and SciPy [29–31].
(2) The usefulness of Python in numerical simulations may be extended by the user with modules enabling transformation of Fortran or C routines to Python code, e.g., F2Py [32] and ctypes [33] modules.
(3) It is possible to access COM libraries using additional modules, e.g., Pywin32 [34,35].
(4) Python enables access to more sophisticated data structures, e.g., XML [36] and HDF5 [37].

(5) There are Python modules integrating this language with geoprocessing software, such as ArcGIS [38] and QGIS [39].

VBA is a compiled language, which makes it faster in comparison with any scripting language in stand-alone applications. However, in the analyzed cases, the programming language is used to run an external application, HEC-RAS. Hence, the efficiency of the external application determines the efficiency of the whole concept, which reduces this advantage of VBA. The above list of Python features is the list of undoubted advantages of this scripting language over standard VBA. In some cases, the application of Python is as easy as VBA, e.g., access to COM libraries or XML data. Other elements, such as libraries for numerical computations, make Python implementation more effective. There are also features that are unique to Python, e.g., access to HDF5 data, and linking with Fortran and C. The undeniable advantage of Python is also clearly visible in the area of geoprocessing. This area is important for any hydrological or hydraulic modeling of flow and transport processes in natural water systems. As mentioned above, the former applications of VBA in older versions of ArcGIS are now replaced by modern scripting languages, such as Python [38], JavaScript [55], and R [56]. Python is the primary scripting language used in commercial Esri software, such as ArcGIS 10.x and ArcGIS Pro. This is also the only scripting language implemented in QGIS, which is the open source and cross-platform equivalent of Esri's packages.

Several Python modules are used in this research, including: (1) Pywin32, (2) NumPy, (3) SciPy with optimization, (4) ElementTree XML API, and (5) HDF 5 for Python. The first of them, the Pywin32 module, is used to access COM objects and COM servers in Python code [34,35]. Brief descriptions of available Pywin32 packages may be found in the PythonCOM Documentation Index [35]. The part of the module win32com.client is used in this paper. To access COM objects in the presented scripts, the Dispatch function is applied.

NumPy is a well-known module including objects and tools for scientific computing [20,57]. The functions and objects from the NumPy module work faster in numerical applications than standard Python functions and variables. The reason is that the main elements of NumPy were translated from Fortran and implemented in Python. This mechanism of method transfer may be continued by the user with the F2PY package [32] or ctypes module [33]. In the present research, only the array constructors available in NumPy were applied. These are array and empty methods. The string, integer, and float element arrays were used. However, extension of some presented examples, e.g., Examples 2 and 3, enables application of one's own methods and opens the way for more sophisticated techniques.

Another important package used here was SciPy [29,31]. The optimization part of SciPy was implemented in Example 2. The scipy.optimization module contains several optimization methods and root finding algorithms [31]. In this study, the most basic and well-known Nelder-Mead downhill simplex algorithm was used. A detailed description of it was provided by Press et al. [58].

The sediment simulation data in HEC-RAS are stored in XML format. It is a text-based markup language, derived from the Standard Generalized Markup Language. A detailed description of XML format and usage may be found in Reference [59]. To access data stored in an XML file, the ElementTree module is applied as described in References [35,60]. This package includes objects of the Element type. They are container objects designed to store hierarchical data structures, such as XML. The Element type is a cross between a list and a dictionary in Python. The module imported is xml.etree.ElementTree.

The results of the sediment simulation are stored in HDF5 format. Large sets of numerical data are easily accessed in HDF5 through the well-described hierarchy. A description of this format may be found in HDF Group [37]. To access these data the h5py module is applied [61].

2.4. General Remarks on Analyzed Examples

Three examples were tested. The first was a code of a basic simulation composed in a way that enabled comparisons with the examples presented by Goodell [20,21]. The comparison revealed the most important differences between VBA and Python, such as initialization of HECRASController, access to controller methods, handling of data and results.

The second example was the calibration of the steady flow model. Although there is a procedure for the calibration of roughness coefficients in HEC-RAS, it works only with the unsteady flow model. However, calibration of the steady flow model is useful in many cases. Additionally, the approach presented in this example could also be applied in sensitivity analysis of the model performance or probabilistic flood risk mapping. The Nelder-Mead simplex algorithm was applied to search for the proper roughness coefficients. This is routine from the optimization part of the SciPy module, namely scipy.optimize. It works for similar purposes as the AHYDRA application [24], but the optimization algorithm is applied instead of manual alteration of roughness coefficients. The code is prepared in such a way that the applied Nelder-Mead method may be replaced with any optimization algorithm, e.g., the Genetic Algorithm as proposed by Leon and Goodell [25] with MATLAB.

The objective of the third example was to automatically change sediment samples in sediment data and run a sediment simulation. Such an approach has not been found by the author in the literature to date. The sediment sample data describe the percentage of sediment fractions in a sample. In HEC-RAS, they are called gradations. This function plays the role of a model parameter. It is necessary for the calculation of the total sediment transport, as well as transport of particular fractions. However, in practical applications it should be a piece of material sampled from the river bed. Collection of sediment samples is not very common. The measurements are local and extrapolated to a region, e.g., the cross-section. Hence, such measurements are highly uncertain, which is an additional reason for presentation of Example 3. This example is very simple, but may be extended for more sophisticated cases. It requires handling of XML and HDF5 data with special Python modules. Although, the control of sediment simulation is a more complex problem, the example is focused on change of the sediment samples, because these data are accessible in the most difficult way. If the sediment fractions may be changed by the Python script, all other elements of the sediment simulation, e.g., sediment transport formula, parameters of chosen formula, method for calculation fall velocity, etc., may be changed too. Hence, the adjustment of sediment sample in the XML file is a good illustrative example of sediment data management in general.

There are two main application areas for the mechanism presented in Example 3. The first is obviously the uncertainty or sensitivity analysis of sediment transport simulation. Considering the complexity of such computations, the analysis of sensitivity is too difficult to perform in the standard manual way. However, it is necessary if the sediment transport models are going to be used for forecast purposes, which sometimes happens today. The sediment, as well as flow data, are too uncertain to neglect this problem. The second application area is related to historical data and their processing. Sometimes data collected in the past include the full longitudinal profiles, as well as cross-sections, for different moments in time. The present author had at his disposal such data for Ner river collected for 1983, 2003, and 2007 [62]. These data were collected by the company BIPROWODMEL, for the analysis of regulation requirements. They illustrate the effects of the sedimentation process along the analyzed channel. Unfortunately, the company taking measurements did not collect sediment samples at the same time. Hence, the modeling of sediment transport in the Ner river is possible only if the model is calibrated with respect to these data. There is one important aspect which should be carefully considered. The calibration should also consider other factors, such as the sediment transport formula. However, the differences in the role and structure of the factors influencing the sediment transport simulation lead to the design of separate computational processes for each factor. One such process should be searching for an optimal sediment sample or samples, with other factors kept as constants. The third example presents the powerful capabilities of Python in this area.

An example with change of data stored in ASCII files of HEC-RAS is not presented here. Such examples have been discussed by other authors [20,25] with VBA and MATLAB. In the author's personal opinion, the handling of the ASCII files with Python is easier than with the other languages mentioned above. Hence, such an example would not bring anything new. The third and more difficult example, with XML and HDF files, may be treated as a help or guide in dealing with ASCII files. If Python enables reading and writing of information stored in such complex formats as XML and

HDF5, and it is quite easy to access the data in the ASCII files, it means that the presented scripting mechanisms could be applied to manipulate all data and results of any HEC-RAS model. This is also a huge advantage of the presented approach.

2.5. Applied Rules of Coding

There are important differences between VBA and Python, which have to be discussed before the particular examples are presented. To explain them, the well-known examples given by Goodell [20,21] are quoted and compared with the devised Python code.

The first difference is declaration of variables. In Python, the variables are created dynamically during running of the program. Their type is recognized as the values that are assigned to them. The assignment operator may change this type, any time in the script. There are exceptions to these rules, e.g., special variables such as NumPy arrays. Their shape and type have to be declared before the variable is used. In VBA, dynamic creation is also available. However, it is more convenient to switch off this mechanism with the command Option Explicit at the beginning of the module. Then the variables have to be declared statically with the Dim statement.

To use the HECRASController object in the VBA code, it has to be declared as a variable [20,21]. In Python, the object is created in a similar way by calling the Dispatch function from the module win32com.client. Then all the functions and subroutines of the HEC-RAS controller are available in Python. The access is similar to the VBA calling of the HECRASController. The scheme is shown in Scripts 1–4.

In Script 1. HECobject is the HECRASController variable in the project, arguments is the list of input parameters and return_values is the list of the values we would like to get from the subroutine. The HECobject has to initialized with the Dispatch function of the win32com.client module. Although it seems to be quite simple, there is one technical problem. The subprograms in VBA and Python have different structures and different mechanisms of data exchange with the main body of the program. There are two types of arguments used by VBA subroutines. These are the "called-by-value" and "called-by-references" arguments. Exchange of information between the subprogram and the main program based on "by-value" and "by-reference" mechanisms are well known. The arguments of the first type are used only as inputs. The second calling mechanism may work as input and output. When the function construction is used, another method for output is available, i.e., the returned value of the function. Although these mechanisms are very common and are applied in many programming languages, e.g., VBA, Pascal, Fortran, C/C++, etc., the subprograms used in Python are constructed in a different way. In general, the Python subprograms are functions. The parameters set in the function head are only inputs. The output is exchanged with the outer code only by returned values of the function. It is important to indicate that the function may return a list of values and each of them can be of different type.

```
import win32com.client
HECobject = win32com.client.Dispatch("RAS503.HECRASController")
# ...
return_values = HECobject.method(arguments)
```

Script 1. Code sample 1. Initialization of HEC-RAS object and schematic use of its methods.

To adopt HECRASController methods in Python, a change in the calling is necessary. Imagine such an example: One of the subroutines determines the value of the parameter "called-by-reference". Let this argument be x. In the case of arguments "called-by-reference", the variable storing the output value must be created before the subroutine is run. In Python, the variable storing the None value must be created, which means the variable without a value assigned. Such a statement is equivalent to the variable declaration in VBA and other programming languages. Then the x variable is set as the argument of the subroutine and repeated as the variable in the return list, as shown in Script 2.

```
import win32com.client
HECobject = win32com.client.Dispatch("RAS503.HECRASController")
# ...
x = None
x = HECobject.method(x)
```

Script 2. Code sample 2. Initialization of the HEC-RAS object and use of its methods requiring passing of the argument by reference.

In more complex cases it should be noted that the return list should follow the sequence of the argument list. An example is presented in Script 3.

```
import win32com.client
HECobject = win32com.client.Dispatch("RAS503.HECRASController")
# ...
x, y, z = None, None, None
x, y, z = HECobject.method( x, y, z)
```

Script 3. Code sample 3. Passing the arguments by reference to the method of the HEC-RAS object playing the role of a subroutine.

If the function is run, the first variable in the return list stores the original return value of the function. It is illustrated in Script 4. The variable RV used there is the return value of the function. This approach is used in all three presented examples.

```
import win32com.client
HECobject = win32com.client.Dispatch("RAS503.HECRASController")
# ...
x, y, z = None, None, None
RV, x, y, z = HECobject.method( x, y, z)
```

Script 4. Code sample 4. Passing the arguments by reference to the method of the HEC-RAS object playing the role of a function with one return value.

The main codes presented in Scripts 5–8 consist of the main elements necessary to access capabilities of the HEC-RAS application in Python code. There are also additional codes written in support.py. They are used for support of the main scripts. The support.py file contains the subroutines for data loading from ASCII files, for handling files and directory names, etc. All these additional subroutines are subject to the coder and may be written in several ways. Hence, they are not discussed here. The additional codes are loaded to the particular scripts, if there is such a need.

3. Results and Discussion

3.1. Case Study—Model of a River Reach

Two HEC-RAS models, called model A and model B, were used to test the presented concepts. Both of them represented a short reach of the Warta river in the central part of Poland. The reach is about 10 km long. The average width and depth of the main channel are 40 m and 1.5 m, respectively. There are two bridges along the reach. The estimated mean flow in the analyzed channel was about 47.45 m^3/s. This value was estimated on the basis of the hydrological data collected at the Sieradz gauge station during 1970–2013. The observed flows for the same period vary from the minimum of 15 m^3/s to the maximum of 408 m^3/s [63].

In both models, the number of cross-sections was 35. The average distance between cross-sections was about 300 m. The differences between the model A and model B include the geometry and configuration of flow conditions. Model A was based on the full geometry, including bridges. The flow conditions were steady with basic discharge of 120 m^3/s for the whole reach. The downstream boundary condition was the simple Normal Depth with bottom slope as a parameter. This model was used for basic simulation in Example 1. It was also applied in Example 2, for the test of calibration.

In model B, the geometry was simplified. The bridges were removed, because the model was used for simulation of sediment transport with different sediment samples. Hence, the configuration of flow conditions was based on a quasi-unsteady flow module. The upstream boundary condition was Flow Series with a constant discharge of 269 m^3/s. In the downstream boundary the Normal Depth was applied once again. The slope equals 0.2‰. The sediment transport intensity was calculated on the basis of the Meyer-Peter and Müller (MPM) formula. The simulation horizon was set to 4 months, with a 6-h time step. The time step guarantees necessary numerical stability and accuracy. The simulated period was rather short. In practice, the changes of the river bed due to sediment deposition and erosion are observed in longer periods, e.g., years. Such a configuration is chosen to avoid too long computational time in this illustrative example.

3.2. Example 1—Basic Simulation

The first example presented here was composed in a way similar to the introductory example discussed by Goodell [20,21] for the reach of Beaver Creek. In this paper, model A described above was used. The main scheme of computations is shown in Figure 1. The code of this example is presented as Script 5.

```
import win32com.client
import os, numpy
from support import sc1_ShowNodes

# Initiate the RAS Controller class
hec = win32com.client.Dispatch("RAS503.HECRASController")
hec.ShowRas()                        # show HEC-RAS window
# full filename of the RAS project
RASProject = os.path.join(os.getcwd(),r'test01\test01.prj')
hec.Project_Open(RASProject)         # opening HEC-RAS
# to be populated: number and list of messages, blocking mode
NMsg,TabMsg,block = None,None,True
# computations of the current plan
v1,NMsg,TabMsg,v2 = hec.Compute_CurrentPlan(NMsg,TabMsg,block)
# ID numbers of the river and the reach
RivID,RchID = 1,1
# to be populated: number of nodes, list of RS and node types
NNod,TabRS,TabNTyp = None,None,None
# reading project nodes: cross-sections, bridges, culverts, etc.
v1,v2,NNod,TabRS,TabNTyp =                                  \
hec.Geometry_GetNodes(RivID,RchID,NNod,TabRS,TabNTyp)
# ID of output variables: WSE, ave velocity
WSE_id,AvVel_id = 2,23
TabWSE = numpy.empty([NNod],dtype=float)    # NumPy array for WSE
TabVel = numpy.empty([NNod],dtype=float)    # NumPy array for velocities
for i in range(0,NNod):              # reading over nodes
    if TabNTyp[i] == "":             # simple cross-section
        # reading single water surface elevation
        TabWSE[i],v1,v2,v3,v4,v5,v6 =                 \
        hec.Output_NodeOutput(RivID,RchID,i+1,0,1,WSE_id)
        # reading single velocity
        TabVel[i],v1,v2,v3,v4,v5,v6 =                 \
        hec.Output_NodeOutput(RivID,RchID,i+1,0,1,AvVel_id)
hec.QuitRas()             # close HEC-RAS
del hec                   # delete HEC-RAS controller

sc1_ShowNodes( NNod, TabRS, TabNTyp, TabWSE, TabVel )
```

Script 5. Example 1. Calculation of a single water profile.

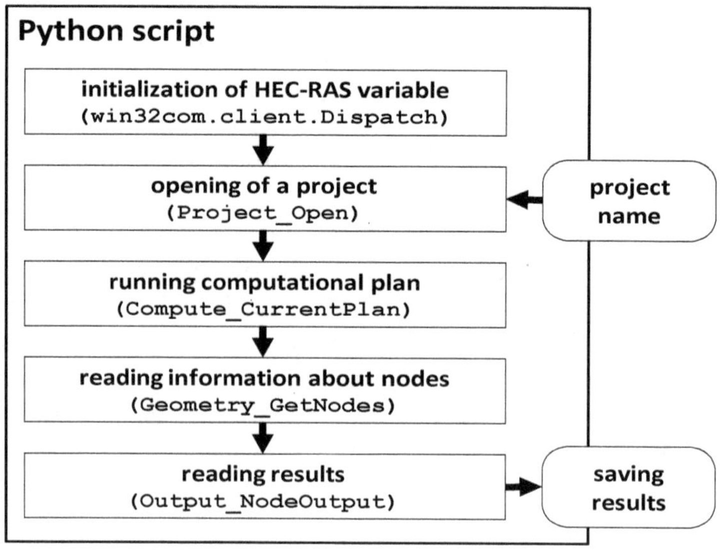

Figure 1. Main elements of computations in Example 1.

As can be seen in the first block of the diagram in Figure 1, the computations have to start with initialization of the variable, which is the HECRASController object. In the brackets, the method used for this purpose is indicated. In the second step, the HEC-RAS project is open. The name of the project is the necessary input. Then the computations are run and the results are processed in two steps. The first is reading information about computational nodes. It is done with one of the "geometric" routines of HECRASController. The second step is reading the output of computations. Then the results may be saved, displayed, or applied in another simulation.

The program presented in Script 5 includes four main steps reflecting the sequence presented in Figure 1. These are: (a) Opening HEC-RAS and loading the project (lines 9–10), (b) running the simulation (lines 12–14), (c) reading the output variables (lines 16–35), and (d) processing the output variables (line 37), e.g., displaying them on the screen. At the beginning, the necessary modules and subroutines are loaded. They are win32com.client for handling COM libraries [34,35], os for management of files and directories and numpy for numerical data. Additionally, the subroutine sc1_ShowNodes is loaded from the script support.py. It is used for displaying results.

The first step of the program is initialization of the HECRASController variable hec (line 6). The Dispatch function of the win32com.client module is applied. The HEC-RAS version used is 5.0.3. When the hec variable is created, the ShowRas and Project_Open functions may be used to open the HEC-RAS windows and load the project. Detailed descriptions of these and other functions used in the paper were provided by Goodell [20]. The next two methods of HECRASCcontroller used are Compute_CurrentPlan (line 14) and Geometry_GetNodes (line 21). The first function runs the current computational plan. The second one reads tables of River Stations, TabRS, and node types, TabNTyp. Both of them need initialization of proper arguments (lines 12 and 18). The approach presented in Code Samples 2-4 is applied. The river and reach ID numbers are also set in line 16. The variables v1, v2, and in general vX, where X is some digit, are used when the return value does not have to be stored.

The next step is reading of the output variables. It starts at line 23, with setting of the ID numbers for water stages and velocities, WSE_id and AvVel_id. Then the NumPy arrays, TabWSE, and TabVel, for storing the results are prepared (lines 24–25). The results are read in a loop over nodes (line 26–33), with calls to the function Output_NodeOutput. The values of water stages and velocities in regular cross-sections are accessed. In such nodes, the node type is an empty string (line 27).

Finally, the HEC-RAS windows is closed with QuitRas. Then the hec variable may be deleted. The last element of the code is the display of results. It is performed by sc1_ShowNode from the support module. A screenshot with the run of Example 1 and display of results is shown in Figure 2.

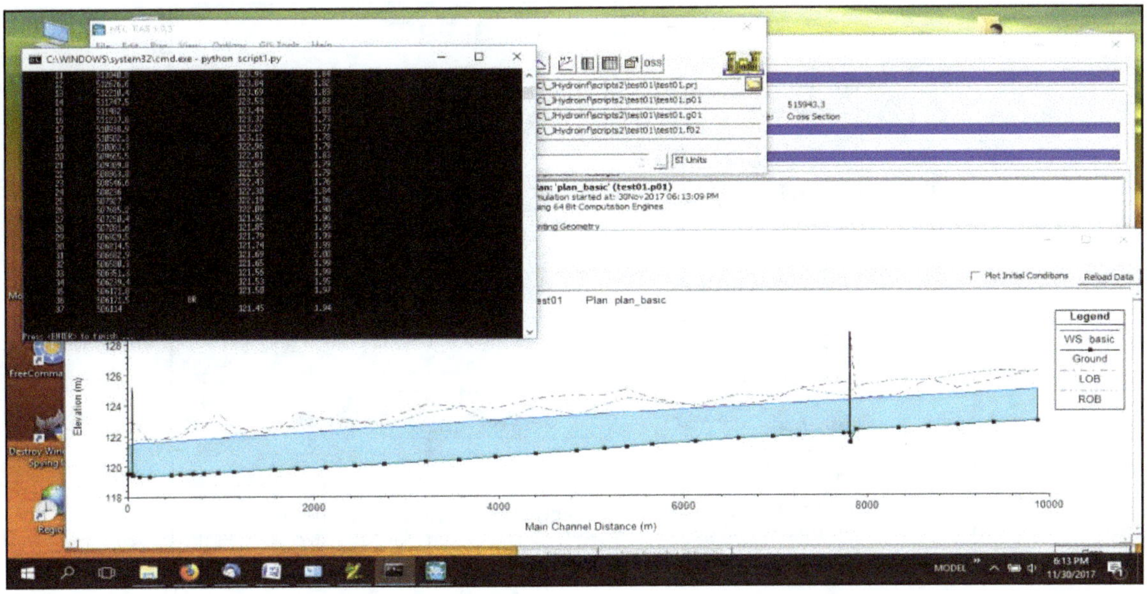

Figure 2. Run of Example 1 and display of results.

3.3. Example 2—Calibration of Roughness Coefficients

The problem of model calibration was considered. Model A described above was used here. The roughness coefficients were sought. The computations were performed in the steady mode. The calculated water surface profile should fit the imposed "observed" values. In fact, the "observed" values were calculated earlier for known roughness coefficients. Very simple distribution of their values was assumed. For the main channel along the whole reach, the value 0.025 was applied. For the left and right floodplains, one value of 0.04 was set in the same way. The purpose of the program was to find the set of optimal coefficients. It was assumed that only three values were necessary. Hence, the dimension of the problem was relatively small, but the generality of this solution is not lost.

The "observed" values were determined for each cross-section along the reach. However, they were applied with different frequencies in the search algorithm. It means that only some values were selected, and the simulated water surface was fitted only in the selected cross-sections. Hence, the objective function is as follows:

$$F(x) = \sqrt{\frac{1}{N}\sum_{i=1}^{N}\left(WSE_i^{(c)} - WSE_i^{(o)}\right)^2} \tag{5}$$

In the above formula, N is number of observations, $WSE_i^{(c)}$ is the computed water surface elevation, and $WSE_i^{(o)}$ is the "observed" one determined in the same i-th cross-section. This function should be minimized to find the optimal set of roughness coefficients x, where $x = [n_{LOB}, n_{Ch}, n_{ROB}]$ for the left floodplain, the main channel, and the right floodplain, respectively.

The algorithm for calculation of the objective function is presented in Figure 3. It starts with processing of the input parameters. At this step, the optimization variable is transformed into computational roughness coefficients. Then the coefficients are set in all model cross-sections. This step requires access to the global variable representing the object of the HECRASController. The next elements of the algorithm are similar to the operations presented in Figure 1. The computations start and the results are read. In both these steps, the HECRASController object is processed. To read the results properly, identification of the nodes where referenced or "observed" values of the water surfaces are located is necessary. This information is taken from another global variable. Then the calculated and referenced water surface elevations are used to determine the final value of the objective function according to Equation (5). The reference water surfaces are also stored in the global variable.

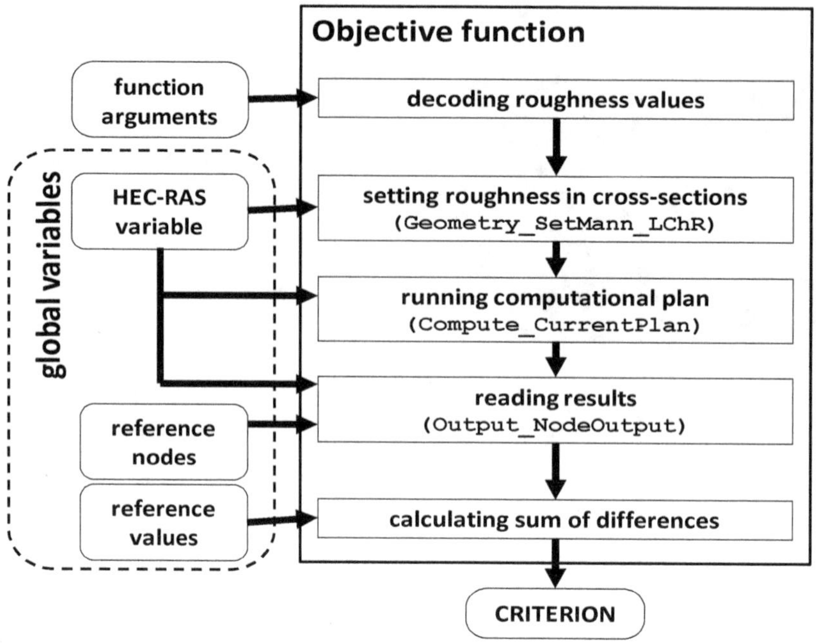

Figure 3. Algorithm for calculation of objective function values.

The main elements of the computations in this example are shown in Figure 4. The computational process starts with initialization of the HECRASController object. Then the HEC-RAS variable is used for opening of the HEC-RAS project, setting the current computational plan and running preliminary computations. These steps are necessary to compare the information about the reference nodes and values with the structure of the computational model. For this purpose, the project name and the reference values must be loaded. After these steps, the object of the HECRASController is ready and the numbers of computational nodes for comparisons with reference nodes are known. After loading of the initial estimate, which is the set of preliminary roughness coefficients, the optimization process starts. It uses the objective function presented in Figure 3.

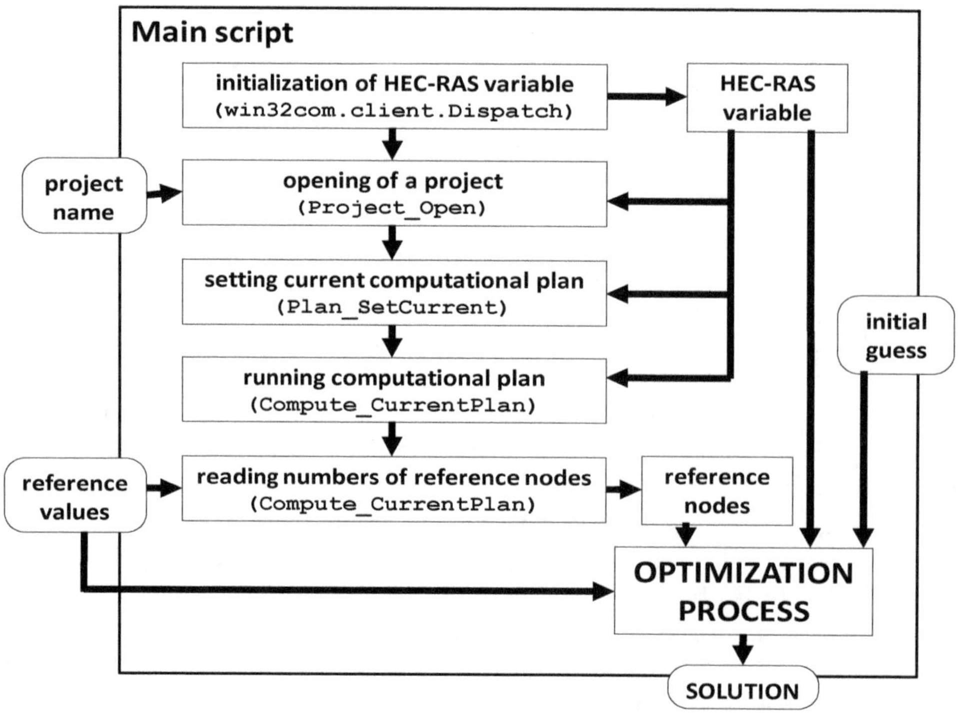

Figure 4. Main computational elements of Example 2.

The starting point is an arbitrarily chosen set of roughness coefficients. The values chosen were 0.08 for the left floodplain, 0.014 for the main channel, and 0.08 for the right floodplain.

In Script 6, such modules as win32com.client, os, math, numpy and scipy.optimize are necessary. They are loaded at the beginning of the script. The function for reading the "observed" values is also loaded from the support module.

```
import win32com.client, os, math
import numpy as np
from scipy import optimize as opt
from support import sc2_ReadObs

def ObjFun(x):                         # objective function
    global hec,RivID,RchID,WSE_ID,ProfID,RivName,RchName,nRS,RScmp,RStyp
    global Nobs,iRSobs,WSEobs
    nLOB,nCh,nROB = x[0],x[1],x[2]
    for i in range(0,nRS):
        ErrMsg    = None       # list of error messages
        v0,v1,v2,v3,v4,v5,v0,ErrMsg =                              \
        hec.Geometry_SetMann_LChR(RivName,RchName,RScmp[i],nLOB,nCh,nROB,ErrMsg)
    NMsg,ListMsg,block = None,None,True          # no. and list of messages, blocking
    v1,NMsg,ListMsg,v2 = hec.Compute_CurrentPlan(NMsg,ListMsg,block)
    TotSum = 0.0        # total sum of square errors
    for i in range(0,Nobs):
        wse,v1,v2,v3,v4,v5,v6 =                                    \
        hec.Output_NodeOutput(RivID,RchID,iRSobs[i],100,ProfID,WSE_ID)
        TotSum += (wse - WSEobs[i])**2         # single square error added
    TotSum = math.sqrt( TotSum / Nobs )        # final value
    return TotSum

def PokazIter(xk):                     # iteration control function
    print 'nLOB=%15.6f, nCh=%15.6f, nROB=%15.6f, ' %            \
    (xk[0],xk[1],xk[2]),
    print '     funk=%15.6f' % (ObjFun(xk))
    return 0

# observed: number, RSes as strings, RSes as floats, WSEs
Nobs,lRSobs,dRSobs,WSEobs = sc2_ReadObs('test01','ObsH','ObsH1x.txt')
# init HEC-RAS Controller
hec = win32com.client.Dispatch('RAS503.HECRASController')
RivID,RchID = 1,1      # ID of river and reach
ProfID,WSE_ID = 1,2    # ID of profile, ID of WSE variable
projekt = os.path.join(os.getcwd(),r'test01\test01.prj')      # project file
hec.Project_Open(projekt)                     # opening HEC-RAS
hec.ShowRas()                                 # show HEC-RAS window
# setting current computational plan
test1 = hec.Plan_SetCurrent('plan_basic')
# river and reach names
RivName,RchName = 'Warta','Dop_Jeziorsko'
# number and list of messages, blocking mode
NMsg,TabMsg,block = None,None,True
# computations of the current plan
v1,NMsg,TabMsg,test2 = hec.Compute_CurrentPlan(NMsg,TabMsg,block)
# numbers of nodes with observations
iRSobs = np.empty([Nobs],dtype=int)
for i in range(0,Nobs):
    iRSobs[i],v1,v2,v3 = hec.Output_GetNode(RivID,RchID,lRSobs[i])
# computational: number, RS and types
nRS,RScmp,RStyp = None,None,None
v1,v2,nRS,RScmp,RStyp = hec.Output_GetNodes(RivID,RchID,nRS,RScmp,RStyp)

print "\nIteration process: Nelder-Mead simplex"
x0 = np.array([0.08,0.014,0.08])              # initial solution
Xopt = opt.fmin(ObjFun,x0,callback=PokazIter)  # optimization

hec.QuitRas()          # close HEC-RAS
del hec                # delete HEC-RAS controller
```

Script 6. Example 2. Calibration of HEC-RAS steady flow model.

Script 6 includes two functions. The first one plays the role of an objective function and is called ObjFun. It is an argument of the optimization algorithm. Its structure is discussed below. The second one, PokazIter, is used by the same algorithm to display results of the single iteration.

The main body of the script starts with the reading of "observed" values in line 31. They are stored as lists of River Stations, lRSobs and dRSobs, and a list of water surface elevations, WSEobs. Then the HEC-RAS variable, hec, is created (line 33). Before the project is opened (line 37), the HEC-RAS window is displayed (line 38) and the current plan is set (line 40), the project constants are set (line 34–35). They are the river and reach IDs, RivID and RchID, profile number, ProfID, and ID of the output variable storing water surfaces, WSE_ID. The names of the river and the reach, RivName and RchName (line 42), are also hard-coded in the script, though they could be read using the methods of the HEC-RAS controller.

The first computations are run in line 46. The necessary variables are prepared earlier (line 44). The only purpose of these preliminary computations is to check the numbers of nodes, at which the "observed" values are determined. The function Output_GetNode is used for this purpose. The important argument of the function is River Station, inserted as a string, lRSobs[i]. The numbers of the "observed" nodes are stored as NumPy array of integers, iRSobs.

The computational nodes, RScmp, and their types, RStyp, are also read from the results of the preliminary computations.

The optimization process with Nelder-Mead simplex starts at line 57. A detailed description of this function may be found at SciPy.org [28]. Before this the initial guess, x0, is prepared as a NumPy array. The main argument of the optimization algorithm is the objective function, ObjFun. It is defined as a separate function (lines 6–22).

The basic argument of the objective function is the set of searched parameters x. It is a 3-element NumPy array. At the very beginning, the parameters x are assigned to the variables, representing roughness coefficients nLOB, nCh, nROB. The roughness coefficients are set for any cross-section of the HEC-RAS project in a loop over nodes with the function Geometry_SetMann_LChR. The subsequent steps are running of the computations (line 15) and reading the results (line 19) in the loop over nodes (lines 17–20). These processes are performed with the HECRASController functions Compute_CurrentPlan and Output_NodeOutput. Both of them were used in the previous example. The difference between the computed, wse, and "observed", WSEobs[i], water surfaces in a single node is calculated in line 20. The final value of the objective function is determined beyond the loop, in line 21. When the optimization process stops, the HEC-RAS window is closed and the controller variable is deleted.

The results obtained in Example 2 are shown in Figure 5. Part (a) presents the convergence of the channel roughness during the run. Part (b) shows changes of the objective function (5) during the computations.

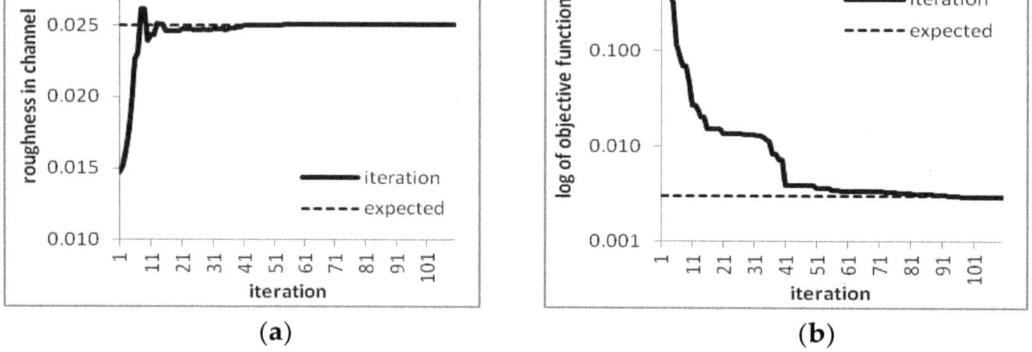

Figure 5. Results of Example 2: (**a**) convergence of channel roughness coefficient, (**b**) convergence of objective function.

3.4. *Example 3—Control of Sediment Simulation*

The purpose of this example is to show that sediment simulation in HEC-RAS may be run several times for different sediment samples. Model B described above is used in this case. The samples are changed automatically by the Python code. The HEC-RAS sediment data file has the XML format. Hence, the xml.etree.ElementTree module [36] is necessary to read and handle data. The results are then read from HDF files with the methods available in the h5py module [59]. The code in this example is very simple. However, the methods presented here may be extended and applied in a more sophisticated code, e.g., calibration of the sediment routing algorithm, analysis of sediment simulation uncertainty, or assessment of flow regime changes caused by sediments.

The third example is too long to present all lines of code in one script. Hence, it is split into Scripts 7 and 8. There are four files with data and results managed in this script. The first is the file storing the tested sediment samples. The format of this file is arbitrary and it will not be discussed here. The only requirement is that the loaded sediment samples should have a form similar to those stored in HEC-RAS. Three other files are more specific. They are the XML file with HEC-RAS sediment data, the HDF file with geometric data, and the HDF file with the results.

```
import win32com.client
import os, h5py, numpy
import xml.etree.ElementTree as ET
from support import sc3_LoadPrb,sc3_SaveRes

# loading samples
pnames,sampA,sampB,sampC,sampD =          \
sc3_LoadPrb('test_sedi','samples','samples.txt')

# init HEC-RAS Controller
hec = win32com.client.Dispatch('RAS503.HECRASController')
projekt = os.path.join(os.getcwd(),r'test_sedi\test_sedi.prj')  # project name
hec.ShowRas()                                               # show HEC-RAS window
# loading River Stations from geometry
hdfname = os.path.join(os.getcwd(),r'test_sedi\test_sedi.g03.hdf')
dane = h5py.File(hdfname,'r')                    # opening of HDF file
# link to the River Stations in HDF file
ListRS = dane.get('Geometry').get('Cross Sections').get('River Stations')
GeomRS = numpy.empty([len(ListRS)],dtype='S15')      # NumPy array of RSes
for i in range(0,len(ListRS)):
    GeomRS[i] = ListRS[i]
dane.close()
NXS = len(GeomRS)                         # number of cross-sections
```

Script 7. Example 3. Automatic control of sediment transport simulation—part 1.

An example of the sediment data file is shown in Figure 6a. It is presented in the XML Viewer. It has a structure similar to a tree. The groups include subgroups and so on. A description of the XML format can be found in the document prepared by the Python Software Foundation [36]. The group with bed gradation is opened in the XML Viewer. The name of the file and the full hierarchy to access these data are shown in Table 1. Particular gradations are stored as the group attribute. This attribute is a dictionary, where the class symbols 'GC1', 'GC2', etc. reflect the gradation in the format '%Finer'. Every XML file is an ASCII file, and it may be opened in Windows Notepad.

The HDF files may be opened in HDFViewer, as presented in Figure 6b. These are simulation results files in binary format. It is not possible to use Notepad to open them, but the HDF also has a structure similar to a tree with successive branches. Two elements are read from the opened HDF file. These are (a) the invert elevations and (b) the water surface elevations. The first is shown in Figure 6b. The hierarchy to access these data is also presented in Table 1.

```

# test of loaded samples
sname = os.path.join(os.getcwd(),r'test_sedi\test_sedi.s01')
ii = 0
for sample in [sampA, sampB, sampC, sampD]:
    plik = ET.parse(sname)  # opening XML file with sediment data
    dane = plik.getroot()   # access to bed gradations
    grad = dane.find('Bed_Gradation').find('Sample').find('Gradation')
    for elm in sample:      # assignment of tested sample
        grad.attrib[elm] = sample[elm]
    plik.write(os.path.join(os.getcwd(),r'test_sedi\test_sedi.s01'))
    del plik
    hec.Project_Open(projekt)                   # open HEC-RAS project
    test1 = hec.Plan_SetCurrent('plan sedi00')  # setting sediment plan
    NMsg,TabMsg,block = None,None,True  # no. and list of messages, blocking
    # computations of the current plan
    v1,NMsg,TabMsg,v2 = hec.Compute_CurrentPlan(NMsg,TabMsg,block)
    del NMsg, TabMsg, block
    hec.Project_Close()             # close HEC-RAS project
    # full path to the HDF file results of the computations
    hdfname = os.path.join(os.getcwd(),r'test_sedi\test_sedi.p03.hdf')
    dane = h5py.File(hdfname,'r')   # opening of HDF file
    # accessing bed elevations
    XSbed = dane.get('Results').get('Sediment').get('Output Blocks')
    XSbed = XSbed.get('Sediment').get('Sediment Time Series')
    XSbed = XSbed.get('Cross Sections').get('Invert Elevation')
    NTS = len(XSbed)
    # accessing WSE
    XSwse = dane.get('Results').get('Sediment').get('Output Blocks')
    XSwse = XSwse.get('Sediment').get('Sediment Time Series')
    XSwse = XSwse.get('Cross Sections').get('Water Surface')
    # NumPy arrays for bed elevations and WSE
    InitBed = numpy.empty([NXS],dtype=float)    # initial bed and WSE
    InitWSE = numpy.empty([NXS],dtype=float)
    LastBed = numpy.empty([NXS],dtype=float)    # final bed and WSE
    LastWSE = numpy.empty([NXS],dtype=float)
    for i in range(0,NXS):          # assigment of bed and WSE form HDF
        InitBed[i] = XSbed[0][i]    # to NumPy array
        InitWSE[i] = XSwse[0][i]
        LastBed[i] = XSbed[NTS-1][i]
        LastWSE[i] = XSwse[NTS-1][i]
    dane.close()
    del dane
    sc3_SaveRes(pnames[ii],sample,GeomRS,InitBed,InitWSE,LastBed,LastWSE)
    ii += 1

hec.QuitRas()           # close HEC-RAS
del hec                 # delete HEC-RAS controller
```

Script 8. Example 3. Automatic control of sediment transport simulation—part 2.

Table 1. Loading data and results from XML and HDF files: type of data, type of file, access hierarchy.

Data	File		Access Hierarchy
	Type	Name	
sediment sample	XML	test_sedi.s01	Data\Bed_Gradation\Sample\Gradation
River Stations	HDF	test_sedi.g03.hdf	Geometry\Cross Sections\River Stations
bed elevations	HDF	test_sedi.p03.hdf	Results\Sediment\Output Blocks\Sediment\Sediment Time Series\Cross Sections\Invert Elevation
water surface elevations	HDF	test_sedi.p03.hdf	Results\Sediment\Output Blocks\Sediment\Sediment Time Series\Cross Sections\Water Surface

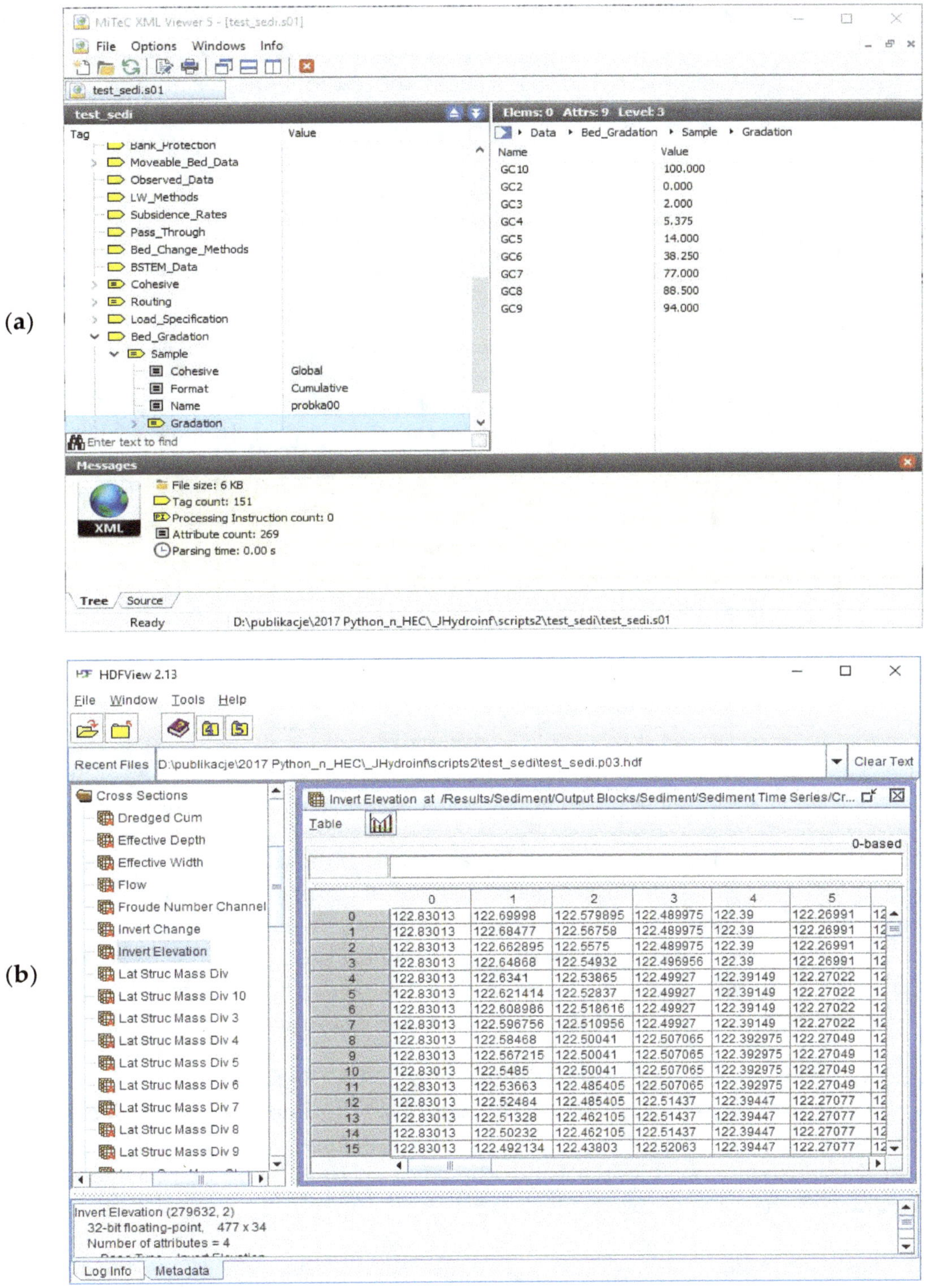

Figure 6. View of data and result files used in Example 3: (**a**) Sediment data file in XML Viewer, (**b**) sediment results in HDFViewer.

The flow chart of the main computations in this example is shown in Figure 7. The main element is a loop over sediment samples. Before the loop starts, the HECRASController object has to be initialized and the name of the project has to be loaded. If the HEC-RAS variable is ready, the information about computational nodes may be read from the HDF5 file storing geometry data. Before the loop starts, the sediment samples have to be loaded to the script. The first step in the loop is access to the XML file with sediment data. The grain fractions are written and the XML file is closed. Then the project can be opened, because any change of the sediment data in the XML file is noticed by the HEC-RAS project only in the phase of the data loading. Further changes of the data in files stored on the disk do not affect the project. Then the computations start and the results are read from the HDF5 file. At the end of each step, the bed elevations and positions of the computational nodes are written to the separate ASCII file, which enables preparation of the bed profiles.

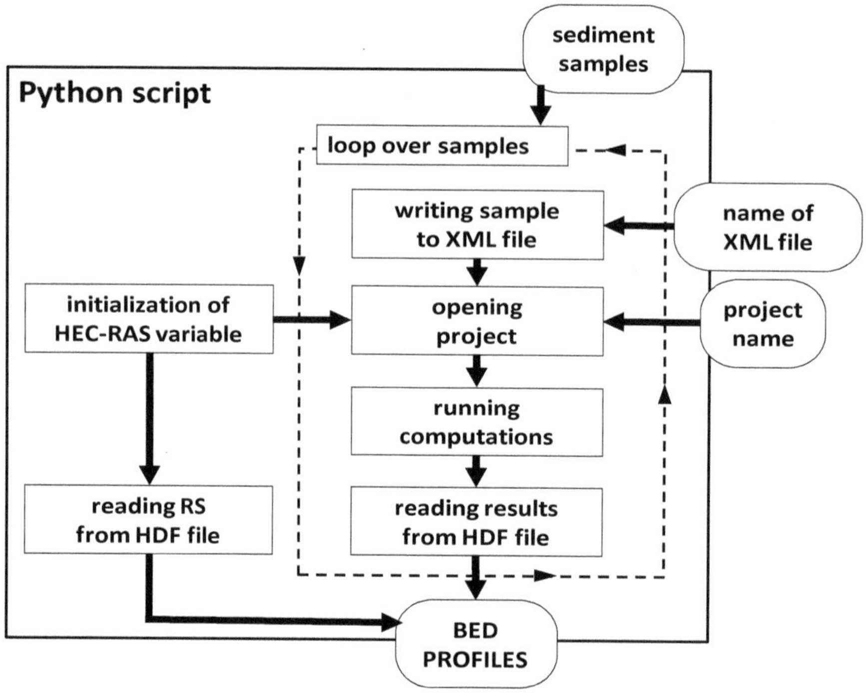

Figure 7. Main computational elements of Example 3.

The program presented in Scripts 7 and 8 starts with import of necessary modules (lines 1–4). The modules win32com.client, os, numpy are also applied in the previous examples. The new elements are xml.etree.ElementTree and h5py. The first one is loaded as ET, and it is used for handling the XML documents. The second one is used for reading of data and results from HDF files. Two functions are loaded from the support module. They are sc3_LoadPrb for loading sediment samples, and sc3_SaveRes for saving results in ASCII files. For the sake of simplicity, no local functions are defined.

The first command is reading of the sediment samples for tests (lines 7–8). The samples are stored as dictionaries sampA, sampB, sampC, and sampD. Their items are composed of sediment class denotation in the format 'GC#' and weighted gradation as '%Finer'. The keys of the dictionary should be the same as those used in the HEC-RAS and XML files with sediment data. The first element of the return list, pnames, is the table with samples' names.

The HECRASController starts at line 11. The HEC-RAS window is opened (line 13) and the full path of the project file is set, but the project is not loaded. Every change in the data requires reloading of the project. Hence, the project is loaded later.

The HDF file including geometry data is opened first (line 16). The link to the River Stations is created (line 18) with the get method of the h5py module. Then the NumPy array for storing these data is prepared (line 19). The data are assigned as value-by-value in a loop over nodes. Then the first HDF file is closed (line 22). The important element taken from the River Station array is the number of cross-sections NXS (line 23). The rest of the commands are shown in Script 8.

The longest part of the script is the loop over tested samples (lines 28–68). At the beginning of each step, the XML file with sediment data is opened (line 29). Then the link to bed gradation grad is created. In the internal loop (lines 32–33) over items of the sample, the bed gradations in the XML file are replaced with bed gradations of the current sample (line 33). Then the file is saved (line 34) and closed by deleting the variable representing this file. At this moment the XML file with sediment data is ready. The HEC-RAS project is opened (line 36) and the computational plan is set properly (line 37). After the current plan is computed (line 40), the project is closed (line 42). The last file of the current iteration is accessed (line 45). This is the HDF file with results of the computations (Table 1). Two elements are read from this file: Bed elevations and water surface elevations. These results are stored as list of successive time steps. Each time step is a list of values stored for each cross-section. The links to the bed and water surface elevations, XSbed and XSwse, are created in lines 47–49 and 52–54. The number of time steps is assigned to the variable NTS.

The NumPy arrays for storing the initial and final bed and surface elevations are created in lines 56–59. The results are assigned to them in another internal loop (lines 60–64) over cross-sections. Later, the HDF file is closed (lines 65–66). Finally, the results are saved in a file specific for each iteration, whose name is the name of the sample. At the end of the script, HEC-RAS is closed and the variable hec is deleted.

The sediment simulation running from the Python script is shown in Figure 8. The display of typical output for such simulation is also shown in the screenshot. The sieve curves of tested sediment samples are plotted in Figure 9a. The variability of assumed sieve curves is typical for a lowland river. The differences in the content of bed sediments are relatively small. However, more complex examples may also be tested in the same way. Figure 9b shows the results of the simulation as bed changes in the selected cross-sections. Considering the time horizon of the simulation, which is 4 months, the bed changes from −20 cm to +60 cm seem to be quite significant. The reason is the relatively huge discharge of 269 m^3/s, kept constant during the whole simulation. The differences between the changes of bed elevations vary from a few to 10 cm. If we compare this result with the range of changes, which is about 80 cm, it transpires that the differences between consecutive simulations are up to 12.5% for a very short simulation period. This stresses the importance of the sediment sample variability for the sensitivity of the sediment transport model.

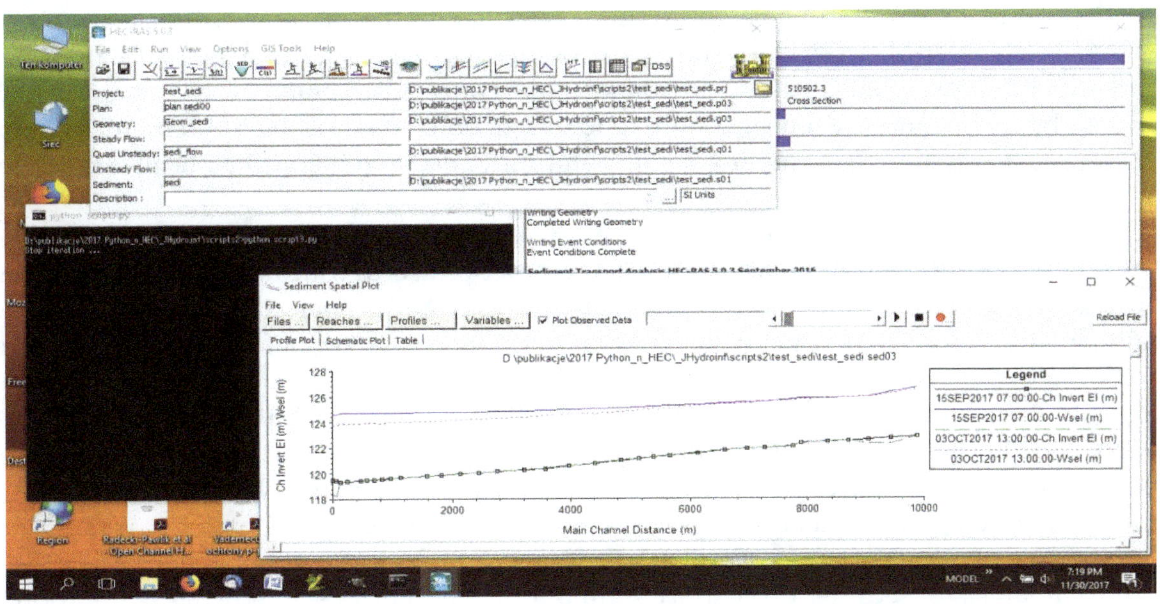

Figure 8. Run of sediment simulation and typical display of sediment output.

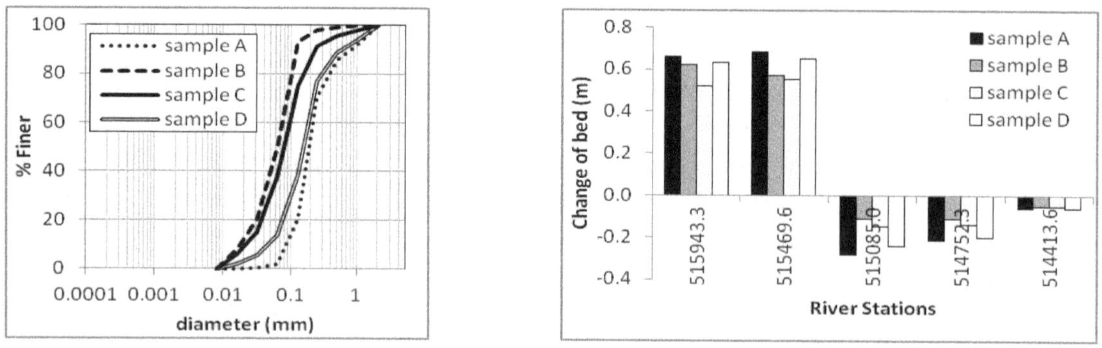

(a) (b)

Figure 9. Sediment simulations: (**a**) Sieve curves of tested samples, (**b**) bed changes in selected River Stations.

4. Conclusions

The examples presented in this paper explain the techniques necessary to control HEC-RAS computations with Python. The main element of the presented methods is the application of Python module Pywin32 for handling COM libraries. This module was used in all three examples. The comparison of the code developed for Example 1, with basic examples discussed by Goodell [20,21] and Leon and Goodell [25], illustrates the ease of Python application for handling HECRASController. The use of Python is not more difficult than VBA or MATLAB. Additionally, Example 1 presents a more general approach for application of Python for control of HEC-RAS computations than described in previous papers [40,41]. This example explains how to control any version of HEC-RAS, including the future developments. It also opens doors for further extension of HECRASController capabilities. Possible further extensions were shown in Examples 2 and 3.

The second example presented the application of HECRASController and Python specific modules, for the automatic calibration of the HEC-RAS model. The roughness coefficients were determined by the optimization method from the SciPy module. This approach was similar to that presented by Leon and Goodell [25], where the Genetic Algorithm from the MATLAB toolbox was applied for control of gate operation. Example 2 was also a more advanced approach for determination of roughness coefficients than that presented in the AHYDRA package [24]. Further development of the techniques presented in Example 2, could focus on application of a faster and/or more sophisticated optimization method than the Nelder-Mead simplex. It is also possible to develop a more general Python module for automatic calibration of HEC-RAS, on the basis of the concept presented here. Example 2 may also be improved by preparation of a user optimization algorithm. It could be implemented in Fortran, which is very fast and aimed at numerical computations, and then it could be translated to Python with the F2PY or ctypes modules.

Example 2 presented the linkage of HEC-RAS computations with numerical routines available in specific Python modules, NumPy and SciPy. The number of useful routines in SciPy is greater than those presented. Other extensions and HEC-RAS applications are also possible. The elements available in NumPy are also very useful for storing and handling of computational results. These features of Python are undoubted advantages of Python over standard VBA.

Example 3 illustrated an extension of HECRASController capabilities to control sediment simulation. To the best of the author's knowledge, such an approach or a similar one has not been tested yet. These examples presented a real advantage of Python, owing to the access to very complex data formats, XML and HDF5. The Python modules enabling such access are ElementTree and H5PY, respectively. There are several applications of the presented approach. The most basic is sensitivity analysis of sediment transport simulation. Another possible use is calibration of the sediment model on the basis of the historical data. These two application areas are discussed in the text. Other applications are also possible in the area of water management. Examples may be the impact of sediment deposition on flood hazard, prediction of morphodynamic changes in the river after regulation, etc. The uncertainty of sediment transport modeling may also be estimated with the techniques used in Example 3.

The illustrated capabilities of Python are not the only advantages of this scripting language over specific VBA applications. Another is access to geoprocessing tools, such as ArcGIS and QGIS. It makes integration of Python with HEC-RAS more necessary. Not only are VBA capabilities too poor to compete with Python in this area; more dedicated numerical software such as MATLAB does not have such broad capabilities for geoprocessing.

Application of Python for control of hydrodynamic simulations with HEC-RAS opens new opportunities in such areas as water management, river engineering, and flood risk management. Good opportunities are available by access to more different data formats from simple text files to more complex XML and HDF5. A huge advantage is the linking of HEC-RAS with other software, such as numerical routines or geoprocessing. More advanced future applications are also possible.

Author Contributions: Everything, T. Dysarz.

References

1. Kachiashvili, K.J.; Melikdzhanian, D.I. Software realization problems of mathematical models of pollutants transport in rivers. *Adv. Eng. Softw.* **2009**, *40*, 1063–1073. [CrossRef]
2. Zischg, A.P.; Mosimann, M.; Bernet, D.B.; Röthlisberger, V. Validation of 2D flood models with insurance claims. *J. Hydrol.* **2018**, *557*, 350–361. [CrossRef]
3. Pinar, E.; Paydas, K.; Seckin, G.; Akilli, H.; Sahin, B.; Cobaner, M.; Kocaman, S.; Akar, M.A. Artificial neural network approaches for prediction of backwater through arched bridge constrictions. *Adv. Eng. Softw.* **2010**, *41*, 627–635. [CrossRef]
4. Shen, D.Y.; Jia, Y.; Altinakar, M.; Bingner, R.L. GIS-based channel flow and sediment transport simulation using CCHE1D coupled with AnnAGNPS. *J. Hydraul. Res.* **2016**, *54*, 567–574. [CrossRef]
5. Gibson, S.; Sánchez, A.; Piper, S.; Brunner, G. New One-Dimensional Sediment Features in HEC-RAS 5.0 and 5.1. In Proceedings of the World Environmental and Water Resources Congress 2017, Sacramento, CA, USA, 21–25 May 2017; pp. 192–206.
6. Horritta, M.S.; Bates, P.D. Evaluation of 1D and 2D numerical models for predicting river flood inundation. *J. Hydrol.* **2002**, *268*, 87–99. [CrossRef]
7. Rodriguez, L.B.; Cello, P.A.; Vionnet, C.A.; Goodrich, D. Fully conservative coupling of HEC-RAS with MODFLOW to simulate stream–aquifer interactions in a drainage basin. *J. Hydrol.* **2008**, *353*, 129–142. [CrossRef]
8. Drake, J.; Bradford, A.; Joy, D. Application of HEC-RAS 4.0 temperature model to estimate groundwater contributions to Swan Creek, Ontario, Canada. *J. Hydrol.* **2010**, *389*, 390–398. [CrossRef]
9. Brunner, G.W. *HEC-RAS River Analysis System Hydraulic Reference Manual*; US Army Corps of Engineers; Report No. CPD-69; Hydrologic Engineering Center (HEC): Davis, CA, USA, 2016.
10. DHI. MIKE 11—A Modeling System for Rivers and Channels User Guide, DHI Software. 2003. Available online: https://www.tu-braunschweig.de/Medien-DB/geooekologie/mike11usersmanual.pdf (accessed on 10 May 2018).
11. Deltares. SOBEK User Manual, Delft, The Netherlands, 2018. Available online: https://content.oss.deltares.nl/delft3d/manuals/SOBEK_User_Manual.pdf (accessed on 10 May 2018).
12. Caponi, F.; Ehrbar, D.; Facchini, M.; Kammerer, S.; Koch, A.; Peter, S.; Vonwiller, L. BASEMENT System Manuals–Reference Manual, VAW–ETH, Zurich, 2017. Available online: http://people.ee.ethz.ch/~basement/baseweb/download/documentation/BMdoc-Reference-Manual-v2-7.pdf (accessed on 10 May 2018).
13. Greimann, B.; Huang, J.V. *SRH-1D 4.0 User's Manual*; Bureau of Reclamation: Denver, CO, USA, 2018. Available online: https://www.usbr.gov/tsc/techreferences/computer%20software/models/srh1d/index.html (accessed on 10 May 2018).
14. Pappenberger, F.; Beven, K.; Horritt, M.; Blazkova, S. Uncertainty in the calibration of effective roughness

parameters in HEC-RAS using inundation and downstream level observations. *J. Hydrol.* **2005**, *302*, 46–69. [CrossRef]

15. Vansteenkiste, T.; Tavakoli, M.; Ntegeka, V.; De Smedt, F.; Batelaan, O.; Pereira, F.; Willems, P. Intercomparison of hydrological model structures and calibration approaches in climate scenario impact projections. *J. Hydrol.* **2014**, *519*, 743–755. [CrossRef]
16. Shrestha, B.; Cochrane, T.A.; Caruso, B.S.; Arias, M.E.; Piman, T. Uncertainty in flow and sediment projections due to future climate scenarios for the 3S Rivers in the Mekong Basin. *J. Hydrol.* **2016**, *540*, 1088–1104. [CrossRef]
17. Guo, A.; Chang, J.; Wang, Y.; Huang, Q.; Zhou, S. Flood risk analysis for flood control and sediment transportation in sandy regions: A case study in the Loess Plateau, China. *J. Hydrol.* **2018**, *560*, 39–55. [CrossRef]
18. Dimitriadis, P.; Tegos, A.; Oikonomou, A.; Pagana, V.; Koukouvinos, A.; Mamassis, N.; Koutsoyiannis, D.; Efstratiadis, A. Comparative evaluation of 1D and quasi-2D hydraulic models based on benchmark and real-world applications for uncertainty assessment in flood mapping. *J. Hydrol.* **2016**, *534*, 478–492. [CrossRef]
19. Afshari, S.; Tavakoly, A.A.; Rajib, M.A.; Zheng, X.; Follum, M.L.; Omranian, E.; Fekete, B.M. Comparison of new generation low-complexity flood inundation mapping tools with a hydrodynamic model. *J. Hydrol.* **2018**, *556*, 539–556. [CrossRef]
20. Goodell, C. *Breaking HEC-RAS Code. A User's Guide to Automating HEC-RAS*; h2ls: Portland, OR, USA, 2014.
21. Goodell, C. Automating HEC-RAS, The RAS Solution. 2014. Available online: http://hecrasmodel.blogspot.com/2014/10/automating-hec-ras.html (accessed on 20 November 2017).
22. Yang, T.-H.; Wang, Y.-C.; Tsung, S.-C.; Guo, W.-D. Applying micro-genetic algorithm in the one-dimensional unsteady hydraulic model for parameter optimization. *J. Hydroinform.* **2014**, *16*, 772–783. [CrossRef]
23. Lacasta, A.; Morales-Hernández, M.; Burguete, J.; Brufau, P.; García-Navarro, P. Calibration of the 1D shallow water equations: A comparison of Monte Carlo and gradient-based optimization methods. *J. Hydroinform.* **2017**, *19*, 282–298. [CrossRef]
24. Vladyman. Automating Hydraulic Analysis (AHYDRA) v. 1.0. 2011. Available online: http://ahydra.yolasite.com/ (accessed on 7 November 2017).
25. Leon, A.S.; Goodell, C. Controlling HEC-RAS using MATLAB. *Environ. Model. Softw.* **2016**, *84*, 339–348. [CrossRef]
26. Python Software Foundation. Python 2.7.14 Documentation. 2017. Available online: https://docs.python.org/2/index.html (accessed on 8 November 2017).
27. Galiano, V.; Migallón, H.; Migallón, V.; Penadés, J. PyPnetCDF: A high level framework for parallel access to netCDF files. *Adv. Eng. Softw.* **2010**, *41*, 92–98. [CrossRef]
28. Patzák, B.; Rypl, D.; Kruis, J. MuPIF—A distributed multi-physics integration tool. *Adv. Eng. Softw.* **2013**, *60–61*, 89–97. [CrossRef]
29. SciPy Community SciPy Reference Guide. 2017. Available online: https://docs.scipy.org/doc/scipy/reference/ (accessed on 24 November 2017).
30. SciPy.org NumPy. 2017. Available online: http://www.numpy.org/ (accessed on 8 November 2017).
31. SciPy.org Optimization and Root Finding (scipy.optimize). 2017. Available online: https://docs.scipy.org/doc/scipy/reference/optimize.html (accessed on 8 November 2017).
32. SciPy.org F2PY Users Guide and Reference Manual. 2017. Available online: https://docs.scipy.org/doc/numpy-dev/f2py/ (accessed on 24 November 2017).
33. Python 2.7.15 Documentation. Available online: https://docs.python.org/2/index.html (accessed on 11 September 2018).
34. Hammond, M.; Robinson, A. *Python Programming on Win32*; O'Reilly Media, Inc.: Sebastopol, CA, USA, 2000.
35. PythonCOM Documentation Index Python and COM. Blowing the Rest Away! 2017. Available online: http://docs.activestate.com/activepython/2.4/pywin32/html/com/win32com/HTML/docindex.html (accessed on 8 November 2017).
36. Python Software Foundation. The ElementTree XML API, in Python Software Foundation. 2017. Available online: https://docs.python.org/2/library/xml.etree.elementtree.html (accessed on 8 November 2017).
37. HDF Group. High Level Introduction to HDF5. 2016. Available online: https://support.hdfgroup.org/HDF5/Tutor/HDF5Intro.pdf (accessed on 24 November 2017).

38. Zandbergen, P.A. *Python Scripting for ArcGIS*; Esri Press: Redlands, CA, USA, 2013.
39. QGIS Project PyQGIS Developer Cookbook Release 2.18. 2017. Available online: http://docs.qgis.org/2.18/pdf/en/QGIS-2.18-PyQGISDeveloperCookbook-en.pdf (accessed on 26 November 2017).
40. Peña-Castellanos, G. PyRAS—Python for River AnalysiS; MIT License. 2015. Available online: https://pypi.python.org/pypi/PyRAS/ (accessed on 7 November 2017).
41. Vimal, S. PyFloods. Python Module for Floods. 2015. Available online: https://github.com/solomonvimal/PyFloods (accessed on 7 November 2017).
42. Chow, V.T. *Open Channel Hydraulics*; McGraw-Hill: New York, NY, USA, 1959.
43. Henderson, F.M. *Open Channel Flow. Macmillan Series in Civil Engineering*; Macmillan Company: New York, NY, USA, 1966.
44. Abbott, M.B.; Minns, A.W. *Computational Hydraulics*; Ashgate Publishing: Farnham, UK, 1979.
45. Cunge, J.A.; Holly, F.M.; Verwey, A. *Practical Aspects of Computational River Hydraulics*; Pitman Advanced Publishing Program: Boston, MA, USA, 1980.
46. Wu, W. *Computational River Dynamics*; Taylor & Francis Group: London, UK, 2007.
47. Szymkiewicz, R. Numerical Modeling in Open Channel Hydraulics. In *Water Science and Technology Library*; Springer: Dordrecht, The Netherlands, 2010; Volume 83.
48. Leonard, B.P. A Stable and Accurate Convective Modelling Procedure Based on Quadratic Upstream Interpolation. In *Computer Methods in Applied Mechanics and Engineering*; North-Holland Publishing Company: Amsterdam, The Netherlands, 1979; Volume 19, pp. 59–98.
49. Brunner, G.W. *HEC-RAS River Analysis System User's Manual Version 5.0*; US Army Corps of Engineers; Report No. CPD-68; Hydrologic Engineering Center (HEC): Davis, CA, USA, 2016.
50. Green, J.; Bullen, S.; Bovey, R.; Alexander, M. *Excel 2007 VBA Programmer's Reference*; Wiley Publishing, Inc.: Indianapolis, IN, USA, 2007.
51. Sandler, R. The 14 Most Popular Programming Languages, According to a Study of 100,000 Developers. Buisness Insider, 2018. Available online: https://www.businessinsider.com/14-most-popular-programming-languages-stack-overflow-developer-survey-2018-4 (accessed on 8 September 2018).
52. Putano, B. Most Popular and Influential Programming Languages of 2018, Stackify. 2017. Available online: https://stackify.com/popular-programming-languages-2018/ (accessed on 8 September 2018).
53. Petkov, A. Here Are the Best Programming Languages to Learn in 2018. freeCodeCamp.org, 2018. Available online: https://medium.freecodecamp.org/best-programming-languages-to-learn-in-2018-ultimate-guide-bfc93e615b35 (accessed on 8 September 2018).
54. Downey, A.; Elkner, J.; Meyers, C. *How to Think Like a Computer Scientist. Learning with Python*; Green Tea Press: Wellesley, MA, USA, 2002; Available online: http://www.greenteapress.com/thinkpython/thinkCSpy.pdf (accessed on 8 November 2017).
55. Rubalcava, R. *Introducing ArcGIS API 4 for JavaScript: Turn Awesome Maps into Awesome Apps*; Apress: New York, NY, USA, 2017.
56. Pobuda, M. Using the R-ArcGIS Bridge: The Arcgisbinding Package. Available online: https://r-arcgis.github.io/assets/arcgisbinding-vignette.html (accessed on 11 September 2018).
57. Tutorials Point NumPy. Tutorials Point (I) Pvt. Ltd., 2016. Available online: https://www.tutorialspoint.com/numpy/ (accessed on 24 November 2017).
58. Press, W.H.; Teukolsky, S.A.; Vetterling, W.T.; Flannery, B.P. Numerical Recipes in Fortran 77: The Art of Scientific Computing, Volume 1 of Fortran Numerical Recipes. Second Edition, University of Cambridge. 1992. Available online: http://numerical.recipes/oldverswitcher.html (accessed on 24 November 2017).
59. Tutorials Point XML. Tutorials Point (I) Pvt. Ltd. 2017. Available online: https://www.tutorialspoint.com/xml/ (accessed on 24 November 2017).
60. Lundh, F. Elements and Element Trees. 2007. Available online: http://effbot.org/zone/element.htm (accessed on 24 November 2017).
61. Collette, A. h5py Documentation. 2017. Available online: http://docs.h5py.org/en/latest/ (accessed on 8 November 2017).
62. Dysarz, T.; Szałkiewicz, E.; Wicher-Dysarz, J. Long-term impact of sediment deposition and erosion on water surface profiles in the Ner river. *Water* **2017**, *9*, 168. [CrossRef]
63. Wicher-Dysarz, J.; Dysarz, T. Analiza procesu akumulacji rumowiska w górnej części zbiornika Jeziorsko. *Gospodarka Wodna* **2016**, *9*, 292–298. (In Polish)

9

Experimental Investigation on Mean Flow Development of a Three-Dimensional Wall Jet Confined by a Vertical Baffle

Ming Chen [1,2,*], **Haijin Huang** [1], **Xingxing Zhang** [1], **Senpeng Lv** [1] **and Rengmin Li** [1]

1 Key Laboratory of Hydraulic and Waterway Engineering of the Ministry of Education, Chongqing Jiaotong University, Chongqing 400074, China; hhj910219@163.com (H.H.); zhangddy@126.com (X.Z.); sinplor@126.com (S.L.); lrengmin@126.com (R.L.)

2 Key Laboratory of Navigation Structure Construction Technology, Ministry of Transport, Nanjing 210029, China

* Correspondence: chenmingjy@126.com.

Abstract: Three-dimensional (3D) confined wall jets have various engineering applications related to efficient energy dissipation. This paper presents experimental measurements of mean flow development for a 3D rectangular wall jet confined by a vertical baffle with a fixed distance (400 mm) from its surface to the nozzle. Experiments were performed at three different Reynolds numbers of 8333, 10,000 and 11,666 based on jet exit velocity and square root of jet exit area (named as B), with water depth of 100 mm. Detailed measurements of current jet were taken using a particle image velocimetry technique. The results indicate that the confined jet seems to behave like an undisturbed jet until 16B downstream. Beyond this position, however, the mean flow development starts to be gradually affected by the baffle confinement. The baffle increases the decay and spreading of the mean flow from 16B to 23B. The decay rate of 1.11 as well as vertical and lateral growth rates of 0.04 and 0.19, respectively, were obtained for the present study, and also fell well within the range of values which correspond to the results in the radial decay region for the unconfined case. In addition, the measurements of the velocity profiles, spreading rates and velocity decay were also found to be independent of Reynolds number. Therefore, the flow field in this region appears to have fully developed at least 4B earlier than the unconfined case. Further downstream (after 23B), the confinement becomes more pronounced. The vertical spreading of current jet shows a distinct increase, while the lateral growth was found to be decreased significantly. It can be also observed that the maximum mean velocity decreases sharply close to the baffle.

Keywords: experiment; particle image velocimetry; 3D confined wall jet; mean flow

---

1. Introduction

It is well understood that three-dimensional (3D) wall jets are typically characterized by the interaction between a turbulent boundary layer and a free jet. Because of their diverse practical engineering applications (e.g., film cooling [1], heat transfer [2], and energy dissipation in hydraulic structure [3–8]), a number of experimental studies on the mean flow development in the jet have been conducted in the past few decades. The experiments of 3D wall jet issuing from rectangular orifices have been firstly performed by Sforza and Herbst [9]. From their results, it is well known that the flow field of jet can be divided into three regions: the potential core region (PC), the characteristic decay region (CD) and the radial decay region (RD). Subsequent to their reviews, considerable investigations were carried out for the unconfined case. For example, Padmanabham and Gowda [10] measured the mean flow characteristics of 3D wall jets using a technique of the total pressure probe and determined the influence of the geometry on the characteristic decay region. Law and Herlina [11] investigated the velocity and concentration characteristics of 3D turbulent circular wall jets using a combined PIV and Planar Laser Induced Fluorescence approach. The results showed that velocity profiles collapsed well in both the longitudinal and lateral directions after 20 nozzle diameters and 25 nozzle diameters, respectively. Additionally, the spreading of the jet and decay of local maximum velocity were critically presented. Agelin-Chaab and Tachie [12] performed PIV (particle image velocimetry) measurements for the 3D wall jet issuing from a square nozzle. The main goal of their measurements was to examine the effect of the rough surface on the flow development of the jet. Later, they presented more detailed laboratory investigations on the jet, including the mean flow in the developing and self-similar regions [13]. Recently, detailed flow structures of a 3D curved wall jet have been reported by Kim et al. [14]. The results indicated that, due to the Coanda effect [15], the jet developed on the cylinder surface after the impingement of the circular jet and self-preserving wall jet profile did not clearly occur in such jet.

In the past, there have been only a few studies on the 3D confined wall jet. In order to determine whether the large lateral growth of the jet is induced by the secondary flow, Després and Hall [16] measured the flow field in a 3D wall jet with and without the grid using hot-wire anemometry and PIV. They found that the grid delayed the lateral growth of the jet and increased its vertical growth. Meanwhile, the grid also decreased the mass entrainment and mixing performance. The grid, however, was placed at the nozzle exit in their studies; thus, the confinement condition differs significantly from that in the present study. Onyshko et al. [17] provided the PIV data for a deflected wall jet. The results showed that the baffle placed on the bed had dramatic impact upon the flow feature in the jet: a wall jet-like flow was observed before reaching the baffle; after the baffle, a plane jet-like was formed and then the jet flow was deflected toward the water surface along a curvilinear trajectory. Successively, experimental studies for a wall jet impinging onto a forward-facing step in a cross-flow were comprehensively carried out by Langer et al. [18] using planar laser induced fluorescence (PLIF). They presented the jet flow regime after the initial impingement and found that the perimeter and aspect ratio of the jet were dependent of jet-step distance, height of the step and Reynolds numbers. Additionally, predictive correlations for the shape and size of the jet after impingement were discussed. In these two investigations, although the baffle or step was positioned away from the jet exit, their heights were relatively low and submerged in water. Consequently, the jet flow development was not highly confined by the baffle.

The present investigation is to focus mainly on the mean flow development of the 3D confined wall jet. In fact, the 3D confined wall jet plays a significantly important role in hydraulic engineering related to its powerful efficacy for enhancing energy dissipation. The filling and emptying system of navigation lock is a typical case of hydraulic structure producing 3D confined wall jet, which issues from rectangular nozzles (side ports) located in longitudinal culvert. In general, the large amount of water energy is dissipated by the interaction among the jet, bounded ambient fluid in the lock chamber and chamber wall so as to provide better mooring conditions for vessels [19,20]. A schematic sketch of the composite flow in current jet is shown in Figure 1, where the deflected streamlines in the lateral and wall-normal directions are clearly illustrated due to the confinement effect. In particular, after the normal impingement of the jet onto the baffle, corner wall jets in the lateral directions [21,22] and upward wall jet are easily generated in the vicinity of the baffle due to the Coanda effect. The figure also servers to define the coordinate system; x, y and z represent the longitudinal, wall-normal and lateral directions, respectively; U_m denotes the local maximum mean velocity; y_m is the wall-normal location where U_m occurs; $y_{m/2}$ is the distance from the bottom wall to the point in the outer layer where the velocity is half of U_m (called half-height); $z_{m/2}$ is the lateral location where the velocity has a half value of U_m (called half-width). It should be noted that the lateral confinement is neglected.

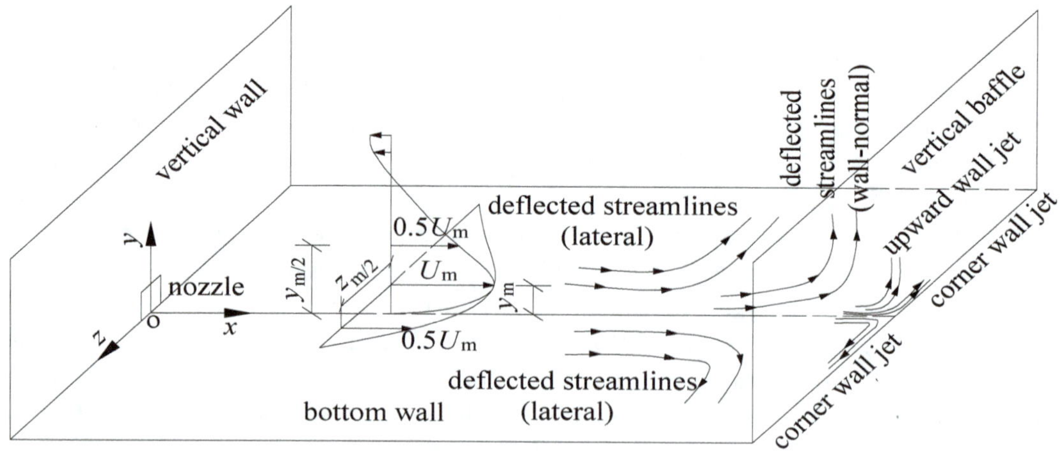

Figure 1. Schematic diagram of a three-dimensional wall jet confined by a vertical baffle.

2. Experimental Setup

The experiments were performed in a test section which is 800 mm long, 400 mm wide and 400 mm deep. Figure 2 shows schematically the experimental arrangement. The side walls and bottom of the test section were made of clear glass to facilitate the PIV measurements. The wall jet was formed by water passing through a long rectangular pipe, which allowed the flow to fully develop. The pipe has a 14 mm × 16 mm (width × height) cross section and was placed to flush the test section floor. The jet exit velocity was conditioned by a flow control valve and an electromagnetic flowmeter. In this experimental facility, two settling basins attached directly to both ends of the test section were specially designed to condition the flow and obtain the various water depths. Moreover, four constant-head skimming weirs, which were used to stabilize the water surface and ensure the overflow to return back to the supply tank, were constructed and installed at each corner of the settling basins. The Cartesian coordinate system, as shown in Figures 1 and 2, was employed. Note that $x = 0$ is at the jet exit plane, $y = 0$ is on the test section floor, and $z = 0$ is at the symmetry plane of the nozzle.

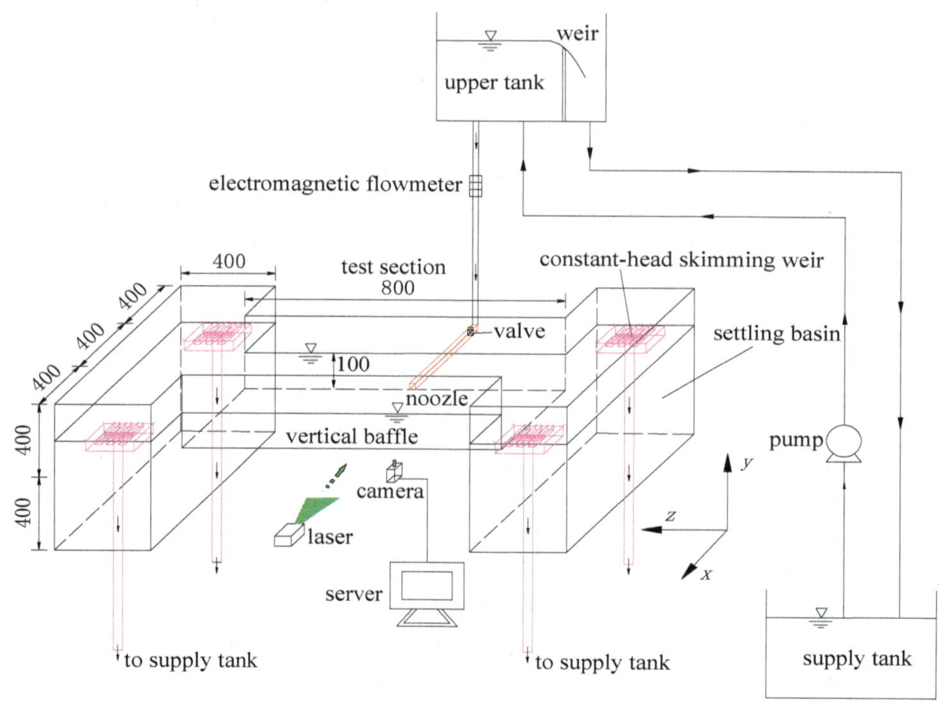

Figure 2. A sketch of experimental setup. (All dimensions in mm).

3. PIV System and Data Analysis

The velocity field for the jet was acquired using a PIV technique. For the present investigation, the water was seeded with 10 μm hollow glass spheres which have a specific gravity of 1.4. This size and density can ensure that the particles follow the flow synchronously [23,24]. A laser with a continuous energy of 10 W was used to illuminate the flow field. The light sheet formed by the laser was set around 1 mm thick and included the central axis of the jet. All the illuminated images were captured by an 8-bit high-resolution digital camera (NX5-S2 series) with a 2560 × 1920 pixels charge-coupled device (CCD). A 50 mm lens (Canon 50 mm f/1.2) was fitted to the camera. All the image pairs were captured at each position with a sampling rate of 1 Hz, a value that is low enough for the images to be uncorrelated [24]. The arrangements of the laser and camera required to be adjusted depending on the plane of measurements. For the x-z plane measurements, the laser and camera were positioned at the side and bottom of the test section, respectively. In case of the x-y plane measurements, the laser was positioned at the bottom of the test section while the camera was positioned at the side of the settling basin. It should be noted that the location of x-z plane measurement depends on the position of U_m, and thus varies with the streamwise distance due to the baffle confinement. It is anticipated that the offset x-z plane measurements at different x-axial locations can describe the vertical variations for mean velocities such as x- and z-axial velocity, U and W, respectively. Regarding x-y plane measurements, Law and Herlian [11] conducted the offset tests and found that the self-similar velocity profiles still occurred at various sections. Therefore, the offset x-y plane measurements from the jet centerline were not performed in the present study.

In the particle image processing, each velocity field involved two consecutive frames, and a time interval between the two frames was critically determined to be 1250 μs such that the maximum particle displacement satisfied the one-quarter rule for PIV correlation analysis [25]. The exposure time for each frame was fixed at 400 μs as a compromise between minimizing image streak and maximizing brightness [26]. The frames presented were divided into numerous small interrogation regions, and the cross-correlation method was used to determine the displacement of the particles in the interrogation window through the peak of the cross-correlation. Subsequently, the local velocity vector of each pair of images was calculated by the displacement and time interval mentioned previously. Detailed information of PIV algorithms is available in the investigation reported by Westerweel et al. [27]. Meanwhile, to improve the computation accuracy for measurements, particle images were processed with the iterative multigrid image deformation method [28]. A three-point Gaussian curve fit was used to determine the peak of displacement with subpixel accuracy. Spurious vectors were removed by the normalized median test method recommended by Westerweel et al. [29] and new vectors were filled by a weighted interpolation approach. The minimum size of the interrogation window of 16 × 16 pixels with 50% overlap was used to process the data. The instantaneous image processing program was developed by Beijing Jiang Yi technology co., LTD (Beijing, China). The mean velocity field was calculated using a MATLAB script developed in our laboratory. Considering the effect of the number of instantaneous image pairs on the calculation accuracy of mean velocity, Hu et al. [30] measured the vertical velocity profiles in a high-precision flume, and obtained mean velocities at various vertical locations based on different sample size of image pairs. The mean velocities obtained were compared with that averaged by expected sample size, thereby gaining the standard deviations between the two mean velocities at various vertical locations. In terms of this analysis method, the convergence test for the experimental data, including three mean velocity components at a typical gauge point location, is shown in Figure 3, where 5000 image pairs were selected as the expected sample size. Small enough deviation for x-axial mean velocity, U, was found after the sample size $N = 2500$, while the corresponding small deviations for y- and z-axial mean velocities, V and W, respectively, were obtained at least after $N = 4000$. To ensure faithful mean flow quantities, 5000 PIV instantaneous image pairs were chosen in the present study. In terms of the curve-fitting algorithm for instantaneous vectors, the size of interrogation window, and the required number of instantaneous image pairs for the calculation of the mean velocity, uncertainties in the mean velocities were estimated to be ±3.4% and ±2.5% for the local velocity close to and away from the wall, respectively.

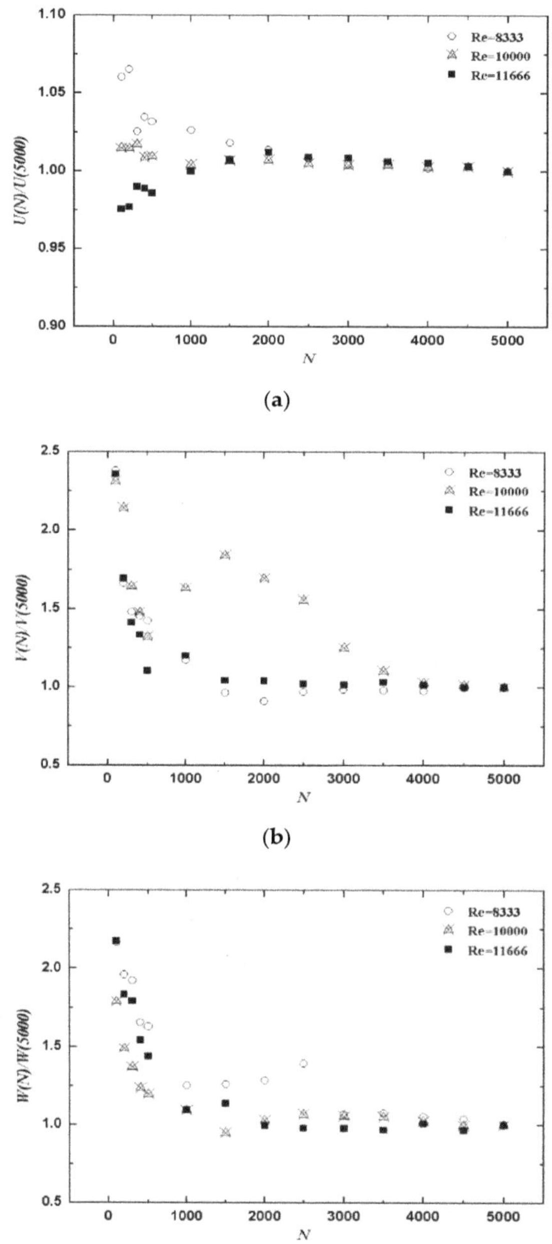

(c)

Figure 3. Convergence test for the experimental data for three different Reynolds numbers (Re) of 8333, 10,000, and 11,666: (**a**) x-axial mean velocity (U) varies with the number of instantaneous image pairs (N), (**b**) y-axial mean velocity (V) varies with the number of instantaneous image pairs (N), and (**c**) z-axial mean velocity (W) varies with the number of instantaneous image pairs (N).

4. Results and Discussion

In the present study, several important mean flow quantities, including velocity profiles and their similarity, growth rate of the half-width and half-height, and decay rate of local maximum velocity, are discussed. Experiments were performed at three Reynolds numbers of 8333, 10,000, and 11,666 (Re = U_0B/v, where U_0 is jet exit velocity, B is square root of jet exit area, and v is kinetic viscosity of water) The jet exit velocities of 0.5 m/s, 0.6 m/s and 0.7 m/s, which correspond to the three Reynolds numbers, were determined from the PIV measurements. The corresponding flow rates (Q) of 0.089 l/s,

0.109 l/s and 0.131 l/s were measured by the electromagnetic flowmeter. A water depth of 100 mm was set for the test section. All measurements in the lateral (x-z) and symmetry (x-y) planes are presented at least for the range of $10 \leq x/B \leq 24$. It should be noted that the measurement locations have not been extended to both the wall and water surface considering the effect of reflection of the laser light on the accuracy of PIV data. The corresponding distances from the measurement edge to the wall and water surface are 1 mm and 5 mm, respectively.

4.1. Spreading Rates

Figure 4 shows the variations of the velocity half-height $y_{m/2}$ and half-width $z_{m/2}$ with downstream distance. More specifically, $y_{m/2}$ and $z_{m/2}$ are the wall-normal and lateral locations where $0.5U_m$ occurs, respectively. In this figure, they were normalized by the square root of jet exit area (B) which is an appropriate scaling parameter as suggested by Padmanabham and Gowda [10] and Agelin-Chaab and Tachie [13]. The results for unconfined case obtained by Law and Herlian [11] are also included for comparison. The half-height increases approximately linearly in the region $16 < x/B < 23$, but after $x/B = 23$ the value of $y_{m/2}$ grows dramatically (Figure 4a). This behavior may be closely related to a clockwise vortex formed in the region $y/y_{m/2} < 0.25$ as mentioned in Section 4.2. Therefore, the position of U_m tends to be deflected away from the bottom wall. The half-width starts to spread after $x/B = 6$ and varies nearly linearly with downstream distance in the region $6 < x/B < 23$ (Figure 4b). However, the $z_{m/2}$ in the region $16 < x/B < 23$ develops more rapidly compared to the early region $6 < x/B < 16$. Beyond $x/B = 23$, although the confinement of the baffle can enhance the development of the jet flow field, the spreading of $z_{m/2}$ tends to significantly decrease, which is contrary to the variation of $y_{m/2}$ in the corresponding region. This is because the value of U close to the baffle gets considerably dropped due to most of the impinged jet fluid moving in both the lateral directions. In general, the variations of $y_{m/2}$ and $z_{m/2}$ are independent of Reynolds number within the present range.

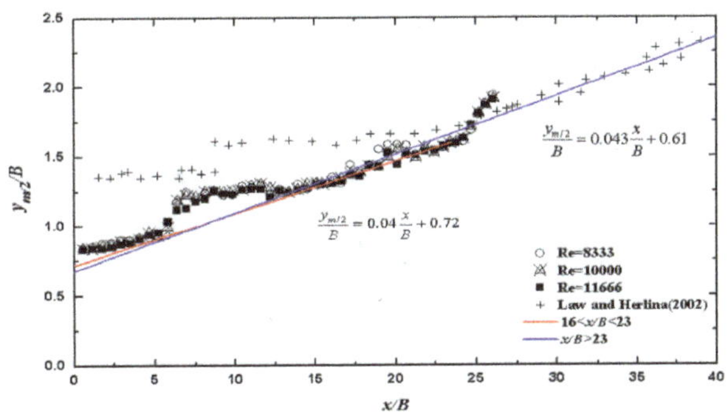

(a)

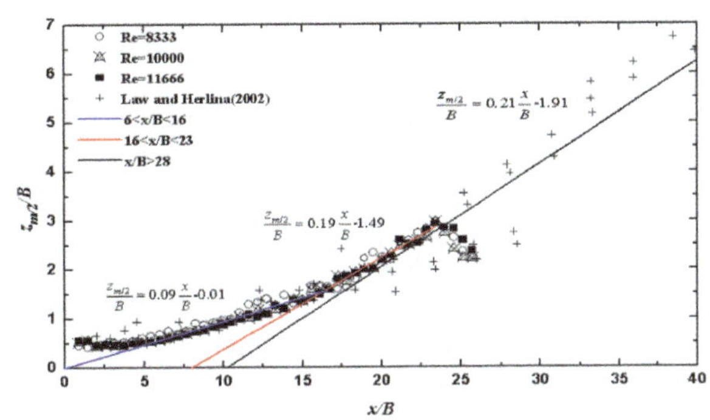

(b)

Figure 4. Mean flow developments of the 3D confined wall jet with Re = 8333, 10,000 and 11,666: (**a**) velocity half-height $y_{m/2}$, and (**b**) velocity half-width $z_{m/2}$.

To estimate the growth rate ($dy_{m/2}/dx$), a linear fit is applied to the data in the region $16 < x/B < 23$ in Figure 4a. The variation of $y_{m/2}$ can be well described by the equation:

$$\frac{y_{m/2}}{B} = 0.04\frac{x}{B} + 0.72 \quad (1)$$

Similarly, the two linear relationships for $z_{m/2}$ (Figure 4b) in the regions $6 < x/B < 16$ and $16 < x/B < 23$ can be well fitted, respectively. These two equations are written as:

$$\frac{z_{m/2}}{B} = 0.09\frac{x}{B} - 0.01 \quad (2)$$

$$\frac{z_{m/2}}{B} = 0.19\frac{x}{B} - 1.49 \quad (3)$$

Therefore, the corresponding spreading rate $dy_{m/2}/dx$ is 0.04. The $dz_{m/2}/dx$ value of 0.09 in the region $6 < x/B < 16$ is comparable to the circular free jet. Further downstream ($16 < x/B < 23$), the lateral spreading becomes to diverge more rapidly and its slope ($dz_{m/2}/dx$) is 0.19. For comparison, some of previous investigations for 3D circular wall jets are illustrated in Table 1. It is interesting to see that, in the region $16 < x/B < 23$, the spreading rates of $y_{m/2}$ and $z_{m/2}$ for the confined wall jet in the present study, respectively, fall well within the ranges of 0.036 [31]–0.045 [10] and 0.17 [32]–0.33 [31] corresponding to the values in RD region for undisturbed jet reported in the literature. It should be pointed out that reaching RD region for the unconfined wall jet requires streamwise distance at least $x/B = 20$, as summarized in Table 1. These results imply that the baffle starts to alter the jet flow development after $x/B = 16$ and the fully developed region for the confined case appears to occur at least $4B$ earlier than the unconfined case.

Table 1. Illustration of previous studies on three-dimensional wall jet.

Authors	Measuring Technique	Re	RD Region	$dz_{m/2}/dx$	$dy_{m/2}/dx$	n
Padmanabham and Gowda [10]	HWA	95,400	>20B	0.216	0.045	1.15
Law and Herlina [11]	PIV	5500, 12,200, 13,700	>23B	0.21	0.042	1.07
Agelin-Chaab and Tachie [12,13]	PIV	5000, 10,000, 20,000	>60B	0.255	0.054	1.15
Després and Hall [16]	PIV	108,000	>45B	0.25	0.047	-
Present data	PIV	8333, 10,000, 11,666	16B–23B	0.19	0.040	1.11

Note: B = square root of jet exit area; RD is the radial decay region.

4.2. Mean Velocity Profiles

The time-averaged velocity profiles at selected x/B locations in both the vertical and lateral planes are summarized in this section and compared with the previous results, including the free jet [33], two-dimensional [34] and three-dimensional [11,13] wall jets. All the velocities were normalized by the local maximum mean velocity, U_m.

Figure 5 shows the streamwise development of the axial (U) and vertical (V) mean velocity profiles measured in the symmetry plane for Re = 11,666. The y coordinate was normalized by the velocity half-height $y_{m/2}$. The profiles of U collapse reasonably well in the region $10 < x/B < 23$, while the quality of collapse at $x/B = 10$ is relatively poor because the exit flow could not fully develop in the vertical direction, as shown in Figure 5a. Additionally, in the region $y/y_{m/2} < 1.5$, the present data are comparable to the unconfined 2D and 3D wall jet results from Verhoff [34] and Agelin-Chaab and Tachie [13], respectively. Some slight fluctuations of the profiles are observed as the jet evolves downstream to $x/B = 24$ near the baffle. This behavior is attributed to the confinement of the baffle. Further downstream, the confinement becomes more noticeable and negative values of U are observed in the region $y/y_{m/2} < 0.25$. This occurs because most of the impinged jet fluid moves in both the lateral directions and resulting low momentum in the vertical direction could not overcome the adverse pressure gradient. As a result, a clockwise vortex is formed in the corner. It also can be seen from Figure 5a, near the baffle ($x/B \geq 24$), there are some significant deviations which occur just approximately from $y/y_{m/2} = 1.5$ to 3 due to the resulting reverse flow in the vicinity of the water surface. The deviations could be supported by profiles of wall-normal mean velocity (V) in the symmetry plane shown in Figure 5b. After the normal impingement, flow separation occurs close to the baffle and the flow is divided into corner jets [21,22] in both the lateral directions and upward wall jet along the baffle surface due to the Coanda effect, followed by most positive values of V beyond $x/B = 23$ shown in Figure 5b. As expected, the reverse flow is formed after impingement of the upward jet onto the water surface. For the region $21 \leq x/B \leq 24$, some negative values of V are observed in the region $y/y_{m/2} > 3.4$ due to the reverse flow. In the range of $x/B \leq 23$, the magnitudes of V are negative over most of the water depth ($y/y_{m/2} < 3$), indicating that the ambient fluid is being drawn towards the bottom wall, owing to the presence of a secondary mean vortex presented by Launder and Rodi [1]. In addition, compared to the measurements made by Law and Herlian [11] and Agelin-Chaab and Tachie [13], similar variations of V with water depth are observed but the values are slightly lower as illustrated in Figure 5b. However, their measurements reported were selected at least after $x/B = 28$ where the jet flow has been fully developed. It should be noted that considerable scattered points are shown in this figure due to the low accuracy of PIV in the wall-normal direction as described by Law and Herlian [11].

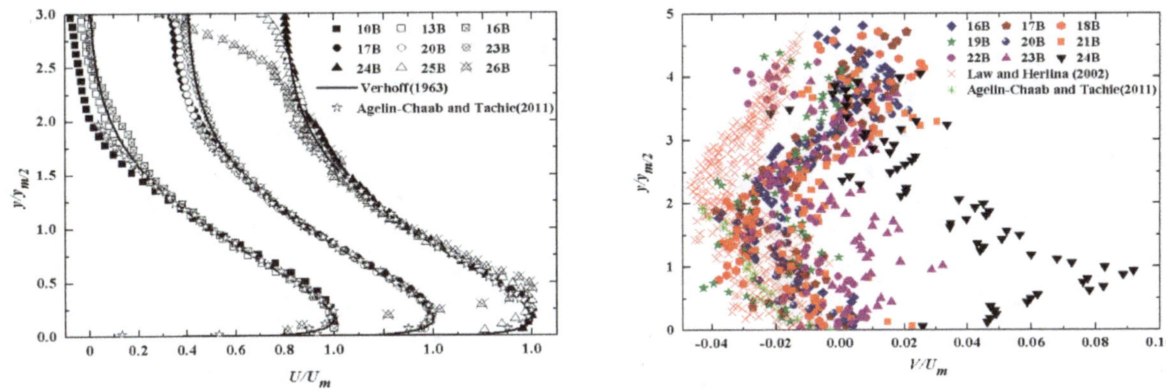

(**a**) (**b**)

Figure 5. Mean velocity profiles measured in x-y plane for Re = 11,666: (**a**) U profiles, and (**b**) V profiles.

To further compare the present observations with analytical and previous results, a quantitative evaluation method with mean absolute relative error (MARE [35]) was used in the present study. The MARE is written as:

$$\text{MARE} = \frac{1}{M}\sum_{i=1}^{M}\left|\frac{U - U_p}{U_p}\right| \times 100 \tag{4}$$

where M is total number of data on one velocity profile in given region, U is present x-axial mean velocities, U_p is analytical or previous x-axial mean velocities. The results of MARE for U profiles in x-y plane at different locations are summarized in Table 2. In the region ($y/y_{m/2} < 1.5$) unconfined by the water surface, the maximum error is substantially within the range of MARE < 5% before $x/B = 24$, while the maximum error increases to 18.7% after $x/B = 24$. The relatively large error between the confined and unconfined cases indicates that the baffle confinement has noticeable impact upon the mean velocity distribution.

The profiles of axial (U) and lateral (W) mean velocities at selected locations in the lateral plane are shown in Figure 6, where the z coordinate was normalized by the half-width ($z_{m/2}$). Being similar to the U velocity distribution in the symmetry plane, measurements ($10 < x/B < 23$) in the lateral plane show reasonable collapse in the region $z/z_{m/2} < 1.2$. However, when the flow evolves downstream ($x/B > 16$), there are some slight differences between the experimental data and previous observations

(Figure 6a). This is critically because the profiles of current jet start to be affected by the confinement of the vertical baffle after $x/B = 16$. Far downstream ($x/B \geq 24$), the confinement increases with increasing longitudinal distance and the velocity profiles across the entire sections seem to be unstable. For example, some significant fluctuations can be observed especially after $x/B = 26$ due to the presence of the baffle. For comparison, the results obtained by Law and Herlian [11] generally agree better with the present data. Similarly, Table 3 gives the MARE values for U profiles in x-z plane at different locations. Except for the region very close to the baffle ($x/B \geq 26$), the maximum error between the present data and analytical and previous results ($z/z_{m/2} < 1.2$) does not exceed the range of MARE < 5%. Figure 6b shows the W distribution at typical x/B locations. The lateral mean velocity (W) increases from zero at the symmetry plane to a peak value which occurs approximately at $z/z_{m/2} = 1.2$ within the region $x/B \leq 24$. Beyond $x/B = 24$, the location of the peak value is gradually delayed (i.e., $z/z_{m/2} = 1.6$ for $x/B = 25$). From this figure, the W profiles in the region $x/B \leq 21$ are relatively lower than those of Law and Herlian [11]. However, as the jet leaves the nozzle, the corresponding W values continuously increase. For example, the present observation at $x/B = 22$ is comparable to those of Law and Herlian [11]. When the jet develops downstream ($x/B = 24$), larger values of W can be observed compared to the results reported by Law and Herlian [11]. Further downstream ($x/B > 24$), the W profiles increase dramatically and are significantly higher as compared to previous results. This indicates that most of the fluid is deflected away from the centerline of the jet in the lateral plane due to the confinement of the vertical baffle.

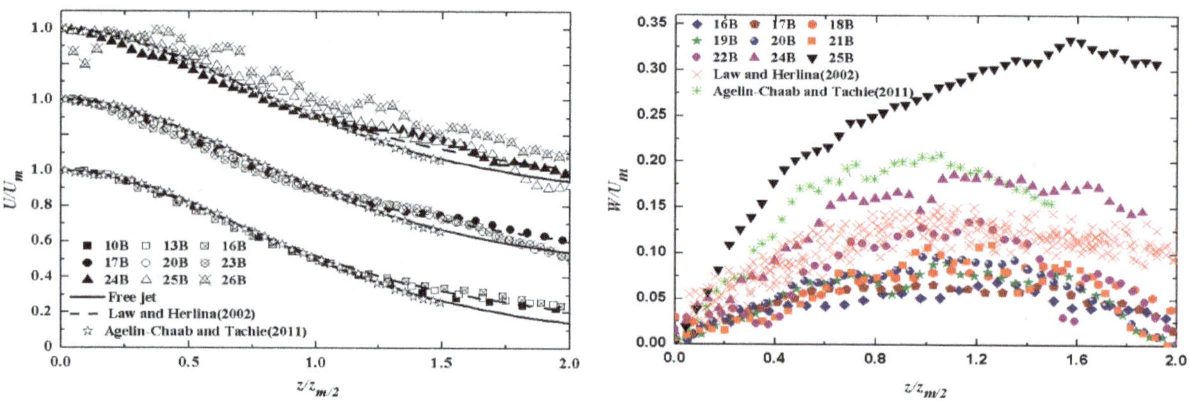

(**a**) (**b**)

Figure 6. Mean velocity profiles measured in x-z plane for Re = 11,666: (**a**) U profiles, and (**b**) W profiles.

In order to examine the effect of the baffle on the mean velocity profiles more closely, detailed velocity distribution with three Reynolds numbers are plotted in Figure 7. Collapsed curves also can be observed both in the lateral and vertical directions especially within the range of $10 \leq x/B \leq 23$. This indicates that the profiles become independent of Reynolds numbers except the regions near the baffle and water surface.

Table 2. MARE (%) values for U profiles in x-y plane at different locations ($y/y_{m/2} < 1.5$).

Location	10*B*	13*B*	16*B*	17*B*	20*B*	23*B*	24*B*	25*B*	26*B*
Analytical solution [34]	4.13	1.45	2.63	1.30	0.92	1.40	2.75	4.65	18.70
Agelin-Chaab and Tachie [13]	4.43	2.54	3.08	1.60	2.24	2.59	3.87	5.12	18.44

Table 3. MARE (%) values for U profiles in x-z plane at different locations ($z/z_{m/2} < 1.2$).

Location	10*B*	13*B*	16*B*	17*B*	20*B*	23*B*	24*B*	25*B*	26*B*
Analytical solution [33]	1.19	1.24	2.85	3.32	3.94	4.72	5.56	6.54	16.22
Law and Herlian [11]	2.06	1.83	2.73	3.11	3.05	3.74	5.47	5.58	15.19
Agelin-Chaab and Tachie [13]	1.44	1.04	2.44	3.00	3.50	4.14	5.38	6.34	15.86

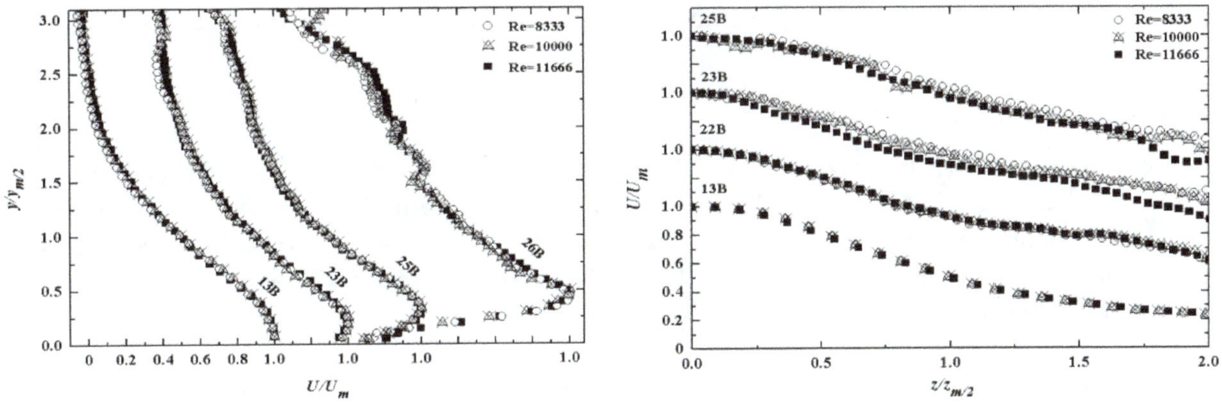

(a) (b)

Figure 7. Variations of U profiles at typical locations with Re = 8333, 10,000 and 11,666: (**a**) x-y plane, and (**b**) x-z plane.

4.3. Decay of Local Maximum Velocity

Further insight into the development of mean velocity can be made by examining how the maximum mean velocity varies with downstream distance, as shown in Figure 8. The value of U_m virtually remains constant in the PC region ($x/B < 3.75$), while it starts to decrease gradually after $x/B = 6$. The U_m decay in both the regions $6 < x/B < 16$ and $16 < x/B < 23$, respectively, can be expressed in power-law forms:

$$\frac{U_m}{U_0} = 2.74\left(\frac{x}{B}\right)^{-0.58} \quad (5)$$

$$\frac{U_m}{U_0} = 11.67\left(\frac{x}{B}\right)^{-1.11} \quad (6)$$

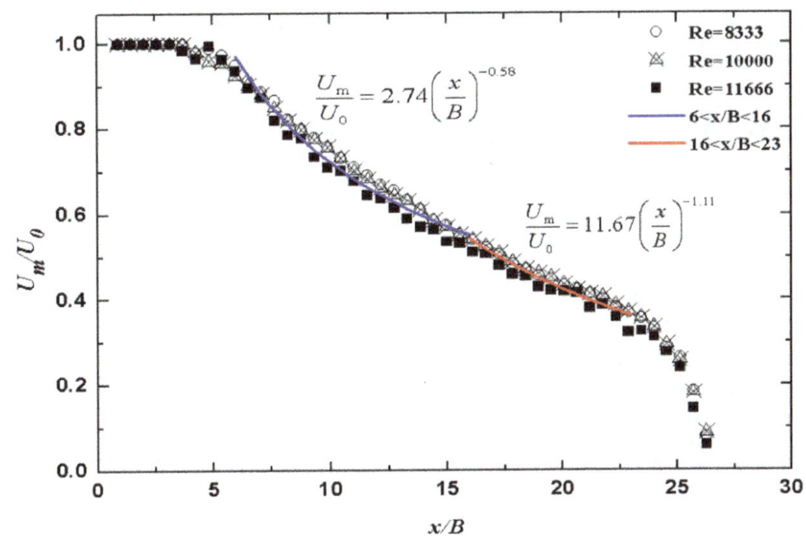

Figure 8. Decay of local maximum velocity U_m.

The decay rate of 0.58 represents the region $6 < x/B < 16$, which is comparable to the values in the CD region for unconfined 3D wall jets given by Padmanabham and Gowda [10]. In the region $16 < x/B < 23$, significant decay of U_m can be clearly seen from Figure 8 and the corresponding decay exponent is 1.11, a value that is compared very well with those of unconfined cases in the RD region summarized in Table 1. These results are consistent with the previous observations mentioned above.

However, it is anticipated that a sharp reduction of U_m can be observed by the strong confinement of the baffle as the jet develops further downstream ($x/B > 24$).

5. Conclusions

An experimental study on a three-dimensional confined wall jet was conducted to examine the effect of a fixed vertical baffle on the mean flow development. Measurements of the flow field were performed using a particle image velocimetry method. The results showed that the confined jet remained unaffected by the baffle until $x/B = 16$, while the other regions of the mean flow development are significantly characterized by the baffle confinement.

In the region ($16 < x/B < 23$), the relatively larger growth rates of 0.04 and 0.19 in the wall-normal and lateral directions, respectively, were obtained for the confined case. These spreading rates fall well within the range of values which correspond to the results in the radial decay region for the undisturbed case. Both the vertical and lateral spreading rates of current jet were found to be independent of Reynolds number. The decay rate was estimated to be 1.11, a value that is consistent with those reported for undisturbed cases in the literature. Similarly, the decay of the maximum mean velocity is independent of Reynolds number. The measurements of mean velocity profiles also exhibit self-similarity and are strongly independent of Reynolds number. Therefore, the fully developed flow of current jet appears to form at least $4B$ earlier than the unconfined case.

As the jet evolves further downstream ($x/B \geq 24$), the confinement becomes more noticeable, and corner jets and upward wall jet are generated close to the baffle. It was found that the jet flow developed more rapidly in the wall-normal direction due to the presence of the baffle. On the contrary, the lateral spreading is significantly reduced. In general, the present mean velocity profiles show reasonable collapse but some differences between the confined and undisturbed jets in the regions very close to the baffle and near the water surface.

It can be concluded that the baffle helps the jet to develop especially beyond $x/B = 16$. This study contributes to a better understanding of the energy dissipation mechanism of current jet. Given the limited measurement cases in the present study, additional tests over considerable distances from nozzle to baffle are needed to fully investigate the mean flow development of the three-dimensional confined wall jet.

Author Contributions: M.C. led the work performance and wrote the article; H.H., X.Z., S.L. and R.L. conducted the experiments and collected data through review of papers.

Notation

B	square root of jet exit area
N	number of instantaneous image pairs
M	total number of data on one velocity profile in given region
n	exponent describing the decay of U_m
Q	flow rate through the jet pipe
Re	jet exit Reynolds number based on jet exit velocity and square root of jet exit area
U	x-axial mean velocity
V	y-axial mean velocity
W	z-axial mean velocity
U_0	jet exit velocity
U_m	local maximum mean velocity
U_p	analytical or previous x-axial mean velocity
x	longitudinal direction in the coordinate system
y	wall-normal direction in the coordinate system
y_m	wall-normal location where U_m occurs

Notation

$y_{m/2}$	wall-normal location where $0.5U_m$ occurs
z	lateral direction in the coordinate system
$z_{m/2}$	lateral location where $0.5U_m$ occurs
v	kinetic viscosity of water

References

1. Launder, B.E.; Rodi, W. The turbulent wall jet-measurements and modeling. *Annu. Rev. Fluid Mech.* **1983**, *15*, 429–459. [CrossRef]
2. Caggese, O.; Gnaegi, G.; Hannema, G.; Terzis, A.; Ott, P. Experimental and numerical investigation of a fully confined impingement round jet. *Int. J. Heat Mass Transf.* **2013**, *65*, 873–882. [CrossRef]
3. Chen, J.G.; Zhang, J.M.; Xu, W.L.; Peng, Y. Characteristics of the velocity distribution in a hydraulic jump stilling basin with five parallel offset jets in a twin-layer configuration. *J. Hydraul. Eng. ASCE* **2014**, *140*, 208–217. [CrossRef]
4. Yang, Y.H.; Chen, M.; Zhang, X.X.; Duan, L.M.; Miao, J.K. 3D numerical simulation of hydraulic characteristics of ditches designed for a navigation lock with high-head and large scale. *Port Waterw. Eng.* **2018**, *4*, 84–90. (In Chinese)
5. Gabl, R.; Righetti, M. Design criteria for a type of asymmetric orifice in a surge tank using CFD. *Eng. Appl. Comput. Fluid Mech.* **2018**, *12*, 397–410. [CrossRef]
6. Adam, N.J.; De Cesare, G.; Schleiss, A.J. Influence of geometrical parameters of chamfered or rounded orifices on head losses. *J. Hydraul. Res.* **2018**. [CrossRef]
7. Wu, J.H.; Ai, W.Z.; Zhou, Q. Head loss coefficient of orifice plate energy dissipator. *J. Hydraul. Res.* **2010**, *48*, 526–530.
8. Zhang, Q.Y.; Chai, B.Q. Hydraulic characteristics of multistage orifice tunnels. *J. Hydraul. Eng. ASCE* **2001**, *127*, 663–668. [CrossRef]
9. Sforza, P.M.; Herbst, G. A Study of Three-Dimensional Incompressible, Turbulent Wall Jets. *AIAA J.* **1970**, *8*, 276–283.
10. Padmanabham, G.; Gowda, B.H.L. Mean and turbulence characteristics of a class of three-dimensional wall jets—Part 1: Mean flow characteristics. *J. Fluids Eng. ASME* **1991**, *113*, 620–628. [CrossRef]
11. Law, A.W.K.; Herlina. An experimental study on turbulent circular wall jets. *J. Hydraul. Eng. ASCE* **2002**, *128*, 161–174. [CrossRef]
12. Agelinchaab, M.; Tachie, M.F. PIV study of three-dimensional wall jet over smooth and rough surfaces. In Proceedings of the FEDSM2007 5th Joint ASME/JSME Fluids Engineering Conference, San Diego, CA, USA, 30 July–2 August 2007.
13. Agelin-Chaab, M.; Tachie, M.F. Characteristics of turbulent three-dimensional wall jets. *J. Fluids Eng. ASME* **2011**, *133*, 021201-1–021201-12. [CrossRef]
14. Kim, M.; Kim, H.D.; Yeom, E.; Kim, K.C. Flow characteristics of three-dimensional curved wall jets on a cylinder. *J. Fluids Eng. ASME* **2018**, *140*, 041201-1–041201-7. [CrossRef]
15. Cîrciu, I.; Boşcoianu, M. An analysis of the efficiency of Coanda-NOTAR anti-torque systems for small helicopters. *INCAS Bull.* **2010**, *2*, 81–88.
16. Després, S.; Hall, J.W. The development of the turbulent three-dimensional wall jet with and without a grid placed over the outlet. In Proceedings of the ASME 2014 International Mechanical Engineering Congress and Exposition, Montreal, QC, Canada, 14–20 November 2014.
17. Onyshko, P.R.; Loewen, M.R.; Rajaratnam, N. Particle image velocimetry applied to a deflected wall jet. In Proceedings of the Hydraulic Measurements and Experimental Methods, Edmonton, AB, Canada, 28 July–1 August 2002.
18. Langer, D.C.; Fleck, B.A.; Wilson, D.J. Measurements of a wall jet impinging onto a forward facing step. *J. Fluids Eng. ASME* **2009**, *131*, 091103-1–091103-9. [CrossRef]
19. Stockstill, R.L. Modeling Hydrodynamic Forces on Vessels during Navigation Lock Operations. ASCE, 2002. Available online: https://doi.org/10.1061/40655(2002)83 (accessed on 8 September 2011).
20. Chen, M.; Liang, Y.C.; Xuan, G.X.; Chen, M.D. Numerical simulation of vessel hawser forces within chamber during navigation lock operations. *J. Ship Mech.* **2015**, *19*, 78–85. (In Chinese)
21. Hogg, S.I.; Launder, B.E. Three dimensional turbulent corner wall jet. *Aeronaut. J.* **1985**, *89*, 167–171.

22. Poole, B.; Hall, J.W. Turbulence measurements in a corner wall jet. *J. Fluids Eng. ASME* **2016**, *138*, 081204-1–081204-8. [CrossRef]
23. Shinneeb, A.M. Confinement Effects in Shallow Water Jets. Ph.D. Thesis, University of Saskatchewan, Saskatoon, SK, Canada, 2006.
24. Shinneeb, A.M.; Balachandar, R.; Bugg, J.D. Confinement Effects in Shallow-Water Jets. *J. Hydraul. Eng. ASCE* **2011**, *137*, 300–314. [CrossRef]
25. Adrian, R.J. Particle-imaging techniques for experimental fluid mechanics. *Annu. Rev. Fluid Mech.* **1991**, *23*, 261–304. [CrossRef]
26. Wang, H.; Zhong, Q.; Wang, X.K.; Li, D.X. Quantitative characterization of streaky structures in open-channel flows. *J. Hydraul. Eng. ASCE* **2017**, *143*, 04017040-1–04017040-10. [CrossRef]
27. Westerweel, J.; Elsinga, G.E.; Adrian, R.J. Particle image velocimetry for complex and turbulent flows. *Annu. Rev. Fluid Mech.* **2013**, *45*, 409–436. [CrossRef]
28. Scarano, F. Iterative image deformation methods in PIV. *Meas. Sci. Technol.* **2002**, *13*, R1–R19. [CrossRef]
29. Westerweel, J.; Scarano, F. Universal outlier detection for PIV data. *Exp. Fluids* **2005**, *39*, 1096–1100. [CrossRef]
30. Hu, J.; Yang, S.F.; Wang, X.K.; Lan, Y.P. Adaptability of PIV based on different working principle to turbulence measurements. *J. Hydroelectr. Eng.* **2013**, *32*, 181–186. (In Chinese)
31. Davis, M.; Winarto, H. Jet diffusion from a circular nozzle above a solid plane. *J. Fluid Mech.* **1980**, *101*, 201–221. [CrossRef]
32. Swamy, N.C.; Bandyopadhyay, P. Mean and turbulence characteristics of three-dimensional wall jets. *J. Fluid Mech.* **1975**, *71*, 541–562. [CrossRef]
33. Schlichting, H. *Boundary-Layer Theory*; Kestin, J., Translator; McGraw-Hill: New York, NY, USA, 1979.
34. Verhoff, A. *The Two-Dimensional Turbulent Wall Jet with and without an External Stream*; Princeton University: Princeton, NJ, USA, 1963.
35. Gumus, V.; Simsek, O.; Soydan, N.G.; Akoz, M.S.; Kirkgoz, M.S. Numerical modeling of submerged hydraulic jump from a sluice gate. *J. Irrig. Drain. Eng. ASCE* **2016**, *142*, 04015037-1–04015037-11. [CrossRef]

10

Numerical and Experimental Comparative Study on the Flow-Induced Vibration of a Plane Gate

Chunying Shen [1,2], Wei Wang [1,*], Shihua He [2] and Yimin Xu [2]

1 State Key Laboratory of Hydraulics and Mountain River Engineering, Sichuan University, Chengdu 610065, China; shenchunying@kmust.edu.cn

2 Faculty of Electrical Power Engineering, Kunming University of Science and Technology, Kunming 650500, China; hoxiwa@aliyun.com (S.H.); yiminxu@sina.com (Y.X.)

* Correspondence: wangwei@scu.edu.cn.

Abstract: A numerical method is applied here to simulate the unstable flow and the vibration of a plane gate. A combination of the large eddy simulation (LES) method and the volume of fluid (VOF) model is used to predict the three-dimensional flow field in the vicinity of a plane gate with submerged discharge. The water surface profile, the streamline diagrams, the distribution of turbulent kinetic energy, the power spectrum density curve of the fluctuating pressure coefficient at typical points underneath the gate, and the complete vortex distribution around the gate are obtained by LES-VOF numerical calculation. The vibration parameters of the gate are calculated by the fluid-structure coupling interface transferring the hydrodynamic load. A simultaneous sampling experiment is performed to verify the validity of the algorithm. The calculated results are then compared with experimental data. The difference between the two is acceptable and the conclusions are consistent. In addition, the influence of the vortex in the slot on the flow field and the vibration of the gate are investigated. It is feasible to replace the experiment with the fluid-structure coupling computational method, which is useful for studying the flow-induced vibration mechanism of plane gates.

Keywords: LES-VOF method; flow-induced vibration; plane gate; numerical analysis

---

1. Introduction

With the wide application of a high head gate in hydraulic engineering, the vibration problem of the plane gate is increasingly prominent. Essentially, the flow-induced vibration of the plane gate is a complex fluid-structure interaction phenomenon. The time varying hydrodynamic load that is caused by the unstable flow around the gate is the main excitation source. The coexistence of the vortices at the gate's bottom edge, in the gate slot and in the gate downstream, further complicates the flow structure of the sluice flow. However, it is difficult to capture these vortices simultaneously and to catch the comprehensive information of the vortex-induced vibration by using an experimental approach. Researchers, such as Hardwick [1], Kolkman [2], Jongeling [3], and Ishii [4], have been making efforts to explore the mechanism of gate vibration. Hardwick conducted a model test investigation on the plane gate vibration. It is believed that this nonlinear vertical vibration response is caused by the resonance between the fluid shear force acting at the bottom edge of the gate and the gate. Kolkman et al. believed that the flow fluctuation and the variation of the pressure due to the inertia effect of sluice flow might be the main reason for the vertical vibration of the gate. Jongeling proved that the streamwise vibration of the gate is induced by the instability of the bottom edge shear layer. Ishii considered that the vortices at the bottom edge of the gate and the interaction with the structure lead to the streamwise vibration of the gate. These research results are mostly limited to the qualitative description of the physical process of the gate vibration, and tend to separate the vertical vibration and the streamwise flow vibration. Therefore, it is especially important to analyze the vibration of the gate combining the vertical and the streamwise vibration. Simultaneously, these researches mainly focus on investigating the vibration mechanism that resulted from the vortex at the bottom edge of the gate.

With the rapid development of computational fluid dynamics (CFD) and computer technology, the accuracy of numerical calculation of flow field and the complexity of solving problems have been greatly improved, and it has been possible to numerically simulate the sluice flow in some extent. The method adopted is mainly based on the finite volume scheme. The SIMPLE series algorithm is used in the calculation of pressure field and velocity field. When dealing with turbulence problems, the turbulence model is employed to solve the Reynolds time-averaged Navier-Stokes equation. Pani et al. [5] used a finite element technique to investigate the fluid-structure interaction effect on the hydrodynamic pressure. Erdbrink et al. [6] also assessed the local velocities and pressures of the two-dimensional field by using the finite element method. Liu et al. [7] adopted the two-dimensional (2-D) mathematical model of Reynolds-averaged equations to study the turbulent flow behind a sluice gate. Kostecki [8–10] predicted the two-dimensional flow field in the vicinity of an underflow vertical lift gate by using a combinative numerical model of the vortex method and the boundary element method. Kazemzadeh [11] applied the smoothed fixed grid finite element method to simulate free surface flow in gated tunnels. In recent years, some novel computational methods and models have been developed. For example, Zhu et al. [12,13] proposed a new slip model, which was used in the study of lid-driven cavity flows in both continuum and transition flow regimes. However, there is still a lack of theories and methods to analyze effectively and deal with the complex vibration phenomenon of plane gates.

The development of parallel algorithms and supercomputers has strongly promoted the progress of direct numerical simulation (DNS) and large eddy simulation (LES) methods. Analyzing the pressure pulsation of turbulent flow using LES method has received more and more attention. However, the application of CFD technology, especially simulating pressure pulsation, in the sluice flow is still preliminary.

Different from the arc gate, the side walls of a plane gate have the gate slots that are used for the lifting movement of gate. In the case of the submerged sluice flow, the random movement of multiple vortices behind the gate, in the gate slot and at the gate's bottom edge will aggravate the pulsation of velocity and the pressure. When water flow crosses the slot, such as that which exists in the hydraulic plane gate, the vortices in the slot will interact with the separated flow near the downstream corner of the slot due to local boundary mutation in the slot area, which can easily lead to flow cavitation and even endanger the safe operation of hydraulic projects [14]. The vortex-induced vibration of a blunt body or other solid boundary caused by vortex shedding has been studied most [15–18], while studies of the gate vibration induced by the vortex in the gate slot has been reported rarely.

Based on the theory of fluid-structure interaction, the present investigation studies the characteristics of the flow-induced vibration of the hydraulic plane gate with a submerged discharge. The three-dimensional numerical simulation method with a large eddy simulation turbulence model can obtain a complete flow structure, including a funnel vortex in the slot. The vibration parameters of the gate are acquired by the fluid-structure coupling interface transferring the hydrodynamic load. The fluid-structure coupling method is validated by the experimental evidence.

2. Computational Modeling

2.1. Turbulence Model

The large eddy simulation (LES) method uses instantaneous flow control equations to directly simulate the large scale vortex in a turbulent flow field. After processing by using the filtering function, the instantaneous Navier-Stokes equation and continuity equation of control fluid flow become:

$$\frac{\partial}{\partial t}(\rho\overline{u}_i) + \frac{\partial}{\partial x_j}(\rho\overline{u}_i\overline{u}_j) = -\frac{\partial\overline{p}}{\partial x_i} + \rho g_i + \frac{\partial}{\partial x_j}\left[\mu\left(\frac{\partial\overline{u}_i}{\partial x_j} + \frac{\partial\overline{u}_j}{\partial x_i}\right) - \rho\tau_{ij}\right] \tag{1}$$

$$\frac{\partial\rho}{\partial t} + \frac{\partial}{\partial x_i}(\rho\overline{u}_i) = 0 \tag{2}$$

where $i, j = 1, 2, 3$; ρ is the specific mass; μ is the dynamic viscosity; g_i is the mass force; and, $\overline{p}$ and $\overline{u}_i$ are the filtered pressure and velocity components, respectively. $\tau_{ij} = \overline{u_i u_j} - \overline{u}_i\overline{u}_j$ are the components of the sub grid scale (SGS) stress tensor representing the effect of the small scale motion on the large scale motion. According to the basic SGS model, Smagorinsky [19] assumes the SGS stress in the following form:

$$\tau_{\text{ij}} - \frac{1}{3}\tau_{kk}\delta_{ij} = -2\mu_t\overline{R}_{ij} \tag{3}$$

The Smagorinsky-Lilly model is used in this paper. In the Smagorinsky-Lilly [20] model, the eddy-viscosity is modeled by:

$$\mu_t = (C_s\Delta)\left|\overline{R}\right| \tag{4}$$

where $\left|\overline{R}\right| = \sqrt{2\overline{R}_{ij}\overline{R}_{ij}}$, $\overline{R}_{ij} = \frac{1}{2}\left(\frac{\partial\overline{u}_i}{\partial x_j} + \frac{\partial\overline{u}_j}{\partial x_i}\right)$, C_s is the Smagorinsky constant; and, Δ is the local grid scale, which is computed according to the volume of the computational cell using $\Delta = V^{1/3}$.

It is more difficult to simulate the free surface accurately, and the volume of fluid (VOF) method is more advantageous to track the complex free surface in the multiphase flow models. Due to the downstream flow directly in contact with the atmosphere, the VOF model is used to track the air-water surface in the present investigation.

The tracking of the interface between the phases is accomplished by solving the continuity equation of the volume fraction of one or more phases. The volume fraction of water is a function of time and space, so the VOF model must be solved by transient. Assuming that α_w denotes the volume fraction for water phase, $\alpha_w = 0$ indicates the exclusion of the water phase in the region, $\alpha_w = 1$ indicates that the region is full of water phase, and $0 < \alpha_w < 1$ indicates the interface for water phase and other phases.

The tracking equations of the interface are as follows:

$$\frac{\partial\alpha_w}{\partial t} + u_i\frac{\partial\alpha_w}{\partial x_i} = 0 \tag{5}$$

2.2. Structural Model

The gate structure is assumed to consist of linear elastic small deformation material, which is made of organic glass. Under the Lagrangian system, structural vibration control equations with a small deformation assumption are as follows:

$$\rho^s\frac{\partial^2 u^s}{\partial t^2} = \text{div}\left(J\sigma^s F^{-\mathrm{T}}\right) + \rho^s g \tag{6}$$

$$\sigma^s = J^{-1}F(\lambda^s(\text{tr}E)I + 2\mu^s E)F^{\mathrm{T}} \tag{7}$$

The F, E, and J are defined as:

$$F = I + \nabla u^s \tag{8}$$

$$E = \frac{1}{2}\left(F^{\mathrm{T}}F - I\right) \tag{9}$$

$$J = \det F \tag{10}$$

where ρ^s is the density of structure, u^s is the vibration displacement, σ^s is Cauchy's stress tensor of the gate structure, $\lambda^s = vE^s/((1+v)(1-2v))$ and $\mu^s = E^s/(2(1+v))$ are the constants of the gate material, E^s is material elastic modulus, and v is Poisson's ratio.

For complete fluid and structure coupling, the geometry coordination condition and the force equilibrium condition on the fluid-structure coupling interface should be satisfied.

The geometry coordination condition on the fluid-structure coupling interface $S^{fs} = \Omega^f \cap \Omega^s$ is:

$$u_i^f = u_i^s \text{ on } S^{fs} \tag{11}$$

The force equilibrium condition on the fluid-structure coupling surface $S^{fs} = \Omega^f \cap \Omega^s$ is:

$$\sigma_{ij}^f n_j^{fs} = \sigma_{ij}^s n_j^{fs} \text{ on } S^{fs} \tag{12}$$

where n_j^{fs} is the normal unit vector on the fluid-structure interface, Ω^f is the fluid region, and Ω^s is the structure region. Due to the water flow effect, the gate structure contacting with the fluid vibrates, which leads to flow field changes. The changing flow field acts on the gate structure again, which forms fluid-structure interaction problems.

2.3. Meshing and Numerical Approach

The direction of flow is the x direction, the depth direction of flow is the y direction, and the direction across the flume is the z direction. Submerged discharge occurs downstream of the gate. The flow inlet is the speed boundary, and the flow outlet is the pressure boundary converted by the head. The top of the downstream flow is the air boundary, and the relative pressure is zero. The flume bottom and the sidewalls are no-slip solid boundary conditions, and the given normal velocity is zero. The method of the wall function is used for processing the near-wall viscous sublayer.

Due to the symmetry of the gate and the flume, half the model is selected to be meshed to reduce the calculating work. In this paper, the cases of the calculation are shown in Table 1, where e is the opening of the gate and H is the upstream water head. The underflow ratio U, is used to access the characteristics of the gate vibration caused by the submerged flow. Comprehensively considering the upstream and the downstream water head, underflow ratio U is defined, as follows:

$$U = \frac{h_t - h_c''}{H_0 - h_c''} \tag{13}$$

where h_t is the downstream water depth, h_c is the contraction section depth, h_c'' is the conjugate water depth of the contraction section, and H_0 is the total head.

Table 1. A list of the cases of calculation.

Case	e/H	U
1	0.049	0.059
2	0.197	0.355
3	0.375	0.730

For U = 0.059, the half grid of fluid with tetrahedral mesh has 252,758 units and a total of 49,112 nodes (Figure 1). The time step length of unsteady flow is 0.001 s, and the maximum number of iterations of each step is 20 times. At the same time, the numerical simulation of turbulent flow under the unslotted condition is carried out to calculate the vortex effect of the gate slot. The half grid of the plane gate is shown in Figure 2a, and the size of the plane gate is shown in Figure 2b.

The ANSYS Worckbench17.0 is used to solve the system of dependent variables. After constructing a three-dimensional fluid-structure coupling model, the flow field in the vicinity of the lift gate with a submerged discharge condition is simulated by the finite volume method. A combination of the $k-\epsilon$ double-equation turbulence model and the VOF model of the multiphase flow is used to calculate the flow field before 10.5 s. With the result as the initial conditions of the Smagorinsky-Lilly subgrid model and the LES method after 10.5 s, the VOF model is still adopted to determine of the water-air interface. Velocity pressure coupling by the PISO algorithm and the momentum equation in discrete form by the second-order windward scheme are used to obtain the hydrodynamic pressure distribution that acts on the gate surface. Under the action of the surface pressure load, the vibration response of the gate structure is analyzed by the finite element method. To analyze the calculation results visually, the magnitude and the materials of the calculation model are consistent with the synchronous experiment model. The density, elastic modulus, and the Poisson's ratio of the gate material are 1400 kg/m^3, 3×10^9 N/m^2, and 0.4, respectively. Constraints are imposed on the top and sides of the gate. The width-to-depth ratio of the gate slot is 1.78.

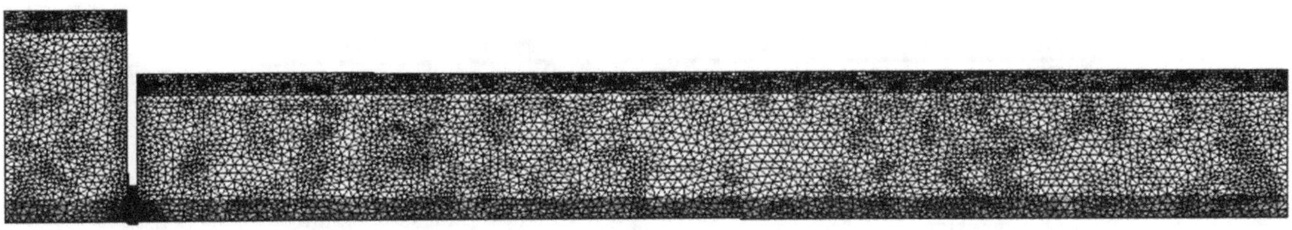

Figure 1. The grid of fluid.

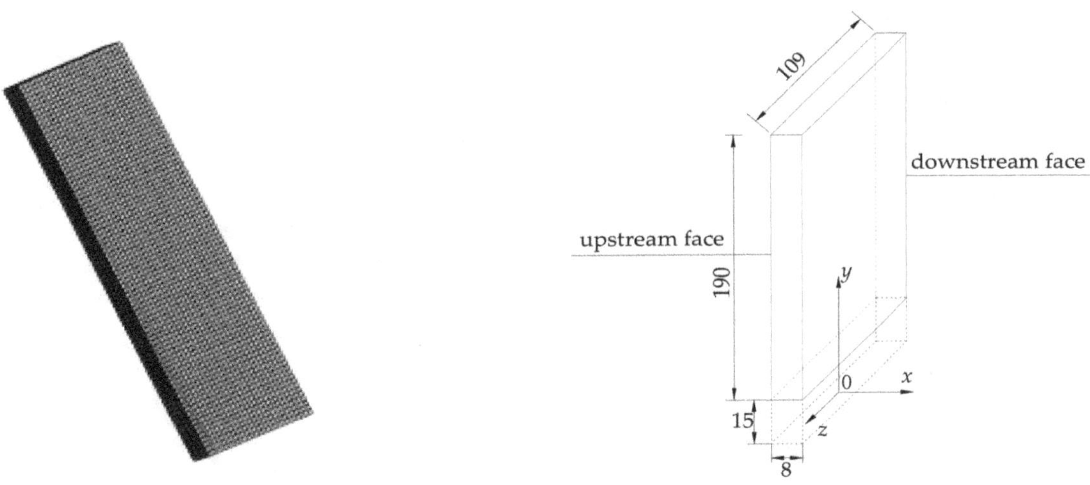

(a) (b)

Figure 2. The plane gate: (**a**) the grid; and, (**b**) the size (unit: mm).

3. The Results of Numerical Simulation

To verify the validity of the numerical simulation method, the numerical results are compared with the experimental data. The flow field behind the plane gate with submerged discharge, the hydrodynamic pressure process at typical points underneath the gate and the vibration parameters of the gate are acquired by the combination experiment [21] of three-dimensional particle image velocimetry (3D-PIV) made by TSI Company, the digital pressure sensor with high precision and the multichannel vibrating data acquisition system made by Lance Measurement Technologies Company, respectively.

The PIV test uses the Insight3G software to start the laser, and simultaneously captures the flow field particle images with two PIV-specific cross-frame CCD cameras. The autocorrelation or cross-correlation principle is used to extract the image of flow characteristics. Finally vector diagram of transient flow velocity in the measured range is obtained by professional post-processing software, such as Tecplot and Matlab. The maximum emission frequency of the laser is set to 14.5 Hz. In this test, a certain concentration of SiO_2 is selected as the tracer particle, and the particle size is 10–15 μm. The three-axis accelerometer is directly attached to the upstream surface of the gate. The accuracy of the pressure sensor and accelerometer both are ±0.1%FS, and the sampling interval of the pulsating pressure and the vibration displacement are 0.01 s and 0.001 s, respectively. Three instruments are tested simultaneously.

The size of the organic glass flume is 4880 mm × 100 mm × 190 mm, and the length behind the plane gate is approximately 3000 mm. The calculations and the experimental cases are consistent with each other. The general arrangement of the experiment is shown in Figure 3.

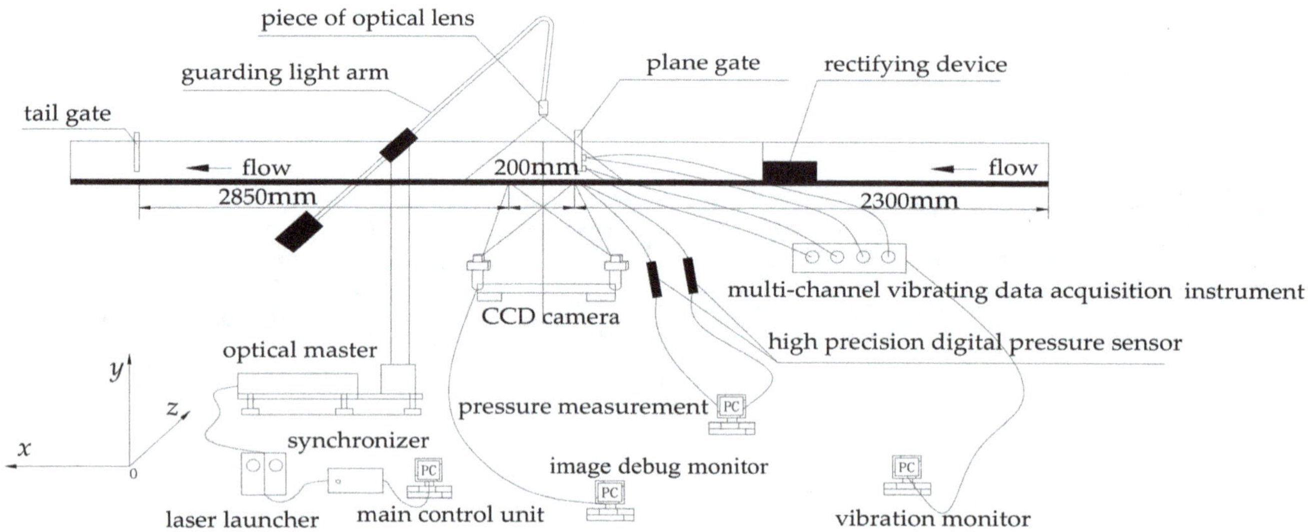

Figure 3. General arrangement of the experiment.

3.1. Flow Field Distribution

The calculation length after the lift gate is 920 mm and width B of the flume is 100 mm. The computation results of the time-averaged surface profiles with the whole flow passage of the modeling after the gate in $U = 0.059$ are shown in Figure 4. Blue represents the water phase and red represents the gas phase.

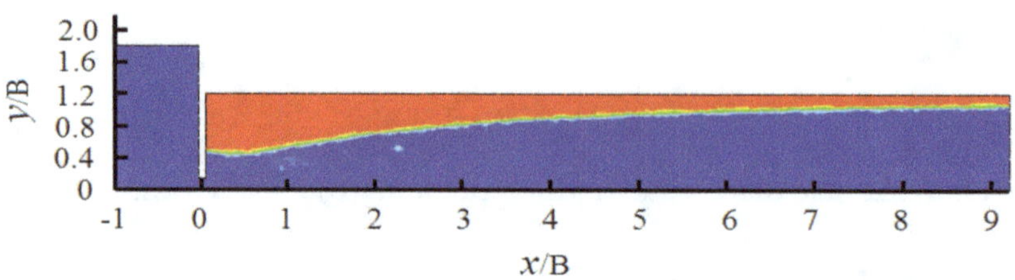

Figure 4. The color nephogram of the time-averaged flow profile.

Figure 5 shows the time-averaged flow profile comparison behind the gate of section $z = 0$ mm for the calculation and experiment of $U = 0.059$. The maximum and minimum of relative difference of the height between the calculation and the test values of the flow profile are 4.8% and 0.16%, respectively, which shows that the VOF method can simulate the water-air interface well.

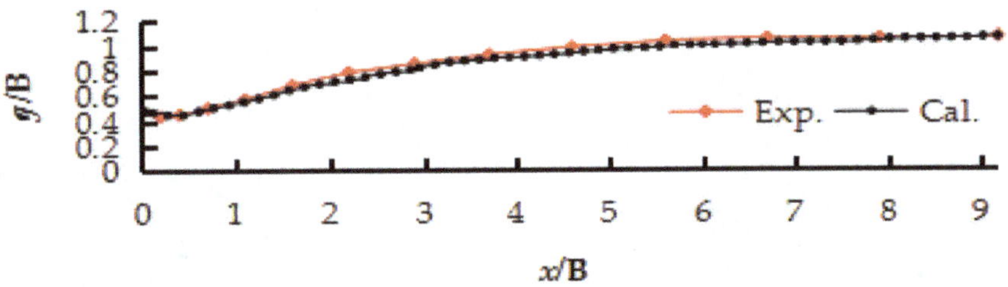

Figure 5. The time-averaged flow profile.

The calculational time-averaged streamlines on sections z = 0 mm and z = 35 mm in different cases are presented in Figure 6. According to (a), (b), and (c) in Figure 6, there is only a vortex on section z = 35 mm in the time-averaged flow field behind the plane gate in the x direction, but there are two vortices on section z = 0 mm that are within the 300 mm range. At different sections in the same case, the vertical distances of the vortex center from the x-axis are approximately the same. With the e/H increasing, the vertical distance of the vortex center from the x-axis increases gradually.

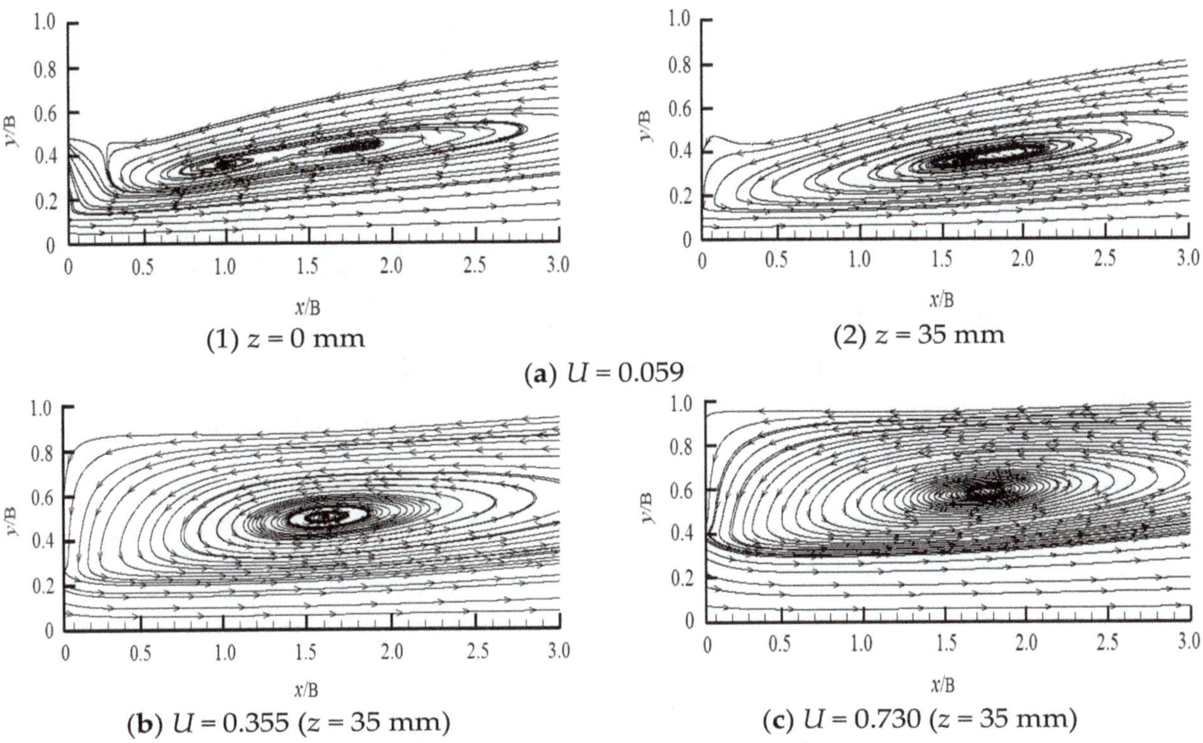

Figure 6. The calculational time-averaged streamlines.

The size of the PIV calibration target is 200 mm × 200 mm. Due to the size limitation, it is impossible to test the vortices in the slot and at the bottom edge of the gate by PIV, so the computational results of only the vortices behind the gate are compared carefully with those of the three-dimensional PIV test. Both the simulated and the experimental velocity vector diagrams are achieved at 20.6 s.

Figure 7 shows the instantaneous streamlines behind the gate on the same section z = 0 mm for the calculation and the experiment. The numerical simulation can make up for the drawback of losing the velocity vector by 3D-PIV positioning. When compared with the experimental results, the vortex center, which is closest to the gate obtained by the simulation, is lower in the y direction under U = 0.059, but this distance is basically same for the other two cases. On the whole, the numerical simulation still obtained the most ideal vortex information.

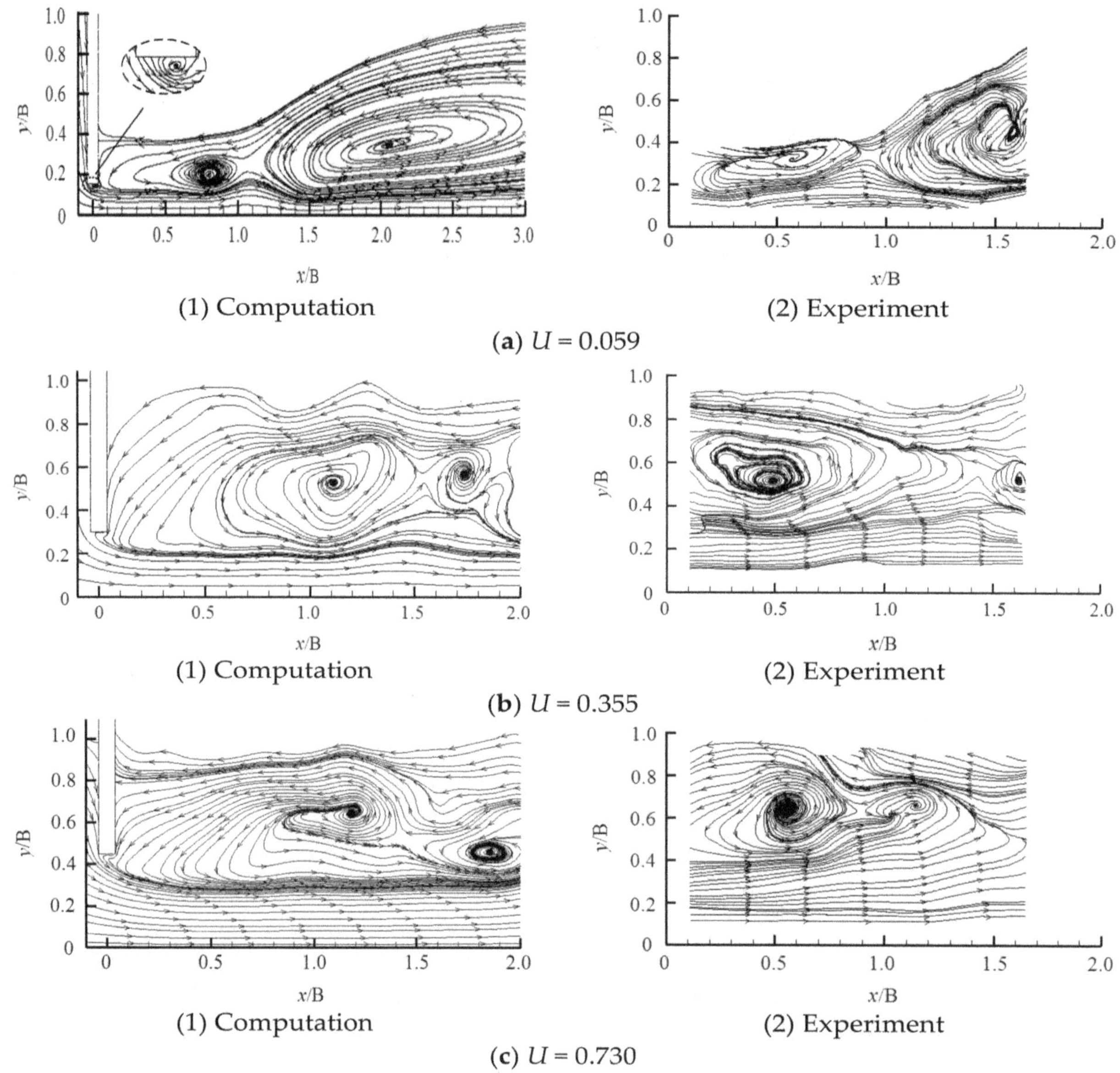

Figure 7. The instantaneous streamlines (z = 0 mm).

It can be seen from the instantaneous streamlines in Figure 7a–c that there are the same number of vortexes on section $z = 0$ mm between the computation and the experiment. Under $U = 0.059$, the numerical simulation captures the phenomenon of the vortex detachment from the gate's bottom edge, but there is no such phenomenon when $U = 0.355$ or $U = 0.730$. Therefore, when the e/H and U reach a certain extent, the vortex shedding phenomenon can be generated from the bottom edge of the gate.

Normalizing the turbulent kinetic energy by the square of the mean flow velocity of the gate downstream section, Figure 8 presents the computational normalized turbulent kinetic energy distribution on section $z = 0$ mm and section $z = 35$ mm. Turbulent kinetic energy under the gate on section z = 0 mm is higher than the value on section $z = 35$ mm. The different magnitude of turbulent kinetic energy is one of the significant factors affecting the gate vibration extremum. The experiment proved that the extremum of vibration displacement on section $z = 0$ mm is larger than that on section $z = 35$ mm.

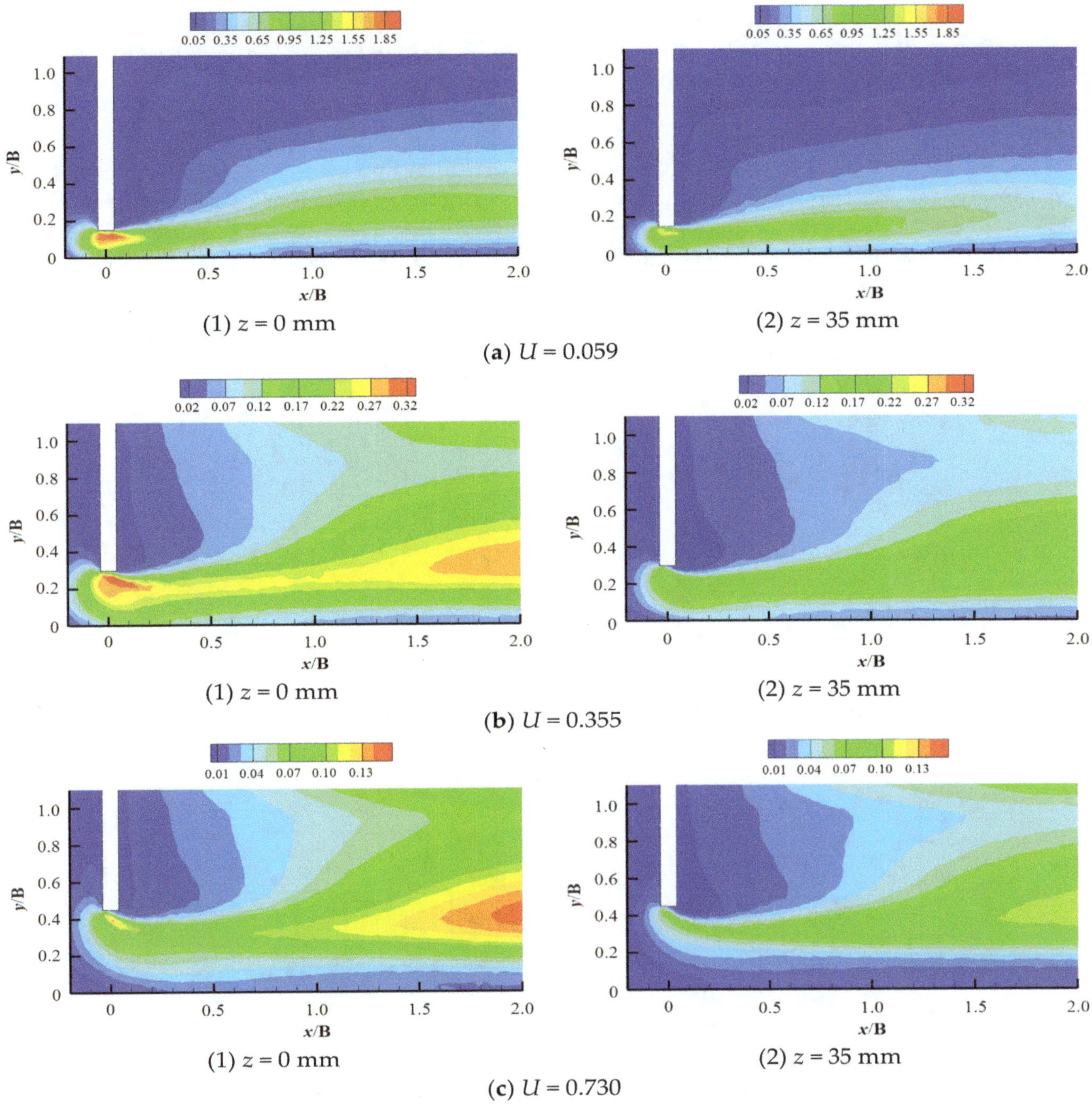

Figure 8. Normalized turbulent kinetic energy distribution.

3.2. *Hydrodynamic Pressure*

The measuring points of hydrodynamic pressure just below the gate are point 1 (0, 0, 0 mm) and point 2 (0, 0, 35 mm). The digital pressure sensor with high precision is used to test the hydrodynamic pressure with a sampling frequency of 0.01 s, while the monitoring points are also set up in the numerical calculation.

Figure 9 shows the time history curves of the fluctuating pressure coefficient of points 1 and 2 in 10 s. $C_p = (p - \overline{p})/0.5\rho v_2$, where C_p is the fluctuating pressure coefficient, p is the instantaneous pressure, $\overline{p}$ is the average pressure, ρ is the water flow density, and v is the average velocity of the exit section. Under the same case, the fluctuating pressure of point 1 is stronger than that of point 2. The time to reach the extreme of the two points is out of synchronization.

Figure 10 shows the normalized power spectrum density curve of the pressure for point 1 in different cases. *St* is the Strouhal number and $St = f \cdot L/\overline{v}$, Where f is the frequency of the pressure pulsating, L is the characteristic length and taken as the gate opening, and $\overline{v}$ is the mean velocity of section just below the gate. With the *e/H* or *U* increasing, the energy of pulsating pressure of point 1 decreases. The computational Strouhal numbers for dominant frequencies of the pressure are 0.332×10^{-3}, 2.217×10^{-3}, 1.545×10^{-3} with the corresponding cases of $U = 0.059, 0.355, 0.730$. Accordingly, the experimental Strouhal numbers for dominant frequencies of the pressure are 0.337×10^{-3}, 2.242×10^{-3}, 1.552×10^{-3} in corresponding cases. The Strouhal number difference for main frequency of the pressure between the computational and the experimental results is very small.

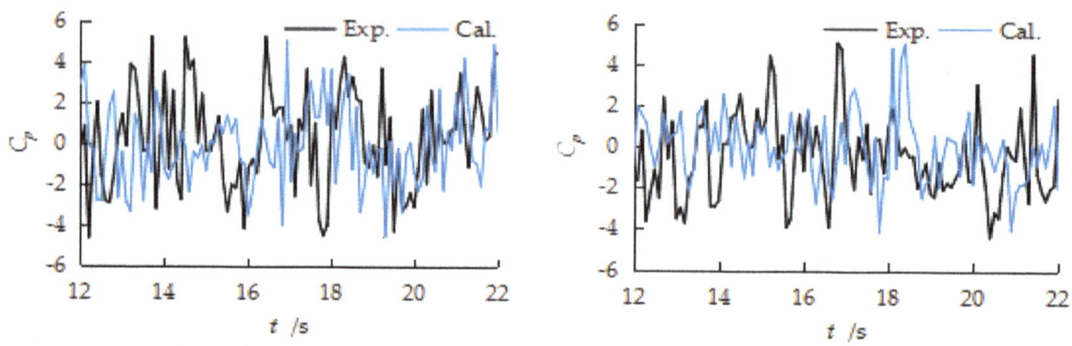

(**a**) Point 1 ($U = 0.059$) (**b**) Point 2 ($U = 0.059$)

Figure 9. The time history curve of the pressure pulsating coefficient for different point.

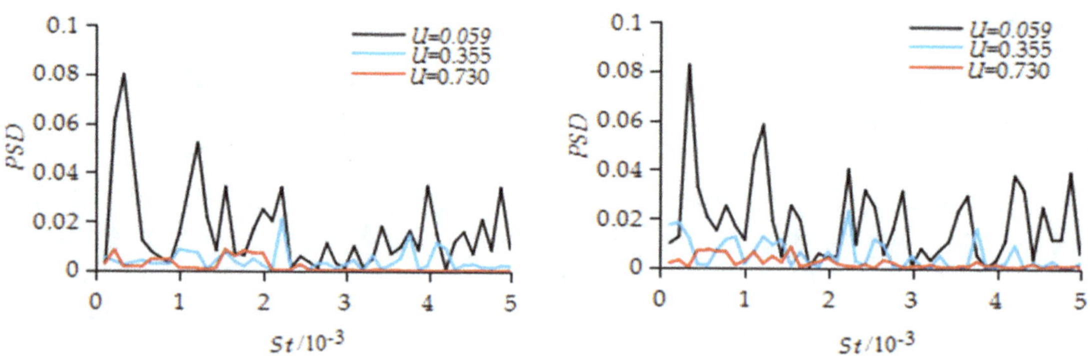

(**a**) Computation (**b**) Experiment

Figure 10. The power spectrum density (PSD) curve of the pressure for point 1 in different cases.

In order to illustrate the influence of the gate slot on the flow field and the gate vibration, both calculation and experiment are carried out on the filling of the slots underneath the gate (unslotted). Figure 11 shows the time history curve of the slotted and unslotted pressure pulsating coefficients in $U = 0.059$. It can be found that the maximum values of the slotted pressure pulsation coefficients for both points 1 and 2 were larger than those of the without a slot. The characteristic values of the pressure pulsating coefficients (slotted and unslotted) for both the experiment and for the calculation in $U = 0.059$ are presented in Figure 12. When comparing the value of the computational and the synchronous experimental results, the relative difference is within 10%, which indicates that the numerical simulation results meet the basic requirements.

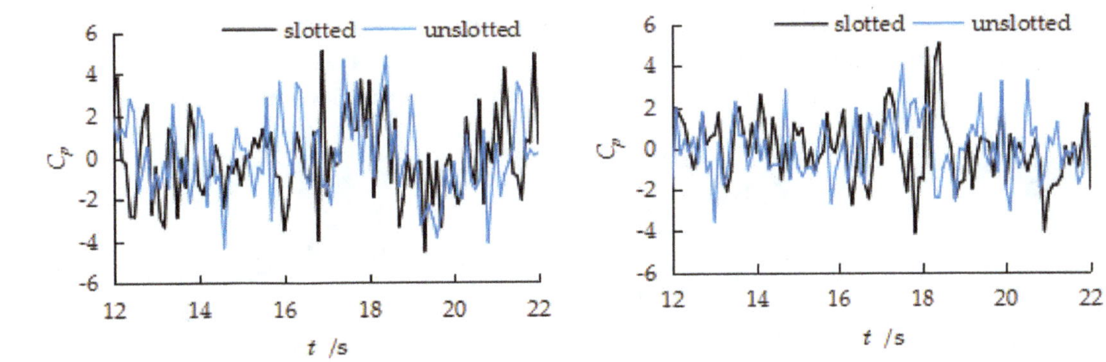

(a) Point 1 (b) Point 2

Figure 11. The time history curve of the pressure pulsating coefficients (slotted and unslotted).

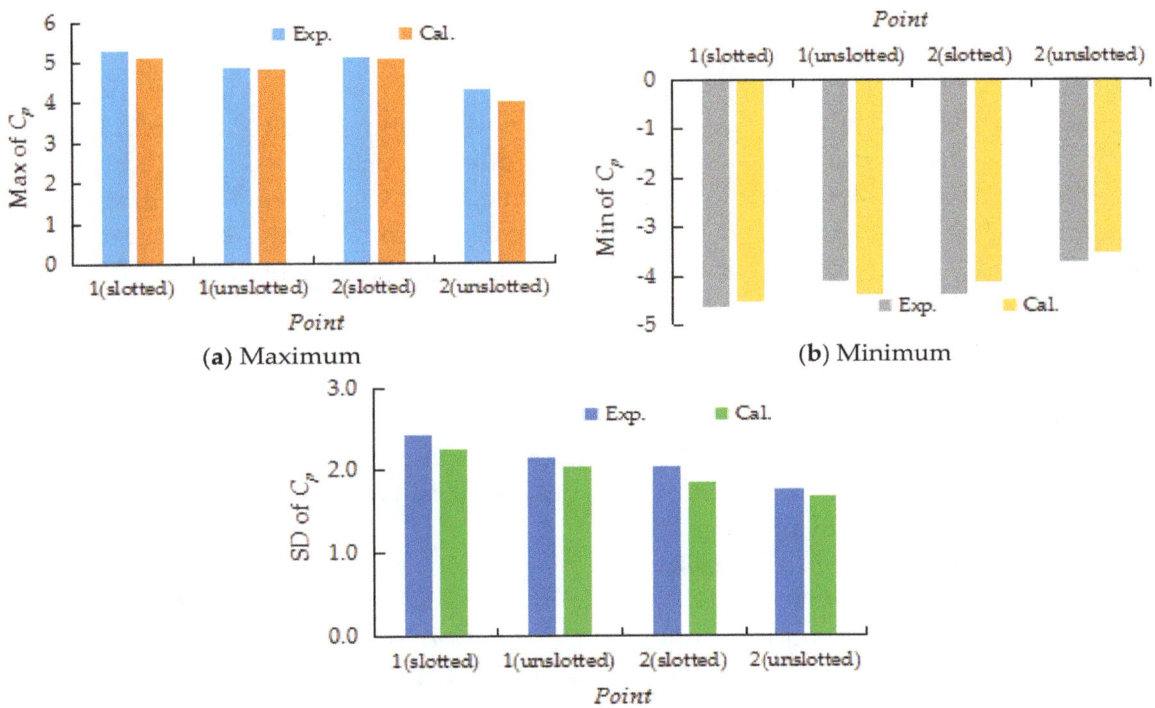

(a) Maximum (b) Minimum

(c) Standard deviation (SD)

Figure 12. The comparison of characteristic value of the pressure pulsating coefficient.

In the experiment, the dominant frequency of the pressure at the measuring points underneath the gate are only 0.025~1.245 Hz, and the dominant frequency of the pressure at the measuring points behind the gate are only 0.183~14.63 Hz. The natural frequencies of the gate vibration by the modal analysis are 247.43 Hz, 529.85 Hz, 1086.6 Hz, 1280.6 Hz, 1605.1 Hz, and 1963.1 Hz, respectively. The natural frequency of the gate vibration is far from the dominant frequency of the flow pressure, so it would not have produced resonance.

Figure 13a presents the computational pressure distribution and streamlines in the gate slot for different level cross sections ($y/e = 0.067$, $y/e = 0.2$, $y/e = 0.333$, $y/e = 0.467$, $y/e = 0.6$, $y/e = 0.733$) from the bottom to the top in y direction for $U = 0.059$. It can be seen from the superimposed vortex diagram that the low-pressure zone of the vortex gradually enlarges from the bottom to the top, which is shaped like a funnel and it shifted to the downstream. The pressure is lowest in the vortex center where it is also most prone to generate cavitation in the gate slot. Due to the water impacting the slot from the left to the right, the pressure is highest at the right of the slot. The closer to the bottom the place is, the higher the pressure is. Under $U = 0.355$ (Figure 13b) and $U = 0.730$ (Figure 13c), the range of the low pressure and the high pressure zones in the gate slot is much smaller than that in $U = 0.059$.

The funnel vortex of $U = 0.355$ still exists, but the center of the vortex is shifted to the upstream from the bottom to the top, and the center pressure of the bottom vortex is relatively large. For $U = 0.730$, the centerline of the vortex is similar to the spiral line, and the funnel vortex is not obvious.

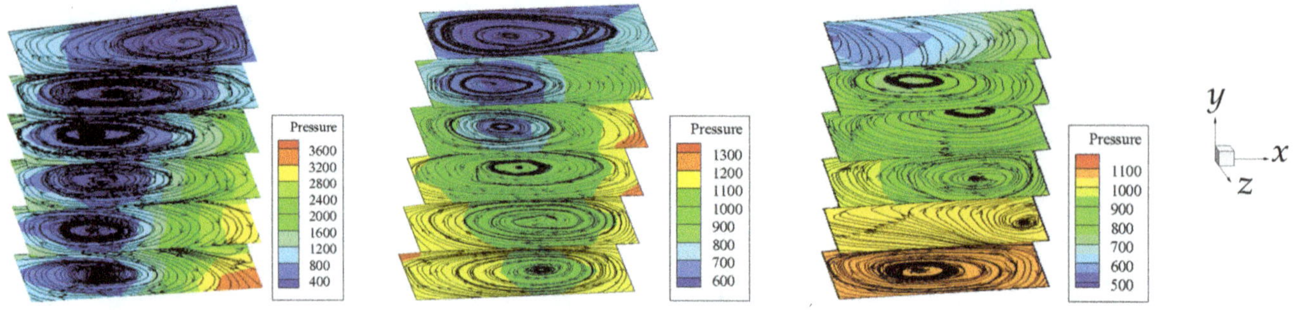

(a) $U = 0.059$ (b) $U = 0.355$ (c) $U = 0.730$

Figure 13. Pressure (pa) distribution and streamlines at 22.0 s in the gate slot.

The maximum vibration displacement of the gate with a submerged condition in the x direction is not due to the resonance, but it is due to the mixed vortex-induced process, including the vortices behind the gate, in the gate slot and at the gate's bottom edge. It can be seen from the above analysis that the numerical simulation method in this paper can obtain not only the pressure fluctuation distribution, but also a more complete flow structure, such as multiple vortices of the sluice flow, and it forms the excitation force of the plane gate vibration.

3.3. Vibration of the Gate

Define the non-dimensional coefficient $K_d = (d - \bar{d})/\sigma_d$, where d is vibration displacement, $\bar{d}$ is the average vibration displacement, and σ_d is the vibration standard deviation. Point A, which is located on the central axis of the upstream surface of the gate and 10 mm away from the bottom edge, is used to study the response of the gate vibration.

The normalized power spectrum density curve of vibration displacement for point A in $U = 0.059$, which is obtained by the computation and the experiment, is shown in Figure 14. In the experiment, the acceleration sensor and the multichannel data acquisition instrument were used to monitor the acceleration of the gate vibration in the x direction and y direction. The vibration displacement is achieved by the second integral of the acceleration. The computational Strouhal number for dominant frequency of vibration displacement in x direction is 2.582 with the corresponding experiment of 2.660. In addition, the computational Strouhal number for dominant frequency of vibration displacement in y direction is 3.724 with the corresponding experiment of 3.768. The Strouhal number for main frequency of the vibration displacement between the computational and the experimental results is closer.

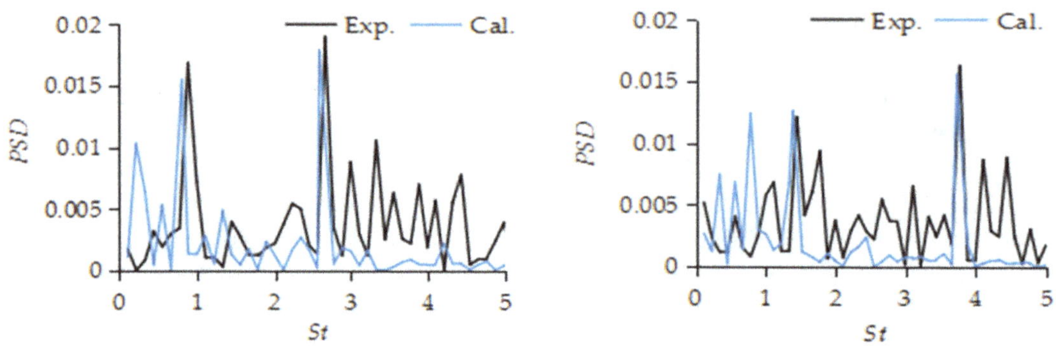

(a) x direction (b) y direction

Figure 14. The power spectrum density (PSD) curve of vibration displacement for point A.

Figure 15 plots the time-history curve of the vibration displacement coefficient K_d in the x direction and in the y direction of point A, with and without slots, by using the numerical simulation. It can be seen that the maximum value of the vibration displacement coefficient of the plane gate with a slot is larger than that of the plane gate without a slot in both the x and y directions. When there is no slot, the maximum value of the x direction vibration displacement coefficient appears at 12.7 s and the minimum value appears at 21.8 s. The maximum value in the y direction appears at 14.7 s and the minimum value appears at 17.6 s. The moment when the vibration displacement coefficient of the x direction and the y direction appears is out of sync.

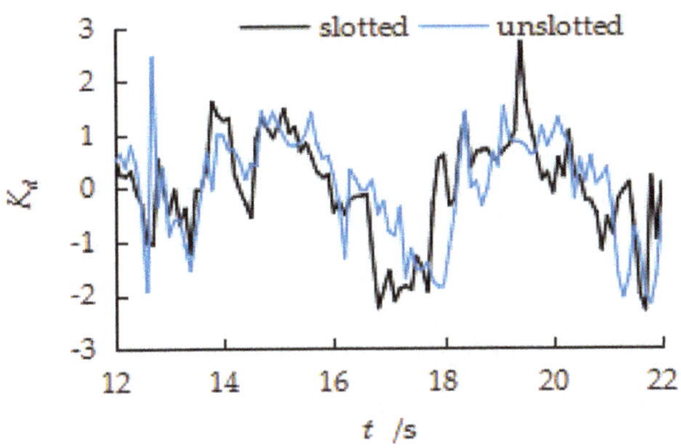

(a) x direction

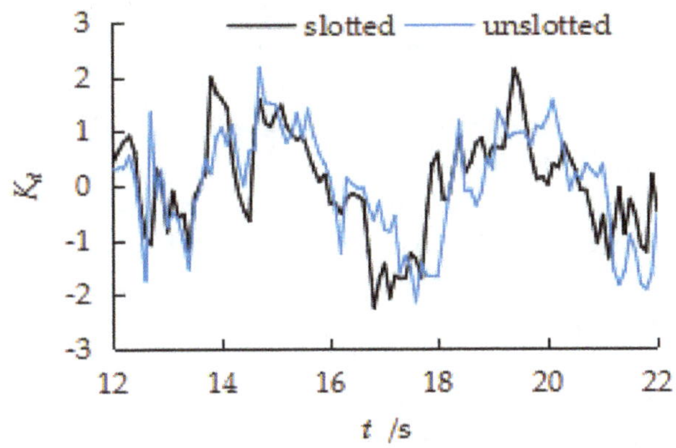

(b) y direction

Figure 15. The time history curve of the vibration displacement coefficient K_d of point A.

4. Conclusions

In this study, numerical simulations were performed to compute the flow field and the water surface profile for a plane gate with a submerged discharge by the combination of the LES and the VOF methods. Overall, the numbers and locations of the vortex are consistent with the synchronous experimental results. When compared with the experimental values, the relative difference of both the standard deviation pressure and the maximum pressure of the calculation is within 10%, which verifies the validity of the numerical simulation methods that were adopted in the present work. Simultaneously, the maximum vibration displacement coefficient was acquired from a calculation of the vibration displacement process of Point A, which was very close to that of the experiment. So, the fluid-structure coupling computational method used in certain conditions can replace the complex experiment and make up for the lack of experiment.

The turbulent kinetic energy under the gate on section z = 0 mm was higher than the value on section z = 35 mm. The change trend of the turbulent kinetic energy was consistent with the amplitude of the gate vibration displacement, which shows that the turbulent kinetic energy under the gate plays an important role in the amplitude of the gate vibration. The maximum value of the vibration displacement coefficient of the plane gate with a slot was larger than that of the plane gate without a slot in both x and y directions. Thus, the vortex in the gate slot aggravates the vibration of the gate. Turbulence pressure pulsation and multiple vortices in the flow field are the main causes of gate vibration. The mixed vortex-induced process includes the vortices behind the gate, in the gate slot, and at the gate's bottom edge.

Author Contributions: C.S. and W.W. conceived and designed the model; C.S. and S.H. set up and debugged the programmes; C.S. analysed the results and wrote the paper; C.S. and Y.X. provided editorial improvements to the paper.

Acknowledgments: Financial support provided by the National Key Research and Development Program (Grant No. 2016YFC0401603) and the National Natural Science Foundation of China (Grant No. 51369013) are gratefully acknowledged.

References

1. Hardwick, J.D. Flow-induced vibration of vertical lift gate. *J. Hydraul. Div.* **1974**, *100*, 631–644.
2. Kolkman, P.A.; Vrijer, A. Gate edge suction as a cause of self-exciting vertical vibrations. In Proceedings of the 17th IAHR Congress, Baden, Germany, 15–19 August 1977; Volume 17, pp. 425–437.
3. Jongeling, T.H.G. Flow-induced Self-excited in Flow Vibrations of Gate Plates. *J. Fluids Struct.* **1988**, *2*, 541–566. [CrossRef]
4. Ishii, N. Flow-induced vibration of long span gates, Part I: Model development. *J. Fluids Struct.* **1992**, *6*, 539–562. [CrossRef]
5. Pani, P.K.; Bhattacharyya, S.K. Hydrodynamic pressure on a vertical gate considering fluid-structure interaction. *Finite Elem. Anal. Des.* **2008**, *44*, 759–766. [CrossRef]
6. Erdbrink, C.D.; Krzhizhanovskaya, V.V.; Sloot, P.M.A. Reducing cross-flow vibrations of underflow gates: Experiments and numerical studies. *J. Fluids Struct.* **2014**, *50*, 25–48. [CrossRef]
7. Liu, S.; Liao, T.; Luo, Q. Numerical simulation of turbulent flow behind sluice gate under submerged discharge conditions. *J. Hydrodyn.* **2015**, *27*, 257–263. [CrossRef]
8. Kotecki, S.W. Numerical modeling of flow through moving water-control gates by vortex method. Part I—Problem formulation. *Arch. Civ. Mech. Eng.* **2008**, *8*, 73–89. [CrossRef]
9. Kotecki, S.W. Numerical modeling of flow through moving water-control gates by vortex method. Part II—Calculation result. *Arch. Civ. Mech. Eng.* **2008**, *8*, 39–49. [CrossRef]
10. Kotecki, S.W. Numerical analysis of hydrodynamic forces due to flow instability at lift gate. *Arch. Civ. Mech. Eng.* **2011**, *11*, 943–961. [CrossRef]
11. Kazemzadeh-Parsi, M.J. Numerical flow simulation in gated hydraulic structures using smoothed fixed grid finite element method. *Appl. Math. Comput.* **2014**, *246*, 447–459. [CrossRef]
12. Zhu, T.; Ye, W. Theoretical and numerical studies of noncontinuum gas-phase heat conduction in micro/nano devices. *Numer. Heat Transf. Part B Fundam.* **2010**, *57*, 203–226. [CrossRef]
13. Liu, H.; Xu, K.; Zhu, T.; Ye, W. Multiple temperature kinetic model and its applications to micro-scale gas flows. *Comput. Fluids* **2012**, *67*, 115–122. [CrossRef]
14. He, S.; Zhang, L. Large Eddy Simulation of Effect of Geometric Parameters on Low Pressure of Turbulent Flow over Gate Slot. *J. Syst. Simul.* **2015**, *27*, 1081–1086.
15. Leclercq, T.; de Langre, E. Vortex-induced vibrations of cylinders bent by the flow. *J. Fluids Struct.* **2018**, *80*, 77–93. [CrossRef]
16. Chizfahm, A.; Yazdi, E.A.; Eghtesad, M. Dynamic modeling of vortex induced vibration wind turbines. *Renew. Energy* **2018**, *121*, 632–643. [CrossRef]
17. Postnikov, A.; Pavlovskaia, E.; Wiercigroch, M. 2DOF CFD calibrated wake oscillator model to investigate vortex-induced vibrations. *Int. J. Mech. Sci.* **2017**, *127*, 176–190. [CrossRef]
18. Bouratsis, P.; Diplas, P.; Dancey, C.L.; Apsilidis, N. Quantitative Spatio-Temporal Characterization of Scour at the Base of a Cylinder. *Water* **2017**, *9*, 227. [CrossRef]
19. Smagorinsky, J. General circulation experiment with primitive equations. *Mon. Weather Rev.* **1963**, *91*, 99–164. [CrossRef]
20. Lilly, D.K. Aproposed modification of the Germa no subgrid-scale closure method. *Phys. Fluid A* **1992**, *4*, 633–635. [CrossRef]
21. Shen, C.; He, S.; Yang, T.; Wang, W. Model tests for synchronous measurement of fluid-structure interaction vibration of a plane vertical lift gate. *J. Vib. Shock* **2016**, *35*, 219–224.

11

Effect of Flaring Gate Piers on Discharge Coefficient for Finite Crest-Length Weirs

Zhong Tian, Wei Wang *, Ruidi Bai and Nan Li

State Key Laboratory of Hydraulics and Mountain River Engineering, Sichuan University, Chengdu 610065, China; tianzhong@scu.edu.cn (Z.T.); bairuidiscu@163.com (R.B.); nancyli1220@163.com (N.L.)

* Correspondence: profwangwei@sina.com

Abstract: The use of flaring gate piers (FGPs) along with finite crest-length weirs changes the shape of plunging jets and increases the efficiency of energy dissipation in some projects; however, the FGPs may affect the discharge capacity. In this study, the flow pattern and discharge coefficient were experimentally investigated under different conditions by varying the weir lengths L_w, contraction ratio β, contraction angle θ, and water heads H. A comparative analysis of the weirs with and without FGPs was performed. For the finite crest-length weirs with FGPs, the water-surface profiles in the flow channel were backwater curves. Moreover, the plunging jets leaving the weir became narrower and then subsequently diffused largely in the transverse and longitudinal directions in air. The discharge coefficients of the weirs with FGPs were approximately equal for various weir lengths. Moreover, following the earlier studies on traditional finite crest-length weirs, a discharge-coefficient equation was developed for the weir with an FGP in this study. The results showed that in the weirs with FGPs, the discharge coefficients clearly increased with the increase in the contraction ratio and water head, but the changes in their values along with the contraction angle were neglected.

Keywords: finite crest length weir; FGP (flaring gate pier); discharge coefficient; subcritical flow

---

1. Introduction

Generally, based on the relative length of the weirs, H/L_w, where H is the head on the weir crest and L_w is the weir length in the direction of flow [1–5], weirs are divided into two groups: the sharp-crested weirs ($H/L_w > 2$) and finite crest-length weirs ($H/L_w < 2$). The finite crest-length weirs can be further classified into three categories: the short-crested ($0.4 \leq H/L_w < 2$), broad-crested ($0.1 \leq H/L_w < 0.4$), and long-crested weirs ($0 < H/L_w < 0.1$). The discharge characteristics of the flows over the finite crest-length weirs with different cross sections, such as rectangular, trapezoidal, and triangular, have been studied. According to a systematic experimental study on the finite crest-length weirs having a rectangular shape with $0 < H/L_w \leq 2$ and $0 < H/P < 1$ (where P is height of the weir), three discharge equations have been developed for the short-crested, broad-crested, and long-crested weirs [6]. Ramamurty et al. [7] studied the influence of upstream rounding on the discharge coefficient. Sarker and Rhodes [8] used computational fluid dynamics (CFD) modeling to model the free-surface profiles of the broad-crested weirs; the experimental results were in good agreement with those of the numerical models. In order to model the free surface flow of real reservoirs, Andersson et al. [9] performed a physical scale model test and three-dimensional simulation of the spilling from a reservoir, which gives a better understand of the flow pattern for the short-crested weir. Chen et al. [10] studied the discharge coefficient of a rectangular short-crested weir with varying slope coefficients and developed an equation for calculation of the discharge coefficient. Recently, Azimi et al. [11] conducted a series of laboratory experiments to study the correlations for nine different types of weirs, which extended the earlier studies on the finite crest-length weirs having a trapezoidal shape, considering sloping crests and either upstream ramps, downstream ramps, or both. In addition, the study was extended to triangular weirs (in which the crest length is zero) with either upstream ramps, downstream ramps, or both. Jan et al. [12] proposed a linear combination of traditional discharge equations of simple triangular and rectangular weirs to develop the discharge equations for the complex broad-crested weirs.

A new concept of energy dissipation by combining the flaring gate pier (FGP) with the weir was proposed in China [13]. The FGP is usually employed on a weir to achieve 3D flow by making the flow emerge from the FGP with lateral contraction and longitudinal dispersion; hence, a shorter length of the stilling basin and a larger energy dissipation ratio could be obtained [14]. FGPs have been extensively applied in roller-compacted concrete dams. During the construction, some sections of the dam serve as gaps for releasing floodwater, and the FGP is integrated in the system to reduce the erosion of downstream channels. Hence, the finite crest-length weirs are integrated with the FGP, and the overflows are largely subcritical when they are constructed to a certain height. The appraisal of the diversion schemes for flooding is related to the construction period, quality, cost, and safety of the whole project; hence, the diversion schemes need to be designed carefully in advance. It is important to study the discharge characteristics of the finite crest-length weirs with the FGP. In previous studies, the FGP was almost integrated on the ogee-overflow weirs (curved short-crested weirs) and the flow over the weir was largely supercritical [15,16]. Li et al. [17] experimentally studied the effects of the FGP on the discharge capacity and flow pattern of the ogee-overflow weirs. The results showed that the effect of the FGP on the discharge capacity is closely related to the flow pattern; four types of flow patterns occurred, depending on the weir head and the contraction ratio. Recently, Li et al. [18] showed that the effect of the variations in the design parameters of the FGP on the discharge capacity of the surface spillway is insignificant when the flow in the channel is supercritical.

Previous studies focused on the hydraulic characteristics of FGP on short-crested weirs. The purpose of this paper is to discuss the flow coefficient after setting an FGP on finite crest-length weirs (including short-crested weirs and broad-crested weirs). A comparative analysis of the weir with different parameters of the FGP was performed.

2. Experimental Setup and Methodology

The experiments were conducted in the State Key Laboratory of Hydraulics and Mountain River Engineering, Sichuan University. The experimental setup comprised a flow measurement section, a large feeding tank (6 m in length and 2 m in width, to ensure steady-state flow condition), a test section, and a flow return system. A right-angled triangle weir is installed upstream of the reservoir, and an electromagnetic flowmeter is installed in the return pipe downstream for flow calibration. Because of the defined relationship between the discharge Q and the head H, the right-angled triangle weir is used as a flow measuring device, for which the discharge calculation formula is

$$Q = 1.33H^{2.465} \tag{1}$$

The head of the right-angled triangle weir is measured by a water level stylus, which is installed 1.0 m upstream of the weir, and the accuracy is 0.1 mm. The water level of the reservoir is also measured by a water level stylus with accuracy of 0.1 mm. The water level gauge point was set at the side of the feeding tank (or reservoir) and is 2 m away from the inlet of the finite-crest length weir. Figure 1 shows the experimental setup.

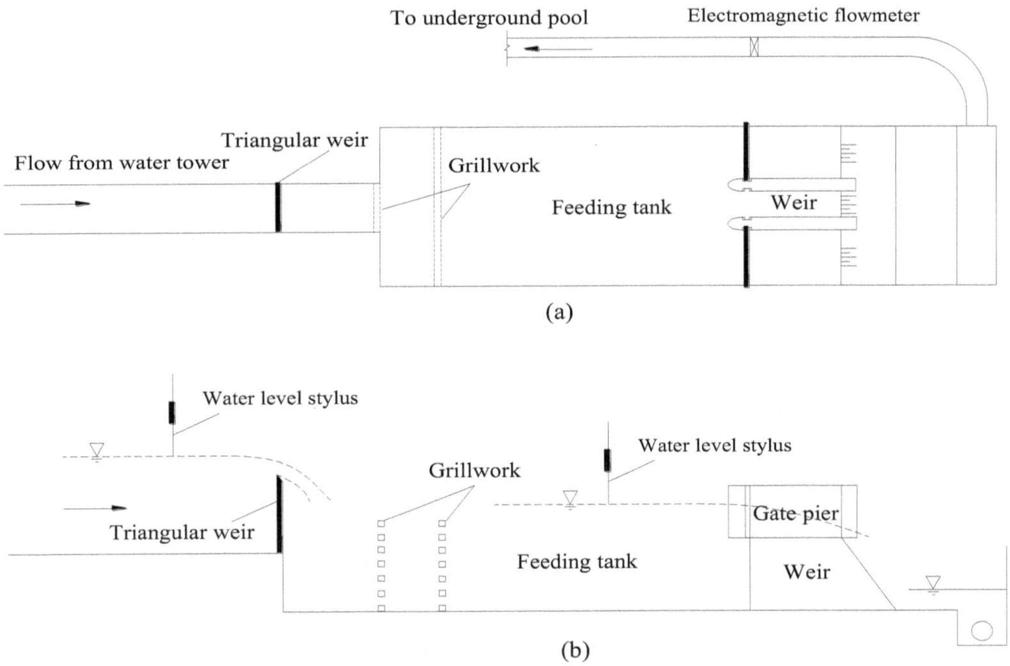

Figure 1. Experimental setup: (**a**) plane view; (**b**) side view.

The head of the right-angled triangle weir and water level of the reservoir were measured every 5 min for each discharge. Comparing discharge between the right-angled triangle weir and the electromagnetic flowmeter shows that the error is less than 1.4%.

The test model, made using Plexiglas, was a finite crest-length weir. The length (L_w), width (B), and height (P) of the type 3 weir were 0.601 m, 0.30 m, and 1.28 m, respectively, wherein the upstream face was upright and the downstream face had a slope of 1.2:1 (i). The channel width B = 0.30 m was reduced to B' = 0.115 m at the edge of the outlet for length L_p = 0.271 m when the FGP was integrated on the weir. The contraction angle was 15°, as shown in Figure 2. In this experiment, three types of finite crest-length weirs were employed, with weir lengths L_w of 0.601 m, 0.519 m, and 0.422 m and weir heights of 1.28 m, 1.378 m, and 1.507 m, respectively. In addition, nine types of FGPs were employed with contraction ratios β = 0.38, 0.51, 0.64, 0.82, and 1.00, and contraction angles θ = 7.0°, 9.6°, 12.7°, 15.0°, and 21.5°, respectively. The length of FGP (L_p in Figure 2) is variable, and the bulkhead slots stay at the same location. The ranges of H and H/L_w were approximately 0.112–0.547 m and 0.249–1.212 m, respectively.

The water-surface profiles were measured along the centerline of the weir. The coordinate system (X–Y) was established to describe the water-surface profiles efficiently, as shown in Figure 2c. The origin was at the center of the weir inlet. The location and average water depth of the measuring section were described by the X and Y values, respectively. X/L_w is the relative location of the measuring section.

To investigate the influence of the FGP on the discharge coefficient of the finite crest-length weir, 13 groups of experiments were performed, as listed in Table 1. The cases were divided into three sets: cases N1–N6 represent the effect of L_w, cases N5–N9 represent the effect of β, and cases N5 and N10–N13 represent the effect of θ. The hydraulic experiments were conducted under different conditions with 5–6 different weir heads (0.112–0.547 m) for each group, including discharge capacity, flow pattern, and water-surface profile in the flow channel.

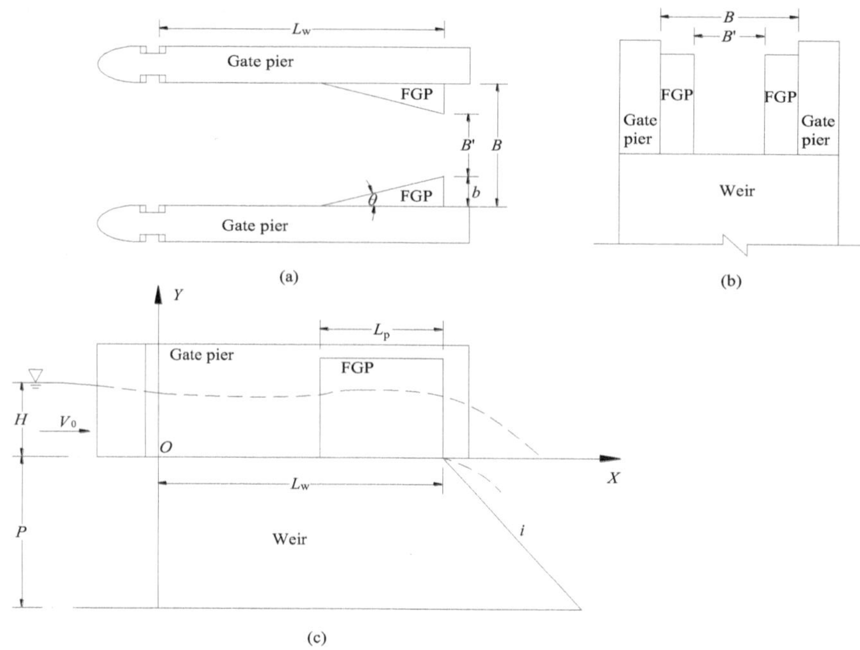

Figure 2. Definition sketch of the finite-crest length weir with FGP (Flaring Gate Pier): (**a**) plane view; (**b**) front view; (**c**) side view (L_w—the length of the weir, L_p—the length of the FGP, B—the width of the weir crest, b—the width of the weir outlet, P—the height of the weir, H—the head of the weir, V_0—approching velocity).

Table 1. Parameters of the finite-crest length weir and FGP.

Weir Type	Weir Height P (m)	Weir Length L_w (m)	Channel Width B (m)	Contraction Ratio $\beta = B'/B$	Contraction Angle θ	Case Name
Type 1	1.493	0.422	0.30	0.51	15	N1
				1	0	N2
Type 2	1.379	0.519	0.30	0.51	15	N3
				1	0	N4
Type 3	1.28	0.601	0.30	0.51	15	N5
				1	0	N6
				0.38	15	N7
				0.64	15	N8
				0.82	15	N9
				0.51	7	N10
				0.51	9.6	N11
				0.51	12.7	N12
				0.51	21.5	N13

3. Results and Discussions

3.1. Flow Pattern Comparison with and without FGP

As shown in Figure 3a, when there was no FGP on the weir, the water depth gradually decreased along the channel and the plunging jet dropped into the stilling basin in a rigorous manner without the longitudinal spreading for a large discharge. In contrast, when the FGP was applied in the weir, the water on the upper side flowed slower than that on the bottom and the plunging jet became narrow just outside the weir; subsequently, the flow spread in the transverse and longitudinal directions in air for the contraction, as shown in Figure 3b,c.

Taking type 3 of the finite crest-length weir as an example, the Froude number (Fr) along the crest with FGP ranges from 0.3~0.5, thus it is a subcritical flow. Fr along the crest without FGP ranges from 0.5~1.2, and flow pattern changing occurs at the ending of the crest ($X/L_w > 0.8$). The water-surface profiles that described the change in depth along the channel are plotted in Figure 4. It was clear that the depth of water in the channel increased significantly after setting the FGP for the same discharge. When the depth of the water in the channel was low, the two water-surface profiles shifted toward the horizontal form. When the depth of the water in the channel was relatively high, the two profiles differed depending on whether the FGP was employed; i.e., the water-surface profile gradually decreased for the channel where the FGP was not installed, as shown in Figure 5a. The water surface upstream of the starting point of FGP is smooth; however, the backwater curve occurred at the beginning of the FGP when setting the FGP in the weir, as shown in Figure 5b, which could be regarded as a surface wave [2].

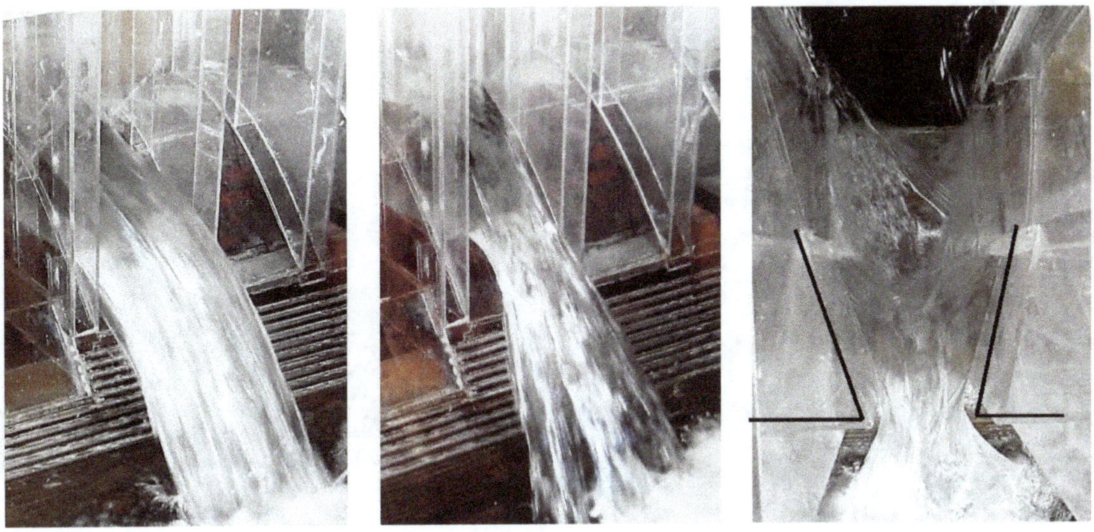

(a) (b) (c)

Figure 3. Photograph of the plunging jets in the present study ($H = 0.3$ m): (**a**) without FGP ($\beta = 1$); (**b**) with FGP ($\beta = 0.51$, $\theta = 15°$); (**c**) with FGP ($\beta = 0.51$, $\theta = 15°$).

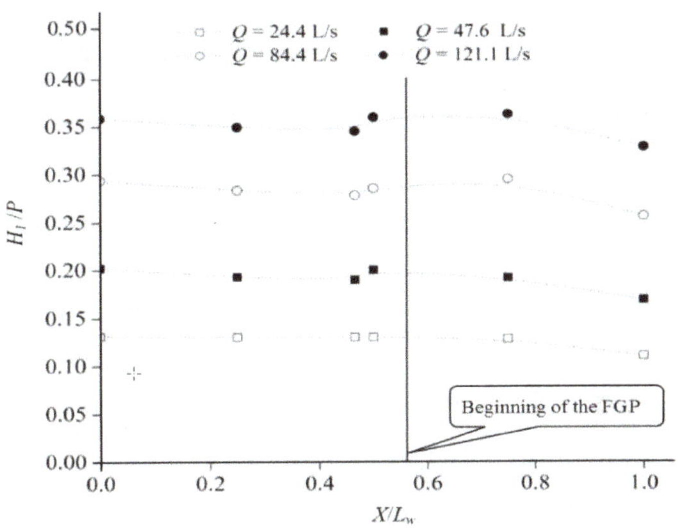

(a)

Figure 4. *Cont.*

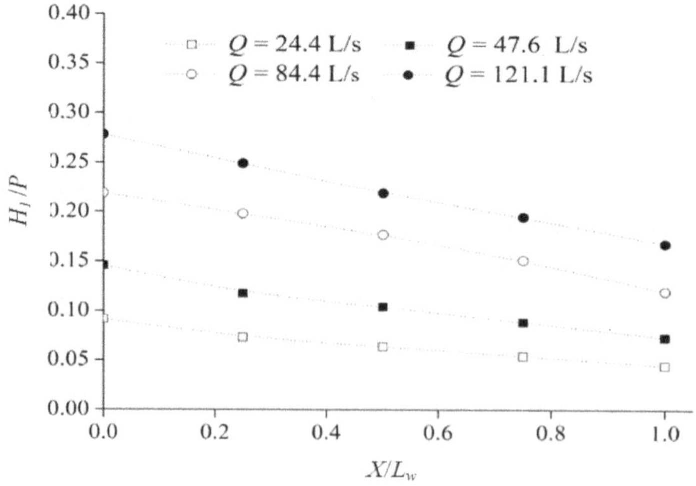

(**b**)

Figure 4. Comparison of water-surface profiles: (**a**) with FGP ($\beta = 0.51$, $\theta = 15°$); (**b**) without FGP ($\beta = 1$) (Q—the discharge of the weir, H_1—water depth of the weir crest).

(**a**) (**b**)

Figure 5. Photograph of water surface: (**a**) without FGP; (**b**) with FGP ($\beta = 0.51$, $\theta = 15°$) (Black arrow stands for the flow direction).

3.2. Discharge Capacity Comparison with and without FGP

Considering type 3 of the finite crest-length weir as an example, Figure 6 shows the head–discharge relationship of the weir for the cases with and without the FGP. The two discharge-stage curves, with slightly similar change rules, were both concave in shape. However, the curves for the weir with the FGP were always below that of the case without the FGP, which indicated that the discharge was reduced when the FGP was installed on the weir. The discharge coefficient C_d was the indicator of the discharge capacity of the weir, which was defined as follows [2]:

$$C_d = \frac{Q}{B\sqrt{2g}H^{\frac{3}{2}}} \tag{2}$$

where Q is the discharge of the finite crest-length weir, which was measured using the right-angled-triangle weir in the model test; B is the width of the channel; g is gravitational acceleration; and H is the head on the weir. Because the height of the finite crest-length weir was considerable ($P/H > 1.33$), the influence of the approaching velocity on the test results could be neglected [19].

Figure 7 shows the relationship between the discharge coefficient (C_d) and the weir head (H). The results showed that the discharge coefficient gradually increased as the head increased, and the spacing of the two curves represented the reduction in the discharge coefficient after setting the FGP on the finite crest-length weir. This might be related to the flow pattern in the channel, because Li et al. [17]

showed that the discharge capacity reduces only if the flow was subcritical when the FGP was applied on the ogee-overflow structure.

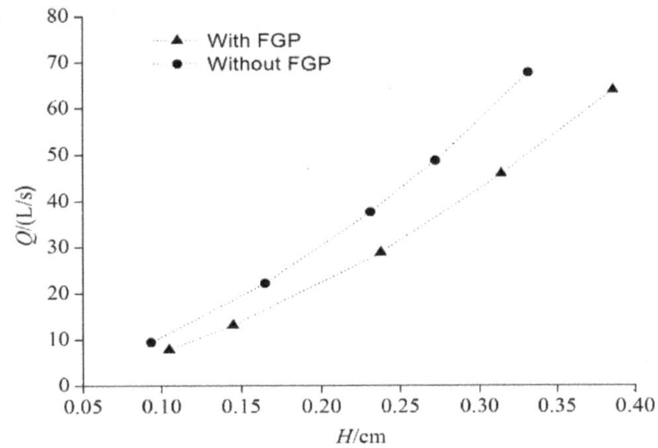

Figure 6. Head–discharge relationship of the weir for the cases with FGP (β = 0.51, θ = 15°) and without FGP.

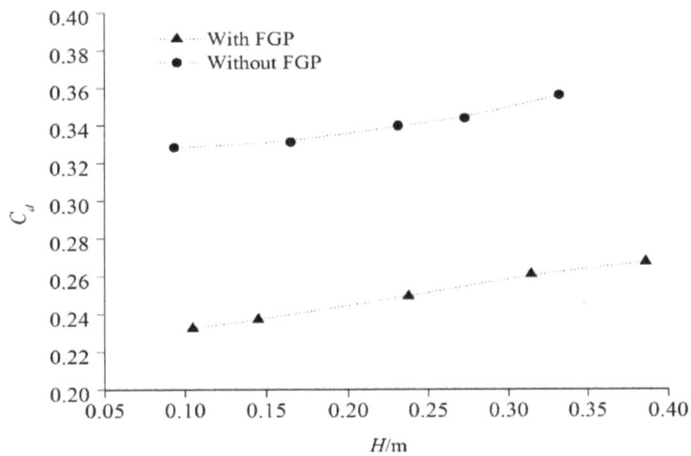

Figure 7. Relationship between the discharge coefficient (C_d) and the weir head (H) for the cases with FGP (β = 0.51, θ = 15°) and without FGP.

3.3. Influence of Weir Length on Discharge Capacity with and without FGP

Figure 8 shows the relationship between the dimensionless head on the weir (H/P) and discharge coefficient (C_d) for different weir lengths L_w, for the cases with and without the FGP. The results showed the effect of the weir length on the discharge coefficient. Moreover, a comparative analysis was performed for the cases with and without the FGP.

As shown in Figure 8a, for the case where the FGP is installed, the three C_d–H curves almost overlap. This implies that the effect of the variations in the weir length on the discharge coefficient of the weir is insignificant when the FGP is installed. In other words, the discharge coefficient of the weir with the FGP was largely unrelated to the weir length in a particular range. The inlet boundary condition could be considered to remain the same, and the boundary condition did not clearly change with the variation in the weir length. However, Rao and Muralidhar [6] and Hager and Schwalt [2] pointed out a particular feature of finite crest-length weirs, referred to as the undular flow on the crest when $H/L_w < 0.1$. Moreover, they showed that the value of C_d was constant, which was similar to the characteristics of the finite crest-length weir with the surface wave; a lower and relatively stable C_d was obtained for various weir lengths when setting the FGP.

It also could be seen that the distribution rule of the three C_d–H curves without the FGP was different from the aforementioned regularity. Two C_d–H curves for L_w = 0.389 m and L_w = 0.317 m almost overlap just above the one for L_w = 0.451 m, as shown in Figure 8b. It indicated that the discharge coefficient of the weir without the FGP increased as the weir length decreased in a particular range, but gradually tended to stabilize when the weir length continued to decrease.

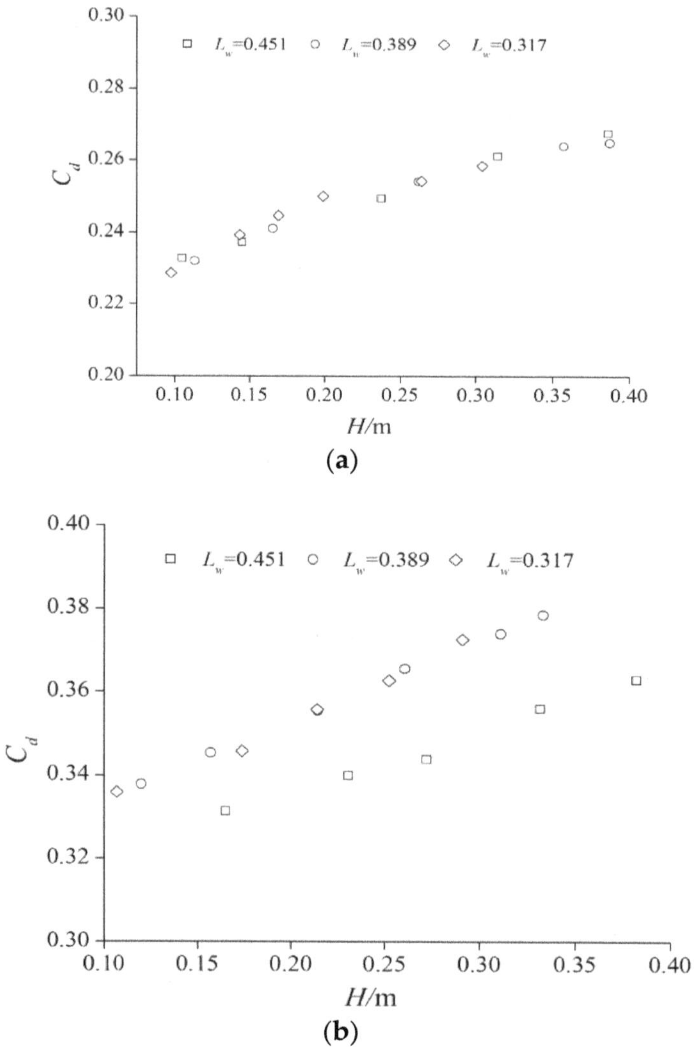

Figure 8. Discharge coefficient (C_d) versus weir head (H) together with different weir lengths (L_w): (**a**) with FGP (β = 0.51, θ = 15°); (**b**) without FGP.

3.4. Discharge Coefficient of the Weir with FGP

Many studies have analyzed the relationship of the discharge coefficient for the finite crest-length weirs of different types, i.e., long-crested weirs, broad-crested weirs, short-crested weirs, and sharp-crested weirs [2,6,9–11]. They all indicated that the nondimensional parameter H/L_w was the dominant factor.

In the present experiment, the range of H/L_w was approximately 0.187–0.909. Keeping the other parameters constant and varying one parameter, the effects of β and θ were shown in Figure 9. Figure 9a shows the variations in C_d with β for each H/L_w at the same contraction angle θ. As indicated in the figure, the discharge coefficient significantly increased as the contraction ratio increases ($0.38 \leq \beta \leq 1$). Figure 9b illustrates the variations in C_d with θ for each H/L_w at the same contraction ratio β, which indicated that the discharge coefficient changes only slightly as the contraction angle increases $7° \leq \theta \leq 21.5°$.

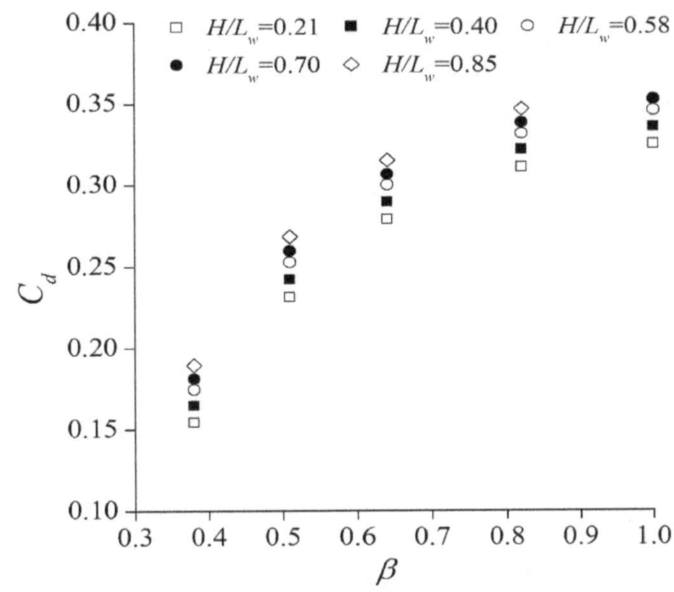

(**a**) Contraction angles $\theta = 7.0°$

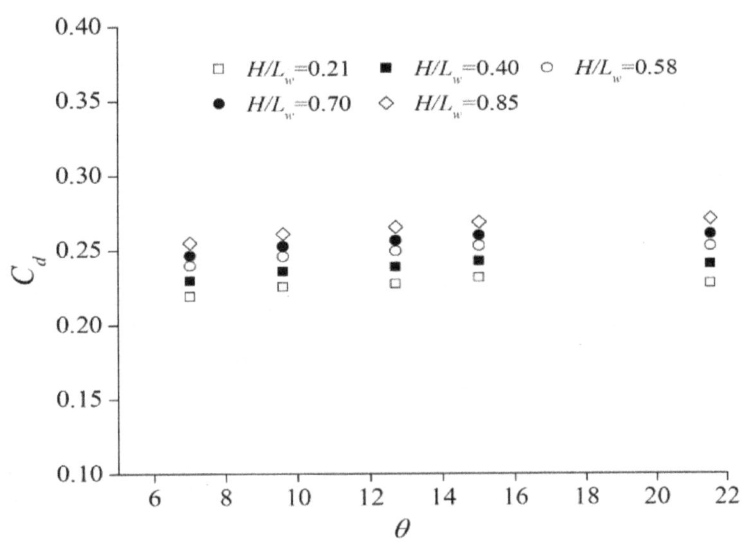

(**b**) Contraction ratio $\beta = 0.51$

Figure 9. Discharge coefficient C_d versus β (**a**) and θ (**b**).

Consequently, H/L_w and β were the dominant parameters of the discharge coefficient of the weir with FGP C_d. By the linear regression method, Equation (3) is obtained. The correlation coefficient R^2 is equal to 0.967, which shows that the accuracy of Equation (3) is higher. The relationship between discharge coefficient C_d and the two parameters described using Equation (3) is shown in Figure 10.

$$C_d = 0.056(H/L_w) - 0.107(\frac{1}{\beta}) + 0.427 \quad (3)$$

with $R^2 = 0.967$.

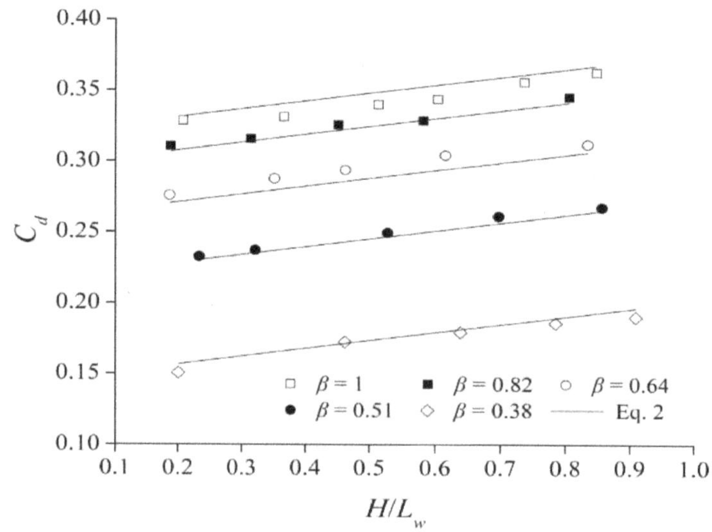

Figure 10. Compared of measured and Equation (3)-computed values.

4. Conclusions

In this study, 65 groups of experiments, including five different flow rates and 13 different geometries of the finite-crest length weir, were conducted to study the variation in the discharge coefficient and flow pattern of a finite crest-length weir with and without an FGP. Unlike the case of the weir without the FGP, for the weir with the FGP, the plunging jet was diffusive in the longitudinal directions in air and the water-surface profile in the flow channel was a backwater curve when the head was relatively high. The discharge capacity was significantly reduced after setting the FGP, which should be considered in engineering design. According to the compared experiments, it was found that the effect of the weir length on the discharge coefficient was different for cases of the weir with and without the FGP. In addition, based on the literature and experimental studies, this study proposed a discharge-coefficient equation for the weir with the FGP.

Author Contributions: Z.T. carried out the experimental investigation and writing; W.W. wrote the methodology and review; R.B. carried out the data curation; N.L. carried out data validation. All four authors reviewed and contributed to the final manuscript.

References

1. Horton, R.E. *Weir Experiments, Coefficients, and Formulas*; Proc. U.S. Geological Survey Water Supply; Government Printing Office: Washington, DC, USA, 1907.
2. Hager, W.H.; Schwalt, M. Broad-crested weir. *J. Irrig. Drain. Eng.* **1994**, *120*, 13–26. [CrossRef]
3. Fritz, H.M.; Hager, W.H. Hydraulics of embankment weirs. *J. Hydraul. Eng.* **1998**, *124*, 963–971. [CrossRef]
4. Bai, R.; Zhang, F.; Liu, S.; Wang, W. Air concentration and bubble characteristics sownstream of a chute aerator. *Int. J. Multiphase Flow* **2016**, *87*, 156–166. [CrossRef]
5. Bai, R.; Liu, S.; Tian, Z.; Wang, W.; Zhang, F. Experimental investigations on air–water flow properties of offset–aerator. *J. Hydraul. Eng.* **2018**, *144*, 04017059. [CrossRef]
6. Govinda Rao, N.S.; Muralidhar, D. Discharge characteristics of weirs of finite crest width. *Houille Blanche* **1963**, *18*, 537–545. [CrossRef]
7. Ramamurty, A.S.; Udoyara, S.T.; Rao, M.V.J. Characteristic of square-edged and round-nosed broad-crested weirs. *J. Irrig. Drain. Eng.* **1988**, *114*, 61–73. [CrossRef]
8. Sarker, M.A.; Rhodes, D.G. Calculation of free-surface profile over a rectangular broad-crested weir. *Flow Meas. Instrum.* **2004**, *15*, 215–219. [CrossRef]
9. Andersson, A.G.; Andreasson, P.; Staffan Lundström, T. CFD-modelling and validation of free surface flow

during spilling of reservoir in down-scale model. *Eng. Appl. Comput. Fluid Mech.* **2013**, *7*, 159–167. [CrossRef]
10. Chen, Y.; Fu, Z.; Chen, Q.; Cui, Z. Discharge Coefficient of Rectangular Short-CrestedWeir with Varying Slope Coefficients. *Water* **2018**, *10*, 204. [CrossRef]
11. Azimi, A.H.; Rajaratnam, N.; Zhu, D.Z. Submerged Flows over Rectangular Weirs of Finite Crest Length. *J. Irrig. Drain. Eng.* **2014**, *140*, 06014001. [CrossRef]
12. Jan, C.D.; Chang, C.J.; Kuo, F.H. Experiments on discharge equations of compound broad-crested weirs. *J. Irrig. Drain. Eng.* **2009**, *135*, 511–515. [CrossRef]
13. Xie, S.Z.; Li, S.Q.; Li, G.F. Development of flaring gate piers in China. *Hongshui River* **1995**, *14*, 3–11. (In Chinese)
14. Guo, J.; Liu, Z.P. Field observations on the RCC stepped spillways with the flaring pier gate on the Dachaoshan Project. In Proceedings of the IAHR XXX International Congress, Theme D, Thessaloniki, Greece, August 2003; pp. 473–478.
15. Zhang, T.; Wu, C.; Liao, H.S. 3D numerical simulation on water and air two-phase flows of the steps and flaring gate pier. *J. Hydrodyn.* **2005**, *17*, 338–343.
16. Wang, B.; Wu, C.; Mo, Z.Y. Relationship of first step height, step slope and cavity in X-shaped flaring gate piers. *J. Hydrodyn.* **2007**, *19*, 349–355. [CrossRef]
17. Futian, L.; Peiqing, L.; Weilin, X.; Zhong, T. Experimental study on effect of flaring piers on weir discharge capacity in high arch dam. *J. Hydraul. Eng.* **2003**, *11*, 43–47. (In Chinese)
18. Li, N.W.; Liu, C.; Deng, J.; Zhang, X.Z. Theoretical and experimental studies of the flaring piers on the surface spillway in a high-arch dam. *J. Hydrodyn.* **2012**, *24*, 496–505. [CrossRef]
19. Ramamurthy, A.S.; Vo, N.D. Characteristics of circular-crested weir. *J. Hydraul. Eng.* **1963**, *119*, 1055–1062. [CrossRef]

12

Modeling the Application Depth and Water Distribution Uniformity of a Linearly Moved Irrigation System

Junping Liu, Xingye Zhu *, Shouqi Yuan and Alexander Fordjour

Research Center of Fluid Machinery Engineering and Technology, Jiangsu University, Zhenjiang 212013, China; liujp@ujs.edu.cn (J.L.); shouqiy@ujs.edu.cn (S.Y.); fordjouralexander27@yahoo.com (A.F.)
* Correspondence: zhuxy@ujs.edu.cn.

Abstract: A model of a linearly moved irrigation system (LMIS) has been developed to calculate the water application depth and coefficient of uniformity (CU), and an experimental sample was used to verify the accuracy of the model. The performance testing of the LMIS equipped with 69-kPa and 138-kPa sprinkler heads was carried out in an indoor laboratory. The LMIS was towed by a winch with a 1.0 cycle/min pulsing frequency while operating at percent-timer settings of 30, 45, 60, 75, and 90%, corresponding to average moving speeds of 1.5, 2.3, 3.3, 4.0, and 4.7 m min^{-1}, respectively. The application depth and CU obtained under various speed conditions were compared between the measured and model-simulated data. The model calculation accuracy was high for both operating pressures of 69 and 138 kPa. The measured application depths were much larger than the triangular-shaped predictions of the simulated application depth and were between the parabolic-shaped predictions and the elliptical-shaped predictions of the simulated application depth. The results also indicate that the operating pressure and moving speed were not significant factors that affected the resulting CU values. For the parabolic- and elliptical-shaped predictions, the deviations between the measured and model-simulated values were within 5%, except for several cases at moving speeds of 2.3 and 4.0 m min^{-1}. The measured water distribution pattern of the individual sprinklers could be represented by both elliptical- and parabolic-shaped predictions, which are accurate and reliable for simulating the application performances of the LMIS. The most innovative aspect of the proposed model is that the water application depths and CU values of the irrigation system can be determined at any moving speed.

Keywords: linearly moved irrigation system; application depth; moving speed; uniformity coefficient

---

1. Introduction

With the development of science and technology and the need for practical production, the rapid development of the linearly moved irrigation system (LMIS) has been popular in various countries. In China, the LMIS was introduced in the 1980s, with increasing use in recent years due to technological innovations, management convenience, economics and water savings [1–3]. Linear systems, also known as mobile side, have structures similar to center pivot; however, circular movement is replaced by linear movement so that the water application rate is constant along the entire length of the lateral move type. The basic elements to determine LMIS operation include the moving speed, V (m/min), and the maximum range of the sprinkler head, R max (m). The moving speed of LMIS directly affects how water is sprayed in the unit area; when the moving speed is slow, the unit area of spraying is greater, and when the moving speed is fast, the unit area of water spraying is lower. Effective irrigation is not the application of water without control or planning, but is the application of the correct amount of water at the right time, especially with uniformity, during the irrigation process. Thus, to achieve water optimization in agriculture, it is important to frequently evaluate the performance of irrigation systems by using some parameters that express and quantify the operation quality. The application depth and water distribution uniformity coefficient are often used as indicators of problems concerning irrigation distribution.

Irrigation uniformity is defined as the variation in irrigation depths over an irrigated area and is an important performance characteristic of the sprinkler irrigation system [4–10]. The importance of sprinkler irrigation uniformity was recognized as early as 1942 [11]. Widespread research has been conducted on the factors affecting sprinkler irrigation uniformity. Such factors include nozzle size and pressure, the type of diffuser device, sprinkler spacing, riser height, field topography, discharge angle, number and configuration of the sprinklers and wind speed and direction, all of which can influence water application uniformity [12–23]. In recent years, several studies have been carried out on irrigation performance by center pivot to identify the main problems of irrigation efficiency [24–28]. However, there have been few studies concerning the application depth and uniformity of the LMIS at different operating speeds [29,30]. In the current studies, none of the existing methods can be conducted to calculate the application depth and uniformity of the LMIS. In some certain special cases, the surface runoff appears in the field as the application of the LMIS. It is difficult to determine a mismatch of the farmland soil infiltration capacity and the hydraulic performance of LMIS. Some targeted solutions for solving the problem are unavailable in the scientific research. Therefore, it is very important to study the calculation method of the application depth and the combination uniformity of the LMIS, which is of great theoretical value and practical significance. The objective of this study is to put forward a calculation method of application depth and uniformity of LMIS and to verify the accuracy of the calculation results through experimental tests.

2. Materials and Methods

2.1. LMIS and Spray Sprinkler

The sprinkler irrigation system used in this study was specifically manufactured as an experimental sample by the Research Center of Fluid Machinery Engineering and Technology (Jiangsu University, Zhenjiang City, Jiangsu Province, China). It was an intermittent linear moving system towed by a winch and installed with a fixed spraying plate package as shown in Figure 1. The fixed spray plate sprinkler studied in this research was the Nelson D 3000 sprayhead (Nelson Irrigation Co., Walla Walla WA, USA). A single-strut pressure regulator (Figure 2a) was part of the package for a low-pressure deflection sprinkler. The shapes of the nozzles were circular (Figure 2b). The nozzle diameter of the fixed spray plate sprinkler was 5.16 mm, corresponding to number #26 specified by the Nelson Company. The sprinkler deflects 36 water streams uniformly, centered from itself, to form a full-circle spray pattern. The pressure regulator and sprinkler were connected immediately with a dedicated screw-type connector (Figure 2c). The reason for using such sprinklers was that prior tests had shown that this package and pressure combination had a good coefficient of uniformity [31–33]. All sprinklers were on flexible drop hoses that set the D 3000 sprayhead approximately 1.0 m above the soil surface, and spaced 3.0 m apart. A pressure regulator of 10 or 20 psi (69 or 138 kPa) was installed just upstream of the spray sprinkler. The discharge of the LMIS was measured with an electrical flow gauge (Model E-mag/DN25, manufactured by Kaifeng Electronic Instrument Company, China) with an accuracy of 0.3%.

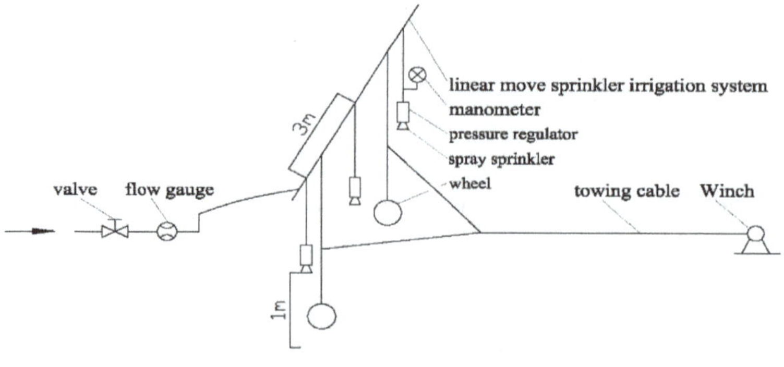

Figure 1. Description of the linearly moved irrigation system (LMIS).

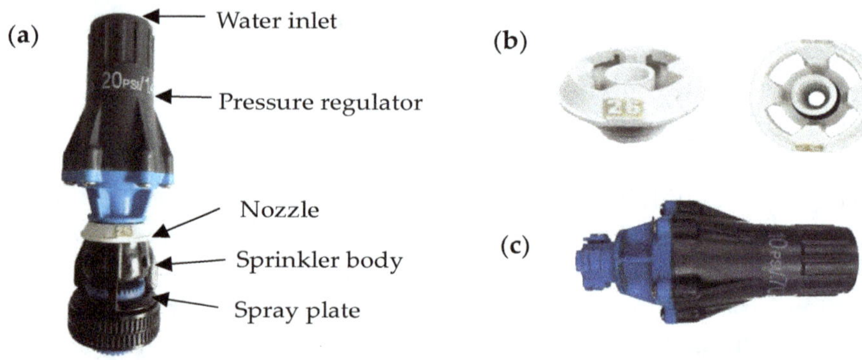

Figure 2. Pressure regulator and sprinkler used in the LMIS (**a**), nozzles used in the sprinkler (**b**), and dedicated screw-type connector of pressure regulator (**c**).

2.2. Modeling for Calculation of Application Depth

As the structure parameters of sprinkler type, installed height and spacing were determined, with the water distribution along the axis direction of the LMIS known. First, the LMIS was maintained at a constant working condition. Then, according to the maximum wetted radius, R max, and the water application rate of the sprinklers, the figure of water application depth could be obtained. The relationship between the water application and spraying distance may be described by a geometric pattern. An elliptical, parabolic or triangular pattern was chosen to represent it as shown in Figure 3, where the curve of abc represents the distribution line of sprinkler irrigation intensity, the vertical axis represents the water application rate p (mm/h), and the horizontal axis represents the spraying distance L (m). In theory, under the conditions of an indoor experiment without any wind effects, the value of L is two times the maximum wetted radius, R max.

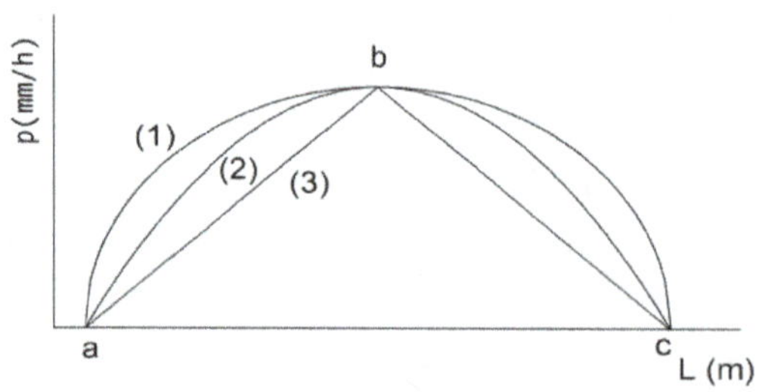

Figure 3. Water application depth: (1) elliptical, (2) parabolic and (3) triangular.

The catch device was supposed to be set at point O, away from the LMIS, with a distance larger than R max, to study the average application depth of the system. The working phenomenon was as following: first, the LMIS approached point O, and point O started to receive the water sprayed out by the LMIS; then, the LMIS gradually receded from point O while moving until the system passed over the catch devices entirely and point O did not receive any water. The LMIS-fulfilled total water application depth was the sum of passing section area of point O. Figure 4 represents the depiction of water application distribution of point O. Combined with Figure 3, when the LMIS was passing through point O at a speed of v, it means that the curve of abc was passing through point O at a speed of v and the passing time could be calculated as $t = L_{ac}/v$.

At this time, the curve of abc also represents the distribution line of sprinkler irrigation intensity, and the vertical axis also represents the water application rate p (mm/h); however, the horizontal axis represents the passing time t (s). When the LMIS was passing through any point of O, the receiving water of point O was the sum of the water application section area passing point O. Figure 4 represents the distribution section S, and S is the superimposed distribution map of all water sprayed to point O. When the travel speed is v, it can be seen that $t = L_{ac}/v$. The relationship between the water application rate ρ and the passing through time T was as follows:

Elliptical shape:

$$\rho = P_k\sqrt{1-\frac{4(T-t/2)^2}{t^2}},\ P_k = \frac{4WDP}{\pi t} \tag{1}$$

Parabolic shape:

$$\rho = -\frac{4P_k}{t^2}(T-t/2)^2 + P_k,\ P_k = \frac{1.5WDP}{t} \tag{2}$$

Triangular shape:

$$\begin{cases} \rho = \frac{2P_k}{t}T & (0 \le T \le t/2) \\ \rho = \frac{2P_k}{t}(t-T) & (t/2 \le T \le t) \end{cases},\ P_k = \frac{2WDP}{t} \tag{3}$$

where

P_k = peak water application rate (mm h^{-1})
t = water application time (h).

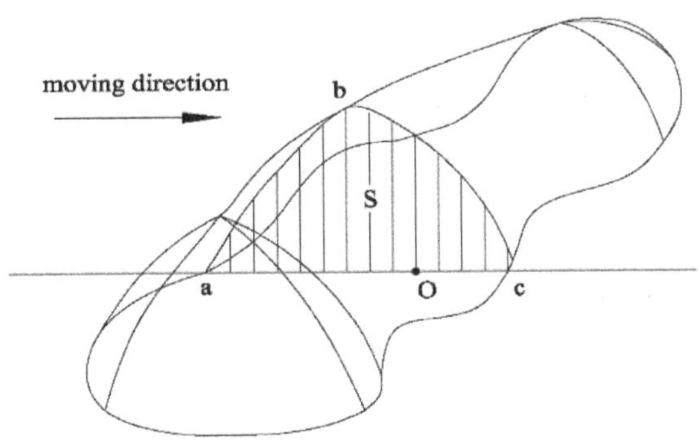

Figure 4. Depiction of water application distribution of point O.

Then, the wetted area S was calculated as follows:

$$S = \int_0^t \rho(T)dt \tag{4}$$

Supposing that the travel speed of LMIS was set at different values of v1, v2, and vn, respectively, t1 = Lac/v1, t2 = Lac/v2, and tn = Lac/vn. The parabolic pattern was selected and was assumed to be representative of LMIS irrigation water, whose pattern for the same water application depth and time has a peak rate between the elliptical and triangular patterns. Figure 5 shows an elliptical application shape. Notice the increase of water application depth (proportional to larger geometric areas) related to high speed, average speed, and low speed, as well as the constant peak water application rate.

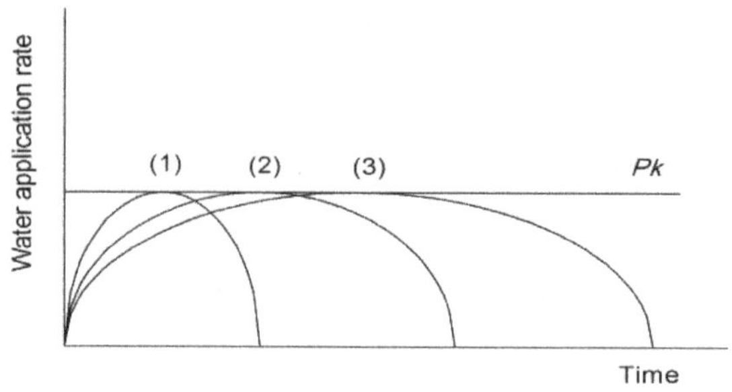

Figure 5. Water application patterns with decreasing speed at the same irrigation point: (1) high speed, (2) average speed, and (3) low speed.

2.3. Modeling for Calculation of Uniformity

The spraying irrigation state of LMIS can be regarded as a limit state of mobile spraying for fixed-type pipe with an infinite narrow width. Supposing that the LMIS traveled at a uniform speed, and the water distribution for every sprinkler was the same and does not change with time, it could be determined that the depth of irrigation water along the direction of the sprinkler is the same. Therefore, only the branch direction sprinkling uniformity degree needed to be considered, which can be representative of the entire area of the spraying uniformity. In this way, the concept of linear spraying uniformity is put forward. The uniformity of the LMIS is equal to the uniformity of the axial direction of the unit. The coefficient of uniformity CU (%), developed by Christiansen (1942), was calculated using the following equation:

$$\mathrm{CU} = 1 - \frac{\Delta\overline{h}}{\overline{h}} \tag{5}$$

When the point was adopted by affirmative grid,

$$\Delta\overline{h} = \frac{\sum_{i=1}^{n}\left|h_i - \overline{h}\right|}{n},\ \overline{h} = \frac{\sum_{i=1}^{n} h_i}{n} \tag{6}$$

where h_i = water depth of calculated point i, mm/h; $\overline{h}$ = mean water depth of all calculated points, mm/h; and n = total number of calculated points used in the evaluation.

Figure 6 represents the calculated sketch of spraying uniformity for LMIS. The K direction is the running direction and the j direction is the axial of the unit. The water depth of h_{jk} was as follows:

$$\begin{gathered} h_{11} = h_{12} = h_{13} = \cdots = h_{1k} = h_1 \\ h_{21} = h_{22} = h_{23} = \cdots = h_{2k} = h_2 \\ \cdots\cdots \\ h_{j1} = h_{j2} = h_{j3} = \cdots = h_{jk} = h_j \\ \text{Then, } \overline{h} = \frac{h_{11}+h_{12}+h_{13}+\ldots h_{1k}+h_{21}+h_{22}+\ldots+h_{2k}+h_{j1}+h_{j2}+\ldots h_{jk}}{n_k \cdot n_j} \\ \text{Then, } \overline{h} = \frac{h_1+h_2+h_3+\ldots+h_j}{n_j} = \sum_{i=1}^{j} h_i/n_j \end{gathered} \tag{7}$$

Here, $\overline{h}$ has been simplified to the average value of the depth of the axial point irrigation water. In the same way:

$$\begin{aligned} \Delta\overline{h} &= \frac{\left|\overline{h}-h_{11}\right|+\left|\overline{h}-h_{12}\right|+\ldots+\left|\overline{h}-h_{1k}\right|+\ldots+\left|\overline{h}-h_{j1}\right|+\ldots+\left|\overline{h}-h_{jk}\right|}{n_k \cdot n_j} \\ &= \frac{n_k\left|\overline{h}-h_1\right|+n_k\left|\overline{h}-h_2\right|+\ldots+n_k\left|\overline{h}-h_j\right|}{n_k \cdot n_j} = \sum_{i=1}^{j} \frac{\left|\overline{h}-h_i\right|}{n_j} \end{aligned} \tag{8}$$

This type has also been turned into an axial $\Delta\bar{h}$. The spray uniformity coefficient can be obtained after substitution of the axial $\bar{h}$ and $\Delta\bar{h}$ into Equation (5).

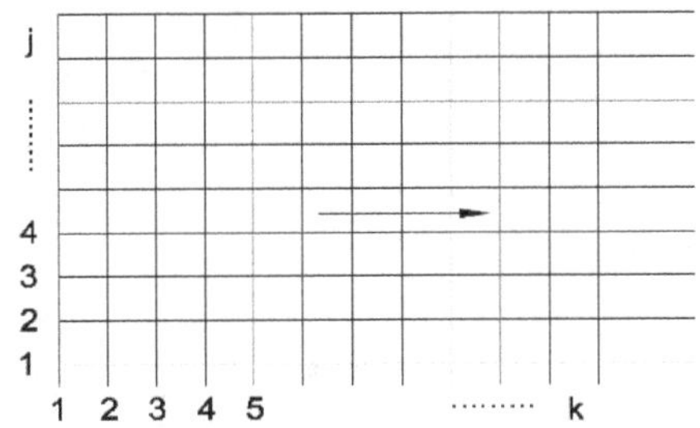

Figure 6. Calculated sketch of spraying uniformity for the LMIS.

2.4. Setup and Procedures of Indoor Experiment

The study site was located at the indoor facilities of the Research Center of Fluid Machinery Engineering and Technology, Jiangsu University (Jiangsu Province, China). Figure 7 shows the test plot for the LMIS mobile water distribution. The structure, built with several metal 40-mm-long and 20-mm-wide rectangular bars, was 12.0 m wide and the spray sprinkler could be adjusted from 0 to 2.5 m above the surface. Because only a short width (12.0 m) was tested, pressure variation through the measured width of the system was assumed to be negligible. The frame corners were equipped with four wheels and a winch was used to tow the system. The water source was a reservoir with a capacity of 60 m^3. A 3.0-kW electric centrifugal pump was connected to the water supplying pipe and a 36-mm external-diameter hose pipe was used to supply water to the spray sprinkler. Manometers and valves were installed as required to control water supply during the experiments (as shown in Figure 1). Catch cans were used to collect the applied water. They were constructed from transparent plastic, with an inverted conical shape. The catch can opening was 200 mm for the inside diameter and the catch can height was 250 mm. Nine rows of catch cans were distributed along the direction of the vertical moving unit, and the distance and spacing of catch cans were 1 m each. The LMIS and the catch cans were installed in a plot with cement flooring. After the spraying test, the weighing method was used to calculate the depth of irrigation water for every catch can. The measured irrigation depth of the point values was taken with an average irrigation depth of the 9 rows of catch cans.

Figure 7. Test plot for the LMIS mobile water distribution.

Experiments were carried out at a constant pressure of 69 and 138 kPa, respectively, maintained by a 10- and 20-psi pressure regulator. The pulsing frequency towed by the winch was 1.0 cycle/min while operating the winch at percent-timer settings of 30, 45, 60, 75%, and 90%, corresponding to the average moving speed of 1.5, 2.3, 3.3, 4.0, and 4.7 m min^{-1}, respectively. These selected five values cover the range of duty cycles that are expected to be used in the field. A value of 100% is the same as normal operating conditions (constant sprinkler discharge) and zero percent is the same as operating the system dry. All operating pressures and moving speeds were within the manufacturer's recommendations. The following standards [34–37] were adopted in the design of the experimental setup and in the experiment itself. The duration of each test was approximately 30 to 45 min. The applied water in the catch cans was read in a measuring cup with a volume of 500 mL and an accuracy of 5 mL. The read data were then converted to average irrigation depth by dividing the cross-sectional area of the catch can. A minimum of three replications were conducted for each pressure and moving speed combination, and data were averaged and used as the final experimental data. Flow rates from the two operating pressures used for the application depth tests were measured three times under the same operating conditions. A metal pipe was positioned over the sprinkler nozzles and the discharge water directed into a bucket for 2 min. Discharge volumes were then weighed with an electronic balance (Otimpa Corp, China) with 1 gram accuracy and converted into flow rate.

The average air temperature during testing was 20.8 °C and ranged from 17.8 °C to 22.4 °C. The average relative humidity was 41% and ranged from 36% to 47%. To verify the accuracy of the model, a moving superposed water quantity verification test was carried out. Taking into account the condition that the work at the beginning of the LMIS may not be stable, the moving procedure of the system was start-up and test data were collected after operating at the working pressure for more than 10 min.

3. Results and Discussion

The stable working state of the LMIS means the spraying speed is stable, and the working pressure of the spray head on the system is stable. The wind speed during the entire tests ranged from 0.0 to 0.12 m s^{-1} and was usually less than 0.1 m s^{-1}. These data are not presented or discussed as the values were less than the lower threshold of the measuring equipment. The potential source of error in this study is any differences in evaporative conditions across the time period of the various tests. Although many studies have been conducted on evaporation during sprinkler irrigation, Schneider [38] noted that no more than 2% losses resulted from evaporation during use of sprinkler irrigation systems. For example, even under a condition of an average temperature of 26 °C, a relative humidity of 64%, and a wind speed of 6.4 m/s, the measured evaporation was only 0.8% of total sprinkler discharge [39]. Therefore, due to the indoor experimental conditions without any wind effects in the study, this particular potential source of error was minimized. An indoor experiment was conducted to verify the accuracy of the model using the Nelson D 3000 spray head with a nozzle diameter of 5.16 mm at working pressures of 69 kPa and 138 kPa and at an elevation of 1.0 m above the soil surface.

3.1. Coefficient of Discharge

The results of measured flow rates of sprinkler irrigation nozzles used in this study are shown in Table 1. Analysis of the measured data were performed to find the influence of the geometrical parameters as well as the operating pressure on discharge of the sprinkler head, which can be expressed in terms of the discharge coefficient, C. The coefficient of discharge of the sprinkler is generally expressed as follows:

$$C = \frac{Q}{d^2 \frac{\pi}{4} \sqrt{2gH}} \tag{9}$$

where C is the coefficient of discharge, Q is the flow rate of the sprinkler (m^3 h^{-1}), d is the diameter of the nozzle (m), g is the gravitational acceleration (m s^{-2}), and H is the pressure head (m).

As shown in Table 1, when using the fixed spray plate sprinkler, the measured nozzle flow rates ranged from 0.77 to 0.84 m^3 h^{-1}, with a mean value of 0.808 m^3 h^{-1}, and 1.15 to 1.22 m^3 h^{-1} with a mean value of 1.183 m^3 h^{-1} for the pressure regulators of specification 69 and 138 kPa, respectively. After calculating using Equation (8), the nozzle coefficients of discharge ranged from 0.880 to 0.960 with a mean value of 0.923, and 0.929 to 0.986 with a mean value of 0.956, for 69 and 138 kPa, respectively. From the aforementioned analysis, it was found that the coefficients of discharge fluctuated within a small acceptable range under the same operating pressure, which could be attributed to the acceptable experimental error. Additionally, the coefficients of discharge obtained using the 138 kPa pressure regulator were higher than those obtained using the 69 kPa pressure regulator, which means that using a pressure regulator with a higher outlet pressure produced higher coefficients of discharge.

Table 1. Measured flow rates of sprinkler irrigation nozzles.

Replications \ Flow Rate, $m^3\ h^{-1}$	69 kPa			138 kPa		
	Nozzle I	Nozzle II	Nozzle III	Nozzle I	Nozzle II	Nozzle III
1	0.8	0.79	0.8	1.16	1.21	1.22
2	0.84	0.83	0.81	1.18	1.17	1.15
3	0.77	0.8	0.83	1.21	1.16	1.19
Mean	0.80	0.81	0.81	1.18	1.18	1.19

3.2. Measured Water Distributions

Figure 8 represents the measured sprinkler intensity and fitting curve of radial points with a nozzle diameter of 5.16 mm and a height of 1.0 m at operating pressures of 69 kPa and 138 kPa, respectively.

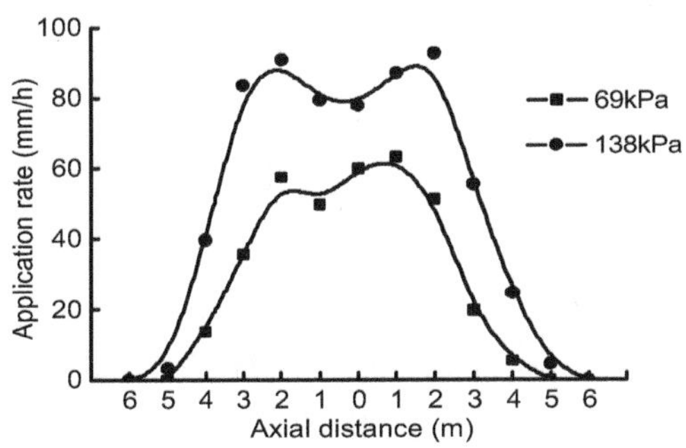

Figure 8. Measured radial application rate and fitting curve for operating pressures of 69 kPa and 138 kPa.

As shown in Figure 8, the sprinkler nozzle wetting radius for 69 kPa and 138 kPa was 4.4 m and 5.7 m, respectively. The sprinkler head produced quite similar water application profiles under different operating pressures. The average values of the sprinkler head application rates varied from 3.2 mm h^{-1} to 92.8 mm h^{-1}. The water application rate increased to a maximum value first and then decreased approximately linearly as the distance from the sprinkler increased. The maximum application rate was determined for the two analyzed pressures: 63.2 mm h^{-1} at distances of 1 m for 69 kPa, and 92.8 mm h^{-1} at 2 m from the sprinkler for 138 kPa. Starting from this distance, the application rate decreased until it reached the minima. For 69 kPa at 4 m, and 138 kPa at 5 m, the minimum values were 5.3 and 4.3 mm h^{-1}, respectively. The application rates had large variability for all measured points. The applied water depths were extremely similar to the radial water depths, indicating that the water distribution pattern of individual sprinklers had a primary influence on the overall water application depth and distribution uniformity. The radial water distribution curve appeared as a saddle type, and had two sprinkler intensity peaks close to the sprinkler. The least squares method was used for curve fitting to simulate the water distribution characteristics, and the fitting degree was as high as 0.89. Table 2 summarizes the measured application depths and corresponding CU values obtained with an LMIS average moving speed of 1.5, 2.3, 3.3, 4.0 and 4.7 m/min under different operating pressures of 69 kPa and 138 kPa. For the various LMIS speeds, the application depth of 69 kPa varied from 1.3 to 4.2 mm with an average of 2.38 mm, the CU varied from 84.5% to 88.9% with an average of 86.58%, and the standard deviation had an average of 0.16; the application depth of 138 kPa varied from 2.6 to 7.7 mm with an average of 4.08 mm, the CU varied from 85.9% to 89.8% with an average of 87.28%, and the standard deviation had an average of 0.68. It appeared that the application depth decreased as the LMIS speed increased, which could be understood easily. The CU value did not linearly relate to the LMIS speed, which was interpreted as adjusting for unimportant variability on a small scale.

Table 2. Measured irrigation application depths and corresponding coefficient of uniformities under different operating pressures and LMIS speeds.

Operating Pressure (kPa)	LMIS Speed (m min^{-1})	Application Depth (mm)	CU (%)	Standard Deviation
69 kPa	1.5	4.2	86.0	0.15
	2.3	2.8	84.5	0.26
	3.3	2.0	88.7	0.12
	4.0	1.6	88.9	0.12
	4.7	1.3	84.8	0.15
	Mean	2.38	86.58	0.16
138 kPa	1.5	7.7	87.0	0.75
	2.3	4.2	89.8	0.82
	3.3	3.1	87.2	1.00
	4.0	2.8	85.9	0.47
	4.7	2.6	86.5	0.36
	Mean	4.08	87.28	0.68

3.3. Comparison of Application Depth and CU Values between Experimental Measured Data and Modeling Simulations

As shown in Sections 2.2 and 2.3, in the modeling for calculation of application depth and CU values of the sprinkler irrigation, the data from Figure 8 were used as the basic foundation, and the fitting curves were supposed to be regulated as elliptical, parabolic, and triangular shapes, respectively.

3.3.1. Application Depth

The simulated application depths were obtained with a constant spacing of 3.0 m for comparison. The data from Table 2 were used as the experimental measured data. As a further illustration of model validation of the application depth calculation, Figure 9 presents a comparison of application depth between measured data and model-simulated results at an operating pressure of 69 kPa and 138 kPa, respectively.

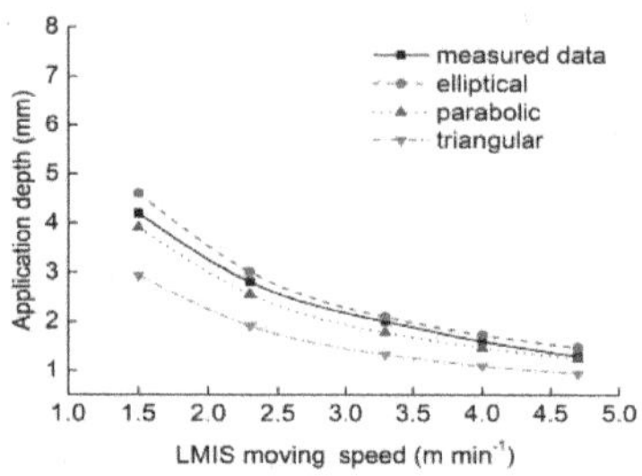

(a)

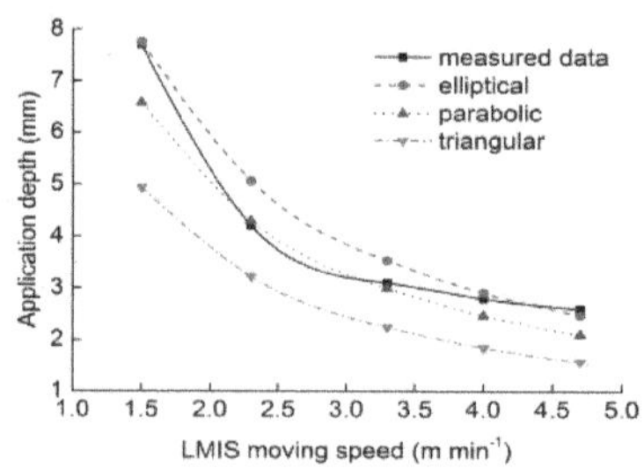

(b)

Figure 9. Comparison of application depth between experimental measured data and model-simulated results: (**a**) 69 kPa, (**b**) 138 kPa.

As seen from Figure 9, there appears agreement between the measured and simulated mean application depths. The measured data and model-simulated application depths for all the elliptical-, parabolic-, and triangular-shaped predictions decreased with an increasing average moving speed, which was in accordance with the analysis that slower moving speeds will result in more spraying water. The measured application depths were much larger than the triangular-shaped prediction simulated application depths and were just between the parabolic-shaped prediction and the elliptical-shaped prediction simulated application depths.

Compared to the experimental measured data at an operating pressure of 69 kPa, the simulated result was from 28.0% to 33.3% lower with an average of 30.9% for the triangular-shaped prediction, from 4.0% to 11.1% lower with an average of 7.8% for the parabolic-shaped prediction, and from 4.7% to 13.1% higher with an average of 8.5% for the elliptical-shaped prediction, respectively. Compared to the experimental measured data at an operating pressure of 138 kPa, the simulated result was from 23.3% to 39.3% lower with an average of 32.0% for the triangular-shaped prediction, from 2.3% to 19.1% lower with an average of 10.2% for the parabolic-shaped prediction, and from 3.9% to 20.4% higher with an average of 8.7% for the elliptical-shaped prediction, respectively. Normally, the measured water depth is generally a slight deviation from the calculated value, caused by the fitting error, the testing error, evaporation and drift during the spraying process [40–42]. Therefore, it was determined that both the parabolic- and elliptical-shaped predictions were the acceptable shapes of the water distribution pattern, as they were closer to the measured application depth.

The measured application depth and model-simulated results based on overlapping the measured water distribution data of individual sprinklers were curved lines. Special attention was given to the development of empirical equations for water application depth with regard to moving speed. The curvilinear Equation (10) was regressed and Table 3 presents the equations of those profiles. The coefficient of determination (R^2) for measured data ranged from 0.9698 to 0.9954 and for all the model simulations was 0.9946.

$$p = Av^2 - Bv + C \tag{10}$$

where p is the application depth (mm) and v is the LMIS speed (m min^{-1}).

Table 3. Equations of water application depth with regard to moving speed.

	69 kPa				138 kPa			
	A	B	C	R^2	A	B	C	R^2
Measured data	0.186	1.814	5.780	0.9954	0.529	4.331	11.26	0.9698
Elliptical prediction	0.231	2.140	6.461	0.9946	0.389	3.605	10.88	0.9946
Parabolic prediction	0.196	1.818	5.487	0.9946	0.330	3.061	9.240	0.9946
Triangular prediction	0.147	1.363	4.115	0.9946	0.248	2.300	6.930	0.9946

3.3.2. Coefficient of Uniformity

The simulated CU values were obtained using MATLAB calculations based on the basic foundation of application depth values from Figure 9. The combined CU values were calculated using the aforementioned method. The data from Table 2 were used as the experimental measured data. As a further illustration of model validation of CU calculations, Figure 10 presents the comparison of CU between experimental measured data and model-simulated results at an operating pressure of 69 kPa and 138 kPa, respectively.

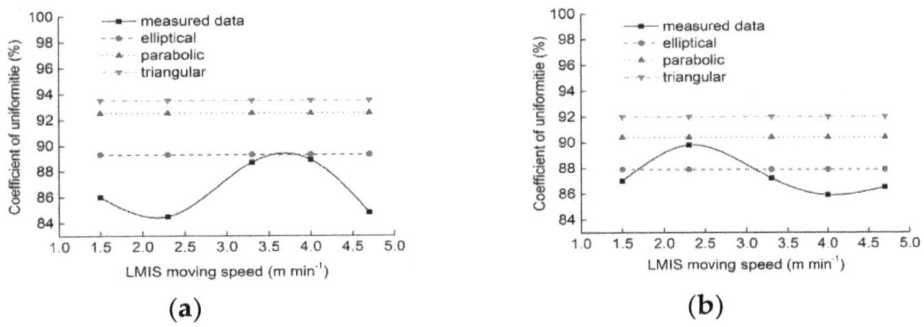

Figure 10. Comparison of CU between experimental measured data and model-simulated results. (**a**) 69 kPa, (**b**) 138 kPa.

As seen in Figure 10, there appears agreement between the measured and simulated CU values. The simulated CU for all the elliptical-, parabolic-, and triangular-shaped predictions maintains a constant value and does not vary with increasing speed. It was determined that a different moving speed has an influence on the depth of irrigation water along the direction of the sprinkler, which varies at a same percentage and does not affect CU values; only the branch direction data affect the sprinkling CU degrees, which can be representative of the entire area of the spraying uniformity. Therefore, when the branch water distribution shape was determined as an elliptical, parabolic, or triangular prediction, the simulated CU keeps a constant value at different moving speeds. However, the measured CUs were variable for all moving speeds. The range of combined CU values at different pressures were as follows: 84.5% at a moving speed of 2.3 m min^{-1} to 88.9% at a moving speed of 4.0 m min^{-1} (69 kPa), and 85.9% at a moving speed of 4.0 m min^{-1} to 89.8% at moving speed of 2.3 m min^{-1} (138 kPa). It was determined that the measured CU values with regard to moving speed fluctuated within a small range, and could be attributed to the acceptable fitting and experimental error.

Compared to the experimental measured data at an operating pressure of 69 kPa, the simulated CU value was 93.5% and of an absolute deviation rate from 5.2% to 10.7% with an average of 8.0% for the triangular-shaped prediction, 92.5% and of an absolute deviation rate from 4.0% to 9.5% with an average of 6.9% for the parabolic-shaped prediction, and 89.3% and of an absolute deviation rate from 0.4% to 5.7% with an average of 3.2% for the elliptical-shaped prediction, respectively. Compared to the experimental measured data at an operating pressure of 138 kPa, the simulated CU value was 92.0% and of an absolute deviation rate from 2.4% to 7.1% with an average of 5.4% for the triangular-shaped prediction, 90.4% and of an absolute deviation rate from 0.7% to 5.2% with an average of 3.6% for the parabolic-shaped prediction, and 87.9% and of an absolute deviation rate from 0.8% to 2.3% with an average of 1.6% for the elliptical-shaped prediction, respectively.

From the aforementioned analysis, it was determined that for the parabolic- and elliptical-shaped predictions, the deviations of the points are within 5% except for several cases at a moving speed of 2.3 and 4.0 m min^{-1}. This deviation may result from inaccurate measurement of the effective wetted widths. Generally, the simplified model has a high computational accuracy, which could be brought into Equation (5) to further calculate the LMIS spraying uniformity. The comparison of measured and model-simulated data indicated that the measured water distribution pattern of individual sprinklers could be represented both as an elliptical- or parabolic-shaped prediction, which was accurate and reliable for simulating the application performance of the LMIS and verified that the calculation of application depth and CU values shown in this work is applicable in practice.

In short, a new model of the LMIS has been developed and a sprinkler irrigation system was specifically manufactured as an experimental sample to verify the accuracy of the model. The differences in model accuracy owing to different operating pressures were not significant, and the moving speed of the LMIS did not appear to influence model accuracy either. Comparisons between experimental data and model simulations revealed that the model can accurately predict water application depth and CU values along the LMIS. Although the model was developed and validated for linear moving systems, it could be readily used for center pivots, which constitute a simplification of the calculation approach adopted.

4. Conclusions

This study presents a model of application depth and uniformity for the LMIS. Based on the obtained results and the conditions in which this trial was carried out, the following can be concluded:

At an operating pressure of 69 kPa and 138 kPa, the sprinkler nozzle wetting radius was 4.4 m and 5.7 m, respectively. The sprinkler head produced quite similar water application profiles, which appeared as a saddle type and had two sprinkler intensity peaks close to the sprinkler.

Compared to the experimental measured data at operating pressures of 69 kPa or 138 kPa, the simulated application depth was on average 30.9% or 32.0% lower for the triangular-shaped prediction, 7.8% or 10.2% lower for the parabolic-shaped prediction, and 8.5% or 8.7% for the elliptical-shaped prediction, respectively. The curvilinear equations for the measured application depth and model-simulated results were regressed and the coefficient of determination (R^2) was from 0.9698 to 0.9954. The simulated CU was an average deviation rate of 8.0% or 5.4% for the triangular-shaped prediction, 6.9% or 3.6% for the parabolic-shaped prediction, and 3.2% or 1.6% for the elliptical-shaped prediction, respectively.

This study indicates that a sprinkler irrigation system for the LMIS that is both elliptical and parabolic in shape could be considered as far as application depths and CU values are concerned. The model can therefore be further developed to provide a useful tool for LMIS design and management.

Author Contributions: J.L. was the supervision of this manuscript; X.Z. was writing the original draft and making all revisions; S.Y. was providing the funding acquisition; and A.F. was doing the literature resources.

Acknowledgments: The authors are greatly indebted to the supports from students of the Research Centre of Fluid Machinery and Engineering, Jiangsu University for their assistances in conducting the experiment.

References

1. Yuan, S.Q.; Li, H.; Wang, X.K. Status, problems, trends and suggestions for water-saving irrigation equipment in China. *J. Drain. Irrig. Mach. Eng. (JDIME)* **2015**, *33*, 78–92.
2. Zhu, X.Y.; Chikangaise, P.; Shi, W.D.; Chen, W.H.; Yuan, S.Q. Review of intelligent sprinkler irrigation technologies for remote autonomous system. *Int. J. Agric. Biol. Eng.* **2018**, *11*, 23–30. [CrossRef]
3. Xu, D.; Li, Y.N.; Gong, S.H.; Zhang, B.Z. Experiment on sweet pepper nitrogen detection based on near infrared reflectivity spectral ridge regression. *J. Drain. Irrig. Mach. Eng. (JDIME)* **2019**, *37*, 63–72.
4. Hart, W.E. Sprinkler distribution analysis with a digital computer. *Trans ASAE* **1963**, *6*, 206–208.
5. Vories, E.D.; Von Bernuth, R.D. Single nozzle sprinkler performance in wind. *Trans. ASAE* **1986**, *29*, 1325–1330.
6. Li, J.; Kawano, H. Sprinkler rotation nonuniformity and water distribution. *Trans. ASAE* **1996**, *39*, 2027–2031. [CrossRef]
7. Zhu, X.Y.; Yuan, S.Q.; Jiang, J.Y.; Liu, J.P.; Liu, X.F. Comparison of fluidic and impact sprinklers based on hydraulic performance. *Irrig. Sci.* **2015**, *33*, 367–374. [CrossRef]
8. Liu, J.P.; Yuan, S.Q.; Li, H.; Zhu, X.Y. Experimental and combined calculation of variable fluidic sprinkler in agriculture irrigation. *AMA-Agr. Mech. Asia Afr. Lat. Am.* **2016**, *47*, 82–88.
9. Xiang, Q.J.; Xu, Z.D.; Chen, C. Experiments on air and water suction capability of 30PY impact sprinkler. *J. Drain. Irrig. Mach. Eng. (JDIME)* **2018**, *36*, 82–87.
10. Hu, G.; Zhu, X.Y.; Yuan, S.Q.; Zhang, L.G.; Li, Y.F. Comparison of ranges of fluidic sprinkler predicted with BP and RBF neural network models. *J. Drain. Irrig. Mach. Eng. (JDIME)* **2019**, *37*, 263–269.
11. Christiansen, J.E. *Irrigation by Sprinkling*; Bulletin 670; California Agricultural Experiment Station, University of California: Berkeley, CA, USA, 1942.
12. Solomon, K. Variability of sprinkler coefficient of uniformity test results. *Trans. ASAE* **1979**, *22*, 1078–1080. [CrossRef]
13. Fukui, Y.; Nakanishi, K.; Okamura, S. Computer evaluation of sprinkler irrigation uniformity. *Irrig. Sci.* **1980**, *2*, 23–32. [CrossRef]
14. Seginer, I.; Kantz, D.; Bernuth, R.D. Indoor measurement of single-radius sprinkler patterns. *Trans. ASAE* **1992**, *35*, 523–533. [CrossRef]
15. Tarjuelo, J.M.; Montero, J.; Valiente, M.; Honrubia, F.T.; Ortiz, J. Irrigation uniformity with medium size sprinklers. Part I: Characterization of water distribution in no-wind conditions. *Trans. ASAE* **1999**, *42*, 665–675. [CrossRef]
16. Michael, J.L.; John, S.S. Sprinkler head maintenance effects on water application uniformity. *J. Irrig. Drain. Eng. ASCE* **2000**, *126*, 142–148.
17. Playán, E.; Zapata, N.; Faci, J.M.; Tolosa, D.; Lacueva, J.L.; Pelegrin, J.; Salvador, R.; Sanchez, I.; Lafita, A. Assessing sprinkler irrigation uniformity using a ballistic simulation model. *Agric. Water Manag.* **2006**, *84*, 89–100. [CrossRef]
18. Zhu, X.Y.; Yuan, S.Q.; Liu, J.P. Effect of sprinkler head geometrical parameters on hydraulic performance of fluidic sprinkler. *J. Irrig. Drain. Eng. ASCE* **2012**, *138*, 1019–1026. [CrossRef]

19. Zhang, L.; Merkley, G.P.; Pinthong, K. Assessing whole-field sprinkler irrigation application uniformity. *Irrig. Sci.* **2013**, *31*, 87–105. [CrossRef]
20. Liu, J.P.; Yuan, S.Q.; Darko, R.O. Characteristics of water and droplet size distributions from fluidic sprinklers. *Irrig. Drain.* **2016**, *65*, 522–529. [CrossRef]
21. Yuan, S.Q.; Darko, R.O.; Zhu, X.Y.; Liu, J.P.; Tian, K. Optimization of movable irrigation system and performance assessment of distribution uniformity under varying conditions. *Int. J. Agric. Biol. Eng.* **2017**, *10*, 72–79.
22. Tang, L.D.; Yuan, S.Q.; Qiu, Z.P. Development and research status of water turbine for hose reel irrigator. *J. Drain. Irrig. Mach. Eng. (JDIME)* **2018**, *36*, 963–968.
23. Lu, M.Y.; Lu, K.J.; Hu, G.; Zhu, X.Y. Experiment on hydraulic performance of type SD-03 pop-up sprinkler. *J. Drain. Irrig. Mach. Eng. (JDIME)* **2018**, *36*, 1120–1124.
24. Buchleiter, G.W. Performance of LEPA equipment on center pivot machines. *Appl. Eng. Agric.* **1992**, *8*, 631–637. [CrossRef]
25. King, B.A.; Kincaid, D.C. Optimal performance from center pivot sprinkler systems. *Trans. ASABE* **1995**, *38*, 1737–1747.
26. Luz, P.B. A graphical solution to estimate potential runoff in center-pivot irrigation. *Trans. ASABE* **2011**, *54*, 81–92. [CrossRef]
27. Martin, D.L.; Kranz, W.L.; Thompson, A.L.; Liang, H. Selecting sprinkler packages for center pivots. *Trans. ASABE* **2012**, *55*, 513–523. [CrossRef]
28. Lu, J.; Cheng, J. Numerical simulation analysis of energy conversion in hydraulic turbine of hose reel irrigator JP75. *J. Drain. Irrig. Mach. Eng. (JDIME)* **2018**, *36*, 448–453.
29. Zhu, X.Y.; Peters, T.; Neibling, H. Hydraulic performance assessment of LESA at low pressure. *Irrig. Drain.* **2016**, *65*, 530–536. [CrossRef]
30. Tian, K.; Zhu, X.Y.; Wan, J.H.; Bao, Y. Development and performance test of lateral move irrigation system. *J. Drain. Irrig. Mach. Eng. (JDIME)* **2017**, *35*, 357–361.
31. Yan, H.J.; Jin, H.Z. Study on the discharge coefficient of nonrotatable sprays for center-pivot system. *J. Irrig. Drain. Eng. ASCE* **2004**, *23*, 55–58.
32. Dogana, E.; Kirnaka, H.; Dogan, Z. Effect of varying the distance of collectors below a sprinkler head and travel speed on measurements of mean water depth and uniformity for a linear move irrigation sprinkler system. *Biosyst. Eng.* **2008**, *99*, 190–195. [CrossRef]
33. Yan, H.J.; Jin, H.Z.; Qian, Y.C. Characterizing center pivot irrigation with fixed spray plate sprinklers. *Sci. Chin.* **2010**, *53*, 1398–1405. [CrossRef]
34. ASAE Standards. *Procedure for Sprinkler Distribution Testing for Research Purposes*, 32nd ed.; S330.1; ASAE: St. Joseph, MI, USA, 1985.
35. ASAE Standards. *Procedure for Sprinkler Testing and Performance Reporting*, 32nd ed.; S398.1; ASAE: St. Joseph, MI, USA, 1985.
36. ASAE. *Test Procedure for Determining the Uniformity of Water Distribution of Center Pivot and Lateral Move Irrigation Machines Equipped with Spray or Sprinkler Nozzles*; ANSI/ASAE S436.1 MAR01; ASAE Standards: St. Joseph, MI, USA, 2001.
37. MOD GB/T 19795.2. *Agricultural Irrigation Equipment—Rotating Sprinklers—Part 2: Uniformity of Distribution and Test Methods*; International Organization for Standardization: Geneva, Switzerland, 2005.
38. Schneider, A.D. Efficiency and uniformity of the LEPA and spray sprinkler methods: A review. *Trans. ASAE* **2000**, *43*, 937–944. [CrossRef]
39. Kohl, K.D.; Kohl, R.A.; DeBoer, D.W. Measurement of low pressure sprinkler evaporation loss. *Trans. ASAE* **1987**, *30*, 1071–1074. [CrossRef]
40. Dechmi, F.; Playán, E.; Cavero, J.; Faci, J.M.; Martínez-Cob, A. Wind effects on solid set sprinkler irrigation depth and yield of maize (*Zea mays*). *Irrig. Sci.* **2003**, *22*, 67–77. [CrossRef]
41. Playán, E.; Salvador, R.; Faci, J.M. Day and night wind drift and evaporation losses in sprinkler solid-sets and moving laterals. *Agric. Water Manag.* **2005**, *76*, 139–159. [CrossRef]
42. Zhang, M.; Zhao, W.X.; Li, J.S.; Li, Y.F. Fertigation uniformity and evaporation drift losses of center pivot irrigation system. *J. Drain. Irrig. Mach. Eng. (JDIME)* **2018**, *36*, 1125–1130.

13

Design of a Streamwise-Lateral Ski-Jump Flow Discharge Spillway

Jun Deng, Wangru Wei *, Zhong Tian and Faxing Zhang

State Key Laboratory of Hydraulics and Mountain River Engineering, Sichuan University, Chengdu 610065, China; djhao2002@scu.edu.cn (J.D.); tianzhong@scu.edu.cn (Z.T.); zhfx@scu.edu.cn (F.Z.)

* Correspondence: weiwangru@scu.edu.cn

Abstract: Spillway outlet design is a major issue in hydraulic engineering with high head and large discharge conditions. A new type of design for a streamwise-lateral spillway is proposed for ski-jump flow discharge and energy dissipation in hydraulic engineering. The water in the spillway outlet is constrained by three solid walls with an inclined floor, a horizontal floor on the bottom and a deflected side wall in the lateral direction. The water flow releases in a lateral direction into the plunge pool along the streamwise direction. It generates a free jet in the shape of "∩" in a limited area, causing the water to fully diffuse and stretch in the air simultaneously, and drop into the plunge pool to avoid excessive impact in the plunge pool. The formation mechanism for the flow pattern is analyzed, and the results show that the optimum inclination is an angle range of 30°~45° for a good performance of free ski-jump jet diffusion shape.

Keywords: ski-jump flow; spillway outlet; jet trajectory; energy dissipation; experimental model

---

1. Introduction

Energy dissipation is always a major issue in hydraulic engineering [1,2], especially for large discharge and complex geological conditions. Ski-jump energy dissipation has prevailed among the hydraulic engineering, considering the construction cost and dissipation performance [3–5]. The characteristics of this ski-jump are the free jets diffusing and aerating in the air [6,7] and then dropping into the water downstream, causing energy dissipation through shearing and disturbance. In the traditional design, water flow discharges in the spillway or steep chute and a flip bucket is set at the end of the spillway to make the water release into the flow [8,9]. The various geometries of flip buckets include a wide range of deflection angles, radius, takeoff angles, etc. [10–14].

In order to improve the energy dissipation rate and reduce the impact on the bottom of the plunge pool, the free jet is always made to widely diffuse in the air in both streamwise and lateral directions, and is dropped far away from the dam for safety [15,16]. Given the enormous flow discharge and velocity, the common ski-jump is not efficient, considering the energy dissipation performance and other damaging problems. In order to make the ski-jump flow diffuse in a certain direction, researchers and engineers developed a series of structure design to drive the water jet. The specific structures [17–19], such as a contract side wall or a bucket radius of flip bucket, can adjust the water jet to a certain direction. However, when the flow velocity exceeds 30–50 m/s, the hydrodynamic pressures on the structure is another important issue, which is extremely dangerous to outlet structure safety. The recent study on the flip bucket proposed a new design principle that is take full advantage of the natural pressure difference at the lateral direction to make more contribution, based on the motion characteristic of high-speed flow. The well performance of long-narrow ski-jump flow on the Jinping-I flood discharge tunnel showed much lower pressure on the structure of a leak-floor flip bucket than that of a lateral contraction of side-wall design. However, there have been limited applications of flip buckets for two reasons: (1) dissipation area and (2) free jet diffusion shape. For some projects, the dissipation area is limited, such as in valleys and in unsatisfactory souring conditions of river courses. The free jet diffusion shape influences the aeration and impact performance directly, and the effect is pivotal in the limited dissipation area.

In the present study, a new spillway design for a ski-jump streamwise-lateral discharge spillway is proposed for the two stated reasons. The spillway side wall close to the plunge pool is removed and an inclined floor is set along the streamwise direction. This makes the water flow release laterally as it moves downstream, and the free jet shape makes the most of the space as it diffuses in the air. The formation mechanism and characteristics of the water jet are presented, based on experiments and numerical simulation.

2. Physical Structure of the Streamwise-Lateral Discharge Spillway

The design of the streamwise-lateral discharge spillway is shown in Figure 1. The spillway is set on the right side of the plunge pool and the axis of the spillway is parallel to the plunge pool in the streamwise direction. The side wall of the spillway closed to the plunge pool (left wall in the present study) is removed (in contrast to the traditional spillway) and an inclined floor is set between the horizontal bottom floor and the right-side wall. An anti-arch design is connected to the straight part of the side wall, which is designed to deflect water into the plunge pool. The intersection part between the horizontal floor and the inclined floor is the diagonal line of the spillway. Here, X is the streamwise coordinate which is perpendicular to the cross-section Y-Z, where Y is the lateral coordinate and Z is the vertical coordinate, respectively.

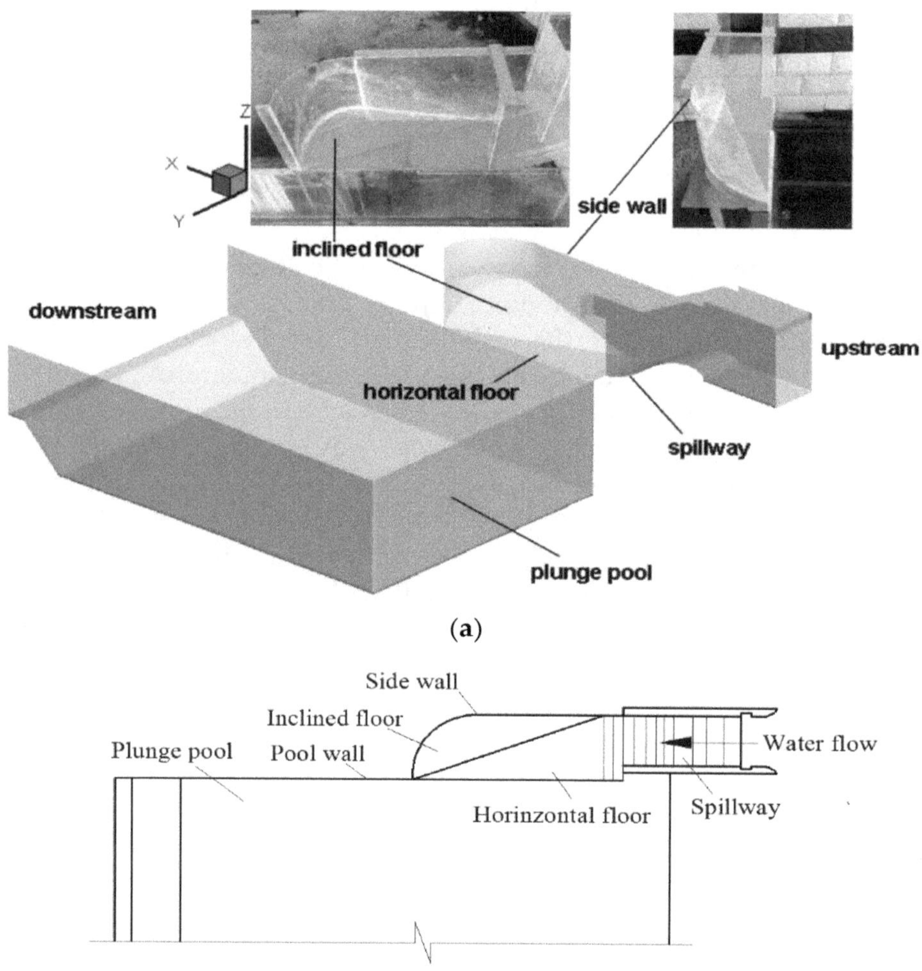

Figure 1. Definition of the streamwise-lateral discharge spillway: (**a**) 3-D model; (**b**) plan view.

The typical flow pattern of the physical model is shown in Figure 2. When the water flow moves downstream in the spillway, the pressure on the left side is reduced to the atmospheric pressure, while the pressure on the right side remains the hydrodynamic pressure. There is a pressure difference due to the specific design, and the water will flow transversely into the plunge pool and the streamwise-lateral discharge will be generated. When the water flow reaches the end of the spillway, the water is restrained by the inclined floor and the anti-arch side wall, and the free jet is then generated and stretches in the air above the plunge pool. The free jet diffuses pronouncedly in the transverse direction above the plunge pool in the very limited area, showing a "curling type" in the air. This will make a beneficial contribution to the energy dissipation through three aspects: (1) The water thickness is thin and the flow discharge per unit width dropping into the plunge pool is small; (2) the aeration of the free jet is fully developed as it diffuses in the air; and (3) the ratio of the shearing dissipation area in the plunge pool increases.

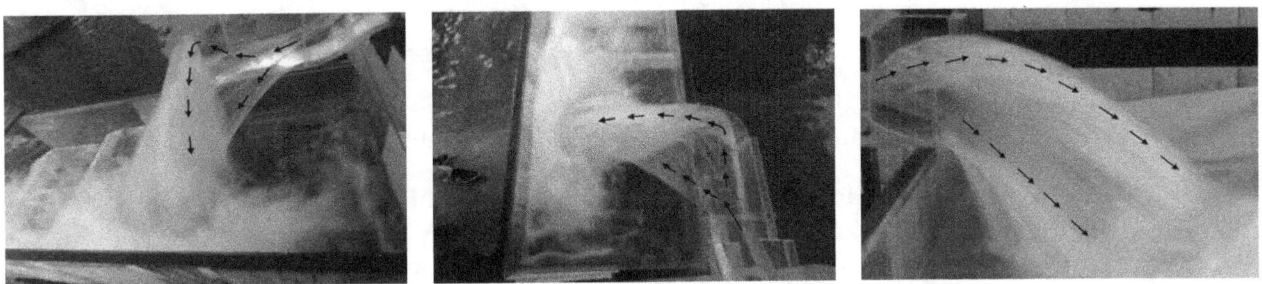

(a) (b) (c)

Figure 2. Typical pattern of the streamwise-lateral discharge spillway: (**a**) physical model from the downstream view; (**b**) physical model from the lateral view; (**c**) ski-jump flow.

To obtain the detained characteristics of configuration and diffusion of the streamwise-lateral ski-jump flow discharge spillway, it is practically impossible to see the jet configuration due to the failure of exiting measurements on the complex 3-D flow. The physical experiments can only provide some jet length and pressure distribution on the floor in order to describe the jet flow pattern. However, the numerical simulation can provide the entire jet field with detailed water profile and velocity distribution in the jet diffusion process. Thus, the physical model test and numerical simulation were used in this paper to systemically study the new design of streamwise-lateral ski-jump flow discharge spillway.

3. Investigation Methods and Definition of the Free Jet Flow Pattern

3.1. Physical Model

The study of the formation of the streamwise-lateral flow discharge spillway is based on energy dissipation in a certain hydropower station. The size of the plunge pool is 88 (length) $\times$ 100 (width) m^2, and the vertical distance from the horizontal floor to the bottom wall of the plunge pool is H_0 = 43 m. The length of the horizontal floor L = 55 m, with W = 21 m in width. The straight length of the side wall is L_b = 36.4 m, with the anti-arch radius R = 19 m and the deflection angle θ = 96° in the anti-arch structure. The angle α of the inclined floor ranges from 0° to 60°, as shown in Figure 3. The flow discharge is 4016 m^3/s, with discharge per unit width q_w = 251 m^3/(s·m). The tail water depth is 18 m deep. The scale of a normal model is 1:60 based on the Froude criterion for experimental study. Considering that the roughness coefficient of the concrete in the prototype is 0.0140 and the roughness scale is $60^{1/6}$ = 1.979, the theoretical value of the roughness coefficient is 0.0071. The physical model is made of transparent plexiglass with a roughness coefficient of 0.0079. The present physical model is approximately satisfactory for the similarity of boundary roughness between the model and prototype.

Considering the 1:60 scale of the physical model, the approach flow Froude number $Fr = V/(gh)^{0.5}$ is about 3.73 with a Reynolds number $Re = Vh/\nu$ of 1.83 $\times$ 10^5, where ν is the kinematic water viscosity,

g is the gravitational acceleration, and $V = 3.2$ m/s and $h = 0.075$ m are the mean velocity and depth of the approach flow on the horizontal floor, respectively. Thus, the present study focuses on the typical pattern of ski-jump flow in a streamwise-lateral discharge spillway, jet trajectory and distribution of mean pressure on the bottom floor of a basin, which satisfies the Froude criterion. The scale effect on air–water properties of supercritical flow cannot be neglected with regards to the surface tension and viscosity effects in high-speed flows [20].

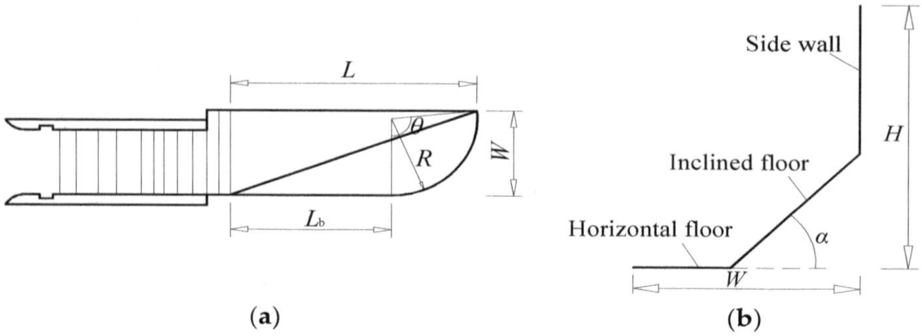

Figure 3. Sketch with relevant parameters: (**a**) plan view; (**b**) sectional view.

A thin plate wire with a graduated cylinder was set downstream for a water flow discharge measurement to the nearest 0.1 L/s, and the water level was read with a conventional point gauge to an accuracy of 0.1 mm. The jet trajectory of the ski-jump flow was measured at about 5 min, ensuring it did not change much with the time between observations. The pressure signal was measured using a series of pressure sensors (CY201, Test, Inc.: Chengdu, China) and an acquisition device (RS485-20, Test, Inc.: Chengdu, China). The accuracy of the pressure sensor was 0.1% of the 100 kPa full scale. The pressure time–domain process was sampled at 100 Hz for 10 s, which was adequate to evaluate the pressure fluctuation under the present hydraulic conditions. The pressure measurement system was calibrated by hydrostatic operation before each test was conducted to ensure measurement reliability. The time mean pressure P was obtained using the time–domain process p, and the maximum mean pressure p_m was determined as $p_m = P - P_0$.

3.2. Numerical Simulation

In the present work, the Fluent computer code (Fluent Inc.: Pittsburgh, PA, USA, 2005) was used to simulate the ski-jump flows of the streamwise-lateral discharge spillway. The Renormalization Group k-ε turbulent model was employed, which is an effective method to simulate the ski-jump flow movement in detail [21]. The governing equations are as follows:

Continuity equation:

$$\frac{\partial \rho}{\partial t} + \frac{\partial \rho u_i}{\partial x_i} = 0 \tag{1}$$

Momentum equation:

$$\frac{\partial \rho u_i}{\partial t} + \frac{\partial (\rho u_i u_j)}{\partial x_j} = -\frac{\partial p}{\partial x_i} + \frac{\partial}{\partial x_j}\left[(\mu + \mu_t) \times \left(\frac{\partial u_i}{\partial x_j} + \frac{\partial u_j}{\partial x_i}\right)\right] \tag{2}$$

k (turbulent kinetic energy) equation:

$$\frac{\partial \rho k}{\partial t} + \frac{\partial (\rho u_i k)}{\partial x_i} = \frac{\partial}{\partial x_i}\left[\left(\mu + \frac{\mu_t}{\sigma_k}\right) \times \frac{\partial k}{\partial x_i}\right] + G_k - \rho\varepsilon \tag{3}$$

ε (dissipation rate of turbulent kinetic energy) equation:

$$\frac{\partial \rho \varepsilon}{\partial t} + \frac{\partial (\rho u_i \varepsilon)}{\partial x_i} = \frac{\partial}{\partial x_i}\left[\left(\mu + \frac{\mu_t}{\sigma_\varepsilon}\right) \times \frac{\partial \varepsilon}{\partial x_i}\right] + C_{1\varepsilon}\rho\frac{\varepsilon}{k}G_k - C_{2\varepsilon}\rho\frac{\varepsilon^2}{k} \tag{4}$$

where t is time. ρ and μ are the average density of the volume fraction and molecular viscosity, respectively. p is the pressure and G_k represents the generation of turbulence kinetic energy due to the mean velocity gradients. μ_t is the turbulent viscosity which can be deduced for the turbulence intensity k and energy dissipation rate ε:

$$\mu_t = C_\mu \frac{k^2}{\varepsilon},\ C_{1\varepsilon} = 1.42 - \frac{\eta\left(1-\frac{\eta}{\eta_0}\right)}{1+\beta\eta^3},\ \eta = \frac{Sk}{\varepsilon},\ S = \sqrt{2\overline{S_{ij}}\,\overline{S_{ij}}},\overline{S_{ij}} = \frac{1}{2}\left(\frac{\partial u_i}{\partial x_j}+\frac{\partial u_j}{\partial x_i}\right) \quad (5)$$

where S is the modulus of the mean strain rate tensor. The detailed values are shown in Table 1.

The Volume of Fluid (VOF) method was used to simulate the free surface and air fraction of ski-jump flow. The governing equation for water fraction α_w is:

$$\frac{\partial \alpha_\mathrm{w}}{\partial t} + u_i \frac{\partial \alpha_\mathrm{w}}{\partial x_i} = 0 \quad (6)$$

$$\rho = \alpha_\mathrm{w}\rho_\mathrm{w} + (1-\alpha_\mathrm{w})\rho_\mathrm{a} \quad (7)$$

$$\mu = \alpha_\mathrm{w}\mu_\mathrm{w} + (1-\alpha_\mathrm{w})\mu_\mathrm{a} \quad (8)$$

where u_i is the velocity components and x_i is the coordinates (i = 1, 2, 3). j is the sum suffix. The air–water interface is tracked by solving the continuity equations. The sum of the water and air volume fraction was one in the controlling body. The ρ_w and ρ_a are the density of water and air, respectively. The μ_w and μ_a are the viscosity of water and air, respectively. When $\alpha_\mathrm{W} = 0$, it means there is full of air; when $\alpha_\mathrm{W} = 1$, it represents the air fraction of fluid is 0. When $0 < \alpha_\mathrm{W} < 1$, it means the certain controlling volume. The total fraction in a certain controlling body is $\alpha_\mathrm{W} + \alpha_\mathrm{a} = 1$. Note that, although the VOF model can give an aeration result in the free surface area, the air concentration distribution and aeration development along the jet trajectory is different to the real situations due to the complex air–water interaction and bubble diffusion process. In the present study, the simulated results will be used for the analysis on formation of ski-jump flow released from the streamwise-lateral spillway, which is little affected by the aeration. The detailed hydraulic characteristics, including jet trajectory and impact pressure on the pool floor, will be analyzed based on the physical model results. The control volume method is introduced to discretize the partial differential equations, and the SIMPLER method, which has good convergence characteristics, is employed for numerical simulation [22,23].

In the present research, both structured and unstructured meshes were used. Unstructured mesh was employed in the streamwise-lateral spillway area due to the complex geometry condition. Structural mesh was used for other regions. Mesh size was refined until the computed results were independent of the size of the scale, based on the fact that the calculated water profile and the pressure on the spillway sidewall did not change more than 3%. The uncertainty analysis of different mesh densities showed that the number of meshes was reasonably set to 1,017,033. The approximate $0.5 \times 0.5\ \mathrm{m}^2$ mesh size was selected to simulate the pattern of ski-jump flow in the air and its diffusion in the still basin. Based on the authors' previous studies [23–25] on the simulation of high-speed flows (flow velocity exceeded over 30 m/s) in hydraulic engineering, these mesh conditions can make sure the maximum uncertainties for the velocity and pressure characteristics in the whole calculated region were approximately smaller 10%, and the uncertainty in most locations was fairly small. The comparison between the experimental and simulated results with regards to the flow characteristics of the streamwise-lateral discharge spillway are shown in Table 2 for $\alpha = 45°$, where $Y = 0$ is the lateral cross-section at the straight side wall. The distributions of pressure on the side wall and water depth indicate that the simulation results are well in agreement with those of experimental measurements. The vector profile of the flow jet velocity in Figure 4 shows similar flow patterns for the experiments. According to the typical flow pattern of the streamwise-lateral discharge spillway, it is difficult to acknowledge the interior characteristics through experimental measurement, due to the irregular "curling" nappe shape and the complicated flow jet movement in the diffusion process. Because the computational water profiles and the pressure on the side wall are well in agreement with the test measurements, the flow patterns are similar between the simulation and the experiment and the complex high-speed fluid can be analyzed with the numerically simulated results. The characteristics of the jet profile and the pressure from the physical model were converted to the prototype according to the length scale. The simulated model was built up based on the real dimensions in the prototype, and the velocity distribution inside the ski-jump flow and stilling basin is the same as the prototype's dimensions.

Table 1. Constants of k-ε equations.

Parameters	η_0	β	C_μ	$C_{1\varepsilon}$	$C_{2\varepsilon}$	σ_k	σ_ε
Value	4.38	0.012	0.0845	1.42	1.68	0.7179	0.7179

Table 2. Comparison of tested and calculated data on the spillway flood discharge ($\alpha = 45°$, $q_w = 251\ m^2/s$).

Pressure on the Sidewall of Spillway (m)						Water Depth along the Side Wall of Spillway (m)		
Y = 15 m			Y = 18 m					
H	Simulation	Test	H	Simulation	Test	X	Simulation	Test
8.9	23.7	23.4	6.7	25.4	24.3	4.1	11.1	10.3
11.3	17.0	17.6	9.2	19.6	19.1	14.1	10.5	10.3
12.4	13.6	12.6	11.7	13.4	13.6	23.8	11.3	11.2
14.8	8.5	9.7	15.3	6.3	5.6	33.1	13.0	13.0
17.1	4.8	4.5	17.8	2.4	2.7	41.5	15.4	14.6
19.5	1.9	1.6	21.5	1.4	0.9	49.5	19.7	19.3
23.0	0.5	0.6	22.8	1.0	0.9	55.1	25.5	25.4

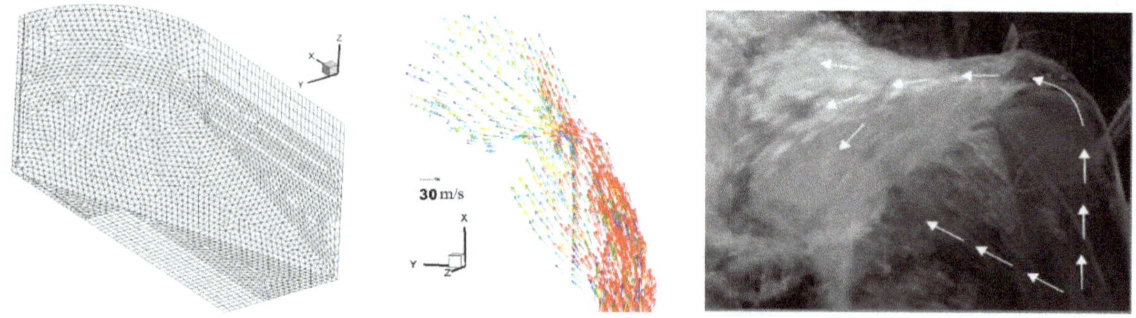

Figure 4. Comparison of numerical simulation with experimental flow: (**a**) a sample of computational mesh in the spillway area; (**b**) velocity vector profile; (**c**) experimental image.

4. Results and Discussion

4.1. Ski-Jump Flow Pattern

Figure 5 shows the simulated results about the water profile and velocity vector distributions of the ski-jump flow released from the streamwise-lateral spillway. At the initial area closed to upstream, the pressure near the left side reduces to atmosphere pressure without the left-side wall, while the water near the right side is constrained by the wall. Thus, the water on the left side moves laterally at the cross-section, and releases from the horizontal floor, flowing towards the plunge pool. Meanwhile, water flow on the right side moves along the streamwise direction and "climbs" along the inclined floor to the side wall of the spillway until it reaches the end of the spillway. Due to the coupling constraint of the inclined floor and the anti-arch wall in both streamwise and lateral directions, a thin and widely stretched nappe shape forms in the air, dropping into the plunge pool. In addition, the water moving laterally continues to move in the streamwise direction. Both of the movements contribute to the formation of three-dimensional ski-jump flow.

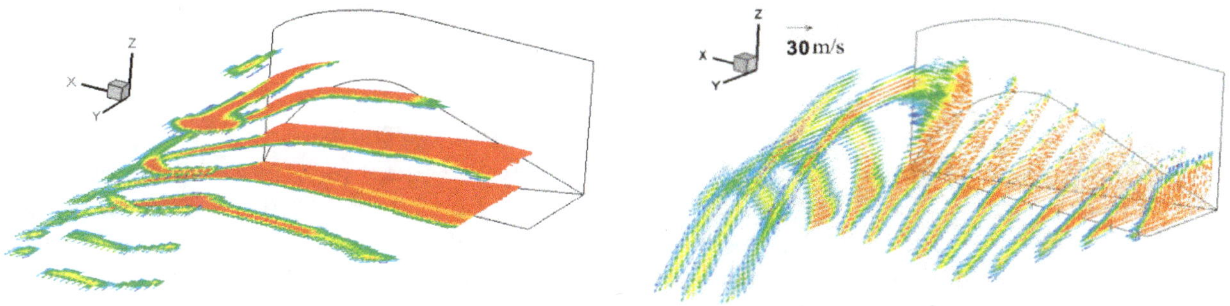

(a) (b)

Figure 5. Jet diffusion of streamwise-lateral ski-jump flow (simulated results for $\alpha = 45°$): (**a**) water profile at different elevations; (**b**) velocity vector profile at different cross-sections.

Figure 6 shows the distributions of the detailed flow velocity vector in the air for various elevations. When the water flow leaves the spillway, the jet flow turns to the transverse direction pronouncedly under the effect of coupling constrained by the inclined floor and the anti-arch wall. In addition, the lateral discharge becomes smaller and the streamwise "curling" discharge increases with the decrease of elevation. The "∩" shape can clearly be seen. The water jet becomes fully stretched in the limited area before dropping into the plunge pool. The thickness of the flow nappe becomes thinner and thinner in the air. This greatly contributes to the aeration and air–water mixture of the flow jet in the diffusion and stretching process, which will accelerate flow velocity attenuation and improve shearing intensity in the plunge pool.

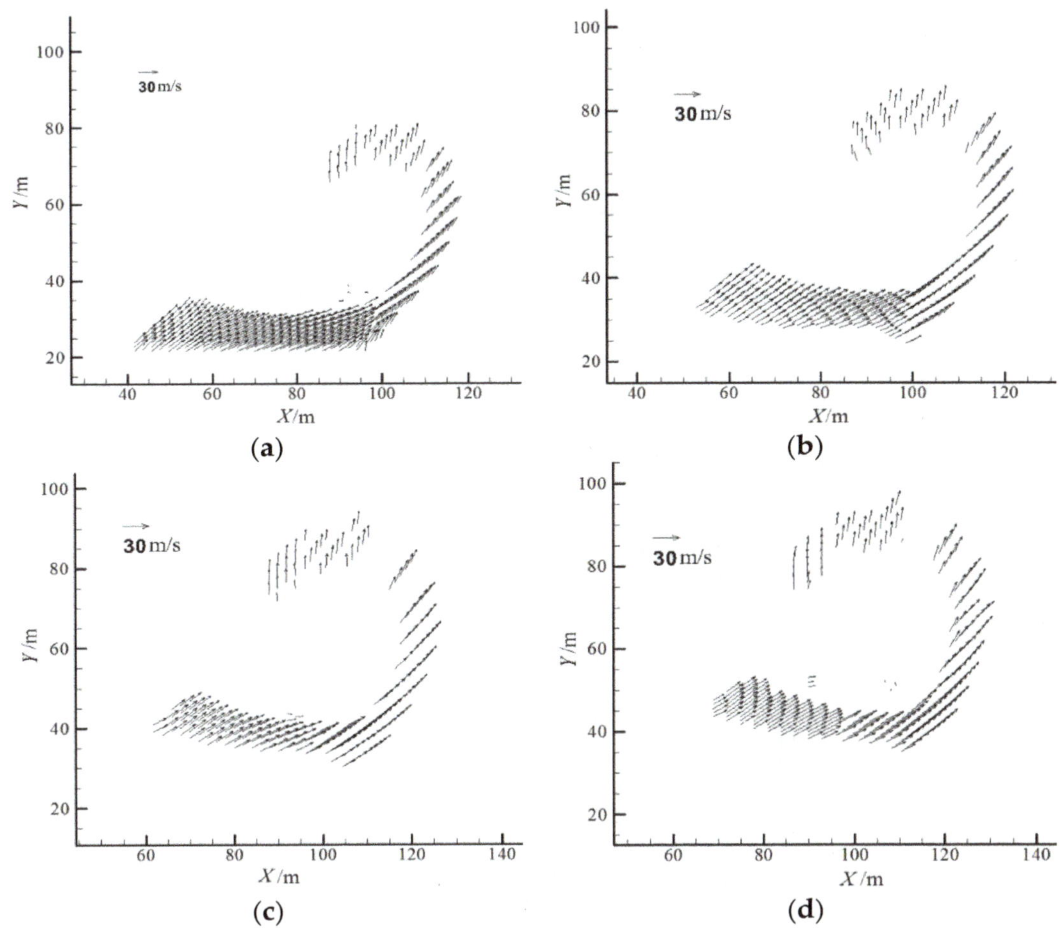

Figure 6. Velocity vector distributions of free jet flow: (**a**) Z = −17 m; (**b**) Z = −22 m; (**c**) Z = −27 m; (**d**) Z = −32 m.

4.2. Effect of Inclined Floor on Jet Length

To describe the free jet trajectory from the experimental results, the nearer jet length L_1 and the further jet length L_2 are introduced, as shown in Figure 7. The further jet length, L_2, is the longest distance in the lateral direction mainly affected by the coupling effect of the inclined floor and the anti-arch side wall of the spillway; and the nearer jet length, L_1, is the lateral jet length distance and is affected mainly by the pressure difference in the horizontal floor. The coefficient $\beta = (L_2 - L_1)/H_0$ describes the free jet diffusion in the air, deflecting from the spillway. The increase of β indicates that the diffusion performance improves and the stretch of water jet in the air space becomes fully developed. The angle α of the inclined floor influences L_1 and L_2 pronouncedly for the otherwise identical conditions. As shown in Figure 8a,b, with the increase of α to 30°, both L_1 and L_2 remain unchanged, which indicates that the effect of the inclined floor on the jet length is not distinct. As α further increases to 45°, L_1 increases while L_2 still remains significantly unchanged. The value of β decreases slightly which indicates that the water flow pattern is stable, as shown in Figure 8c. For $\alpha > 45°$, L_1 further increases while L_2 decreases pronouncedly. This results in a significant reduction in β, which indicates that the diffusion of the jet in the air is not sufficient in the lateral direction. Besides, the further jet length is around five times greater than the nearer jet length for $\alpha < 30°$, and when $\alpha = 60°$, the further jet length is around four times greater than the nearer jet length, indicating that the diffusion of the ski-jump flow nappe remained stable. Based on the effect of the inclined angle on the jet length, it is recommended that α is 30~45° for a good performance of free ski-jump jet diffusion.

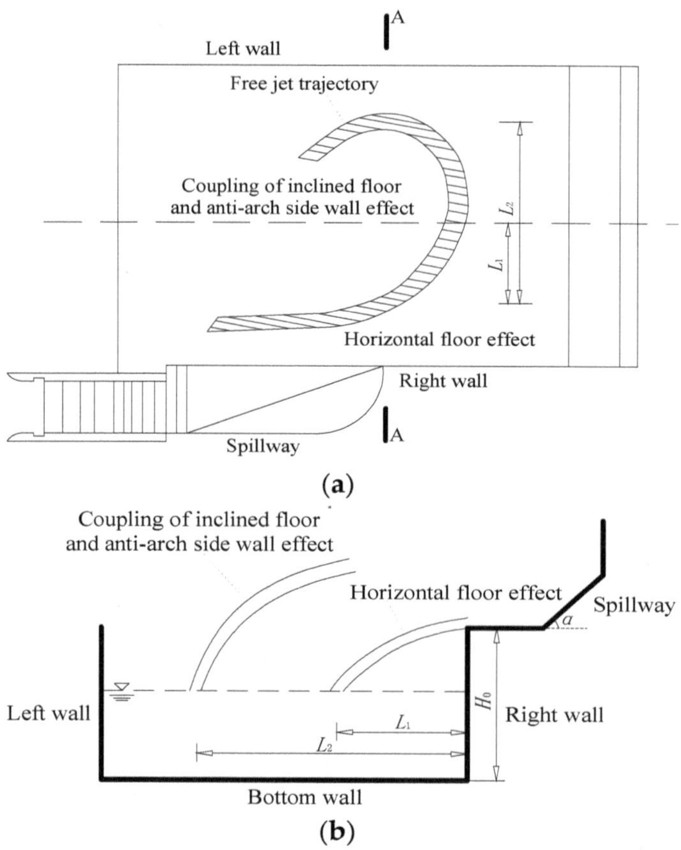

Figure 7. Sketch of free jet flow deflecting from the streamwise-lateral discharge spillway: (**a**) plane view; (**b**) lateral view (A–A).

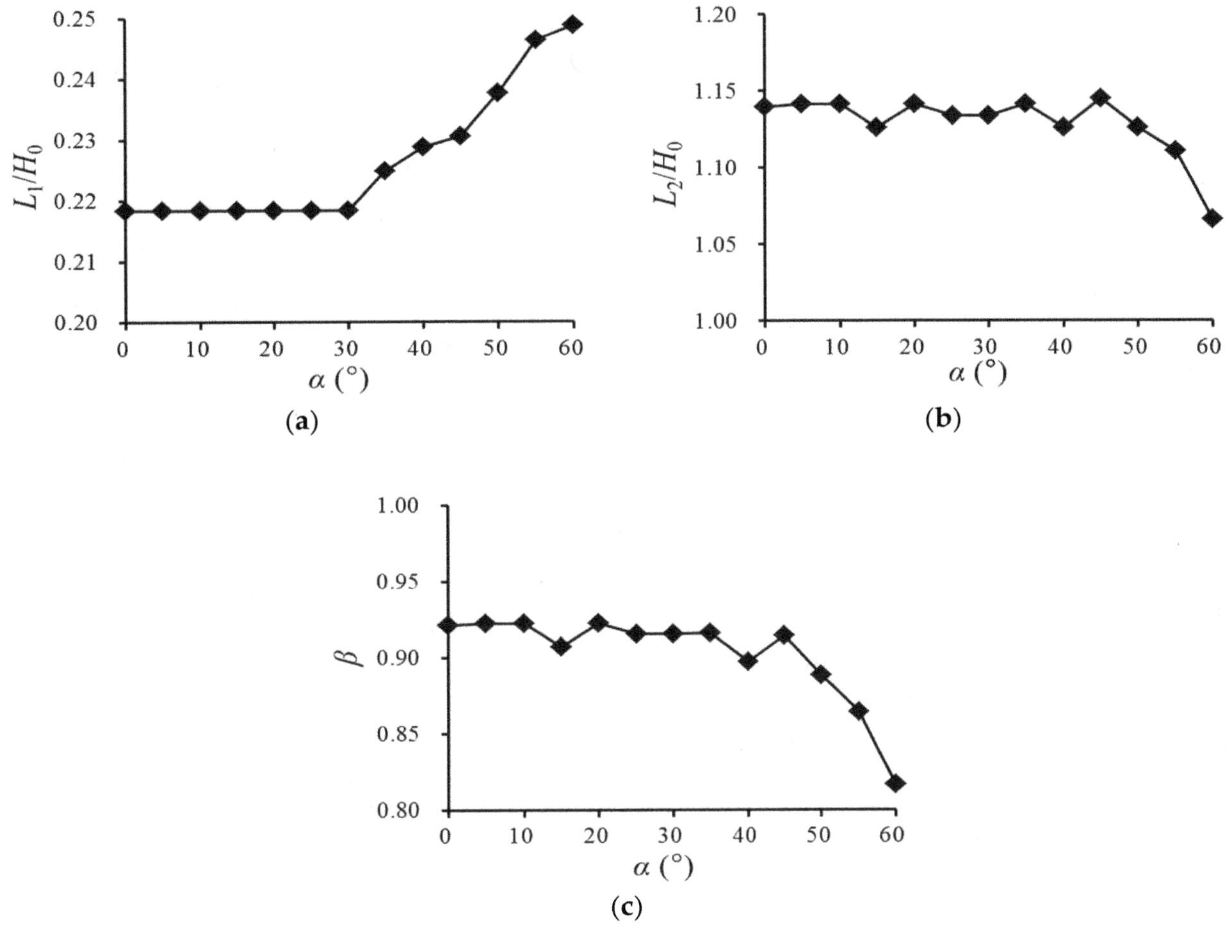

Figure 8. Effect of inclined angle α on free jet length (**a**): L_1; (**b**): L_2; (**c**): β.

4.3. Effect of Inclined Floor on Jet Impact Pressure

The pressure on the bottom floor of the plunge pool is measured experimentally in six areas, Area 1 to Area 6, based on the flow plane-shape, as shown in Figure 9. The maximum impact pressure p_m (defined as the difference between the measured and the mean static pressure in the plunge pool), scaled by the tail water depth P_0, is strongly affected by the inclined floor angle. When α = 0°~30°, the increase of p_m/P_0 is not obvious. While for α > 30°, the p_m/P_0 becomes larger with an increase of α. This is mainly due to the poor effect of the inclined floor constraint on the water flow streamline for a small α condition. Considering that the flow impinging on the anti-arch side wall results in a large amount of energy dissipation, the flow jet's impact pressure on the bottom floor of the plunge pool is relatively low. For a large α, the water flow is well coupling-constrained by both the inclined floor and the anti-arch side wall, and the streamline varies relatively smoothly when it becomes totally free from the streamwise-lateral discharge spillway. Larger α causes a lower energy dissipation in the spillway part. Consequently, the flow energy is mainly transported into the plunge pool, resulting in a high impact pressure on the bottom floor.

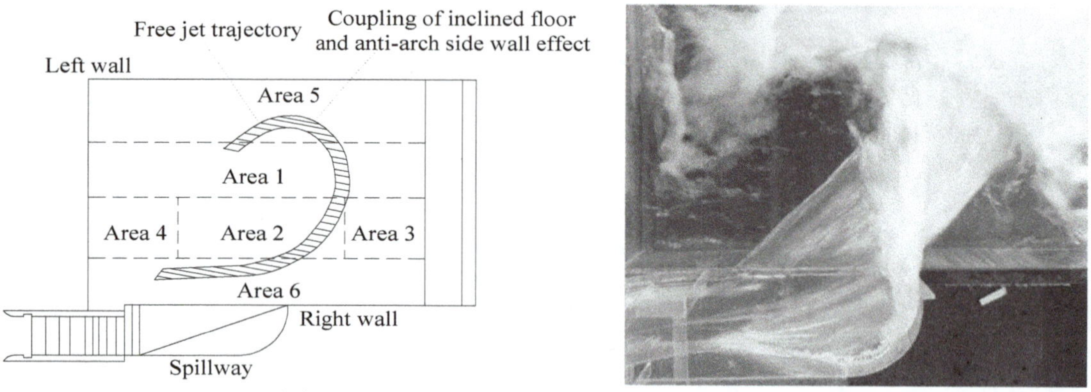

(a) (b)

Figure 9. Flow jet impact in the plunge pool: (**a**) sketch plane view; (**b**) experimental image ($\alpha = 60°$).

The specific maximum impact pressure, p_{m}, in each area of the plunge pool describes the pressure and water discharge rate distributions through the stretched flow jet. For different inclined floor angles, the pressure distribution is obviously different, owing to the non-uniform flow jet stretching, as shown in Figure 10. For $\alpha < 30°$, the pressure profiles distribute smoothly in the whole plunge pool without obvious impact pressure on the bottom floor. The flow jets mainly drop into Area 1, Area 2 and Area 3, which are located in the center of the plunge pool. When α increases to 45°, the flow dropping areas move to Area 1, Area 2 and Area 6, with an obvious increase of impact pressure. This indicates that more water discharge moves in a lateral direction, which is affected by the inclined floor. For $\alpha > 45°$, the pressure is extremely non-evenly distributed, with greater impact pressure in Area 6. When $\alpha = 50°$~$60°$, the local impact pressure is near 1.5~2.0 times greater than the mean static pressure in the plunge pool, and this will cause the bottom floor structure to be affected by negative unbalanced force conditions and may cause structural damage with flood discharge.

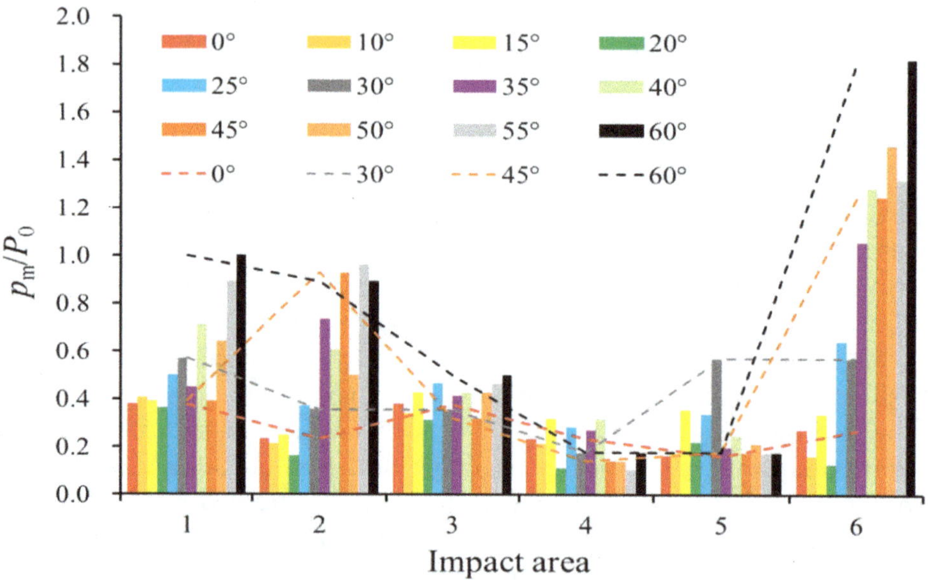

Figure 10. Pressure distributions in the plunge pool affected by the inclined floor angle α.

The ski-jump flow released from the streamwise-lateral spillway can be classified basically into three types. The classification is on the basis of the jet length and impact pressure corresponding to the effect of inclined floor and includes the following: (a) Weak effect: $\alpha < 30°$. The restraint of inclined floor on the ski-jump flow is weak and the jet is mainly affected by the anti-arch radius, which results in the unchanged of jet length (including both nearer and further jet length) and the distribution of impact pressure in the plunge pool. (b) Moderate effect: $30° < \alpha < 45°$. The restraint of the inclined floor becomes obvious and the jet diffusion is affected by both the inclined floor and anti-arch radius of side-wall. The nearer jet length increases with the distribution of impact pressure in the plunge pool is moderate. (c) Strong effect: $\alpha > 45°$. The restraint of inclined floor becomes significant and the stretch of water jet in the air becomes undeveloped. The distribution of impact pressure is extremely asymmetrical with high pressure in the region closed to the spillway.

Figure 11 shows the velocity vector distributions in the plunge pool when the jet flow drops into the pool ($\alpha = 45°$). It is clearly visible that two stable vortexes generate at the nearer and further flow dropping locations in the central part of the plunge pool. With the decrease of elevation, the vortexes' locations move slightly to the left side and the intensive velocity of the vortex attenuates (Figure 11a–c), dispersing near the pool bottom (Figure 11d). Moreover, the velocity attenuates quickly in both the X and Y directions. In terms of structural safety, there is no standing vortex penetrating the bottom floor of the plunge pool for the large flood discharge operation, although the flow pattern dropping into the pool is irregular with the velocity magnitude and direction gradient non-uniformly distributed. The average velocity of the tail-ridge area is about 3.41 m/s, indicating that the streamwise-lateral ski-jump flow is beneficial for energy dissipation due to the turbulent shearing and aeration without the risk of damage to the plunge pool structure.

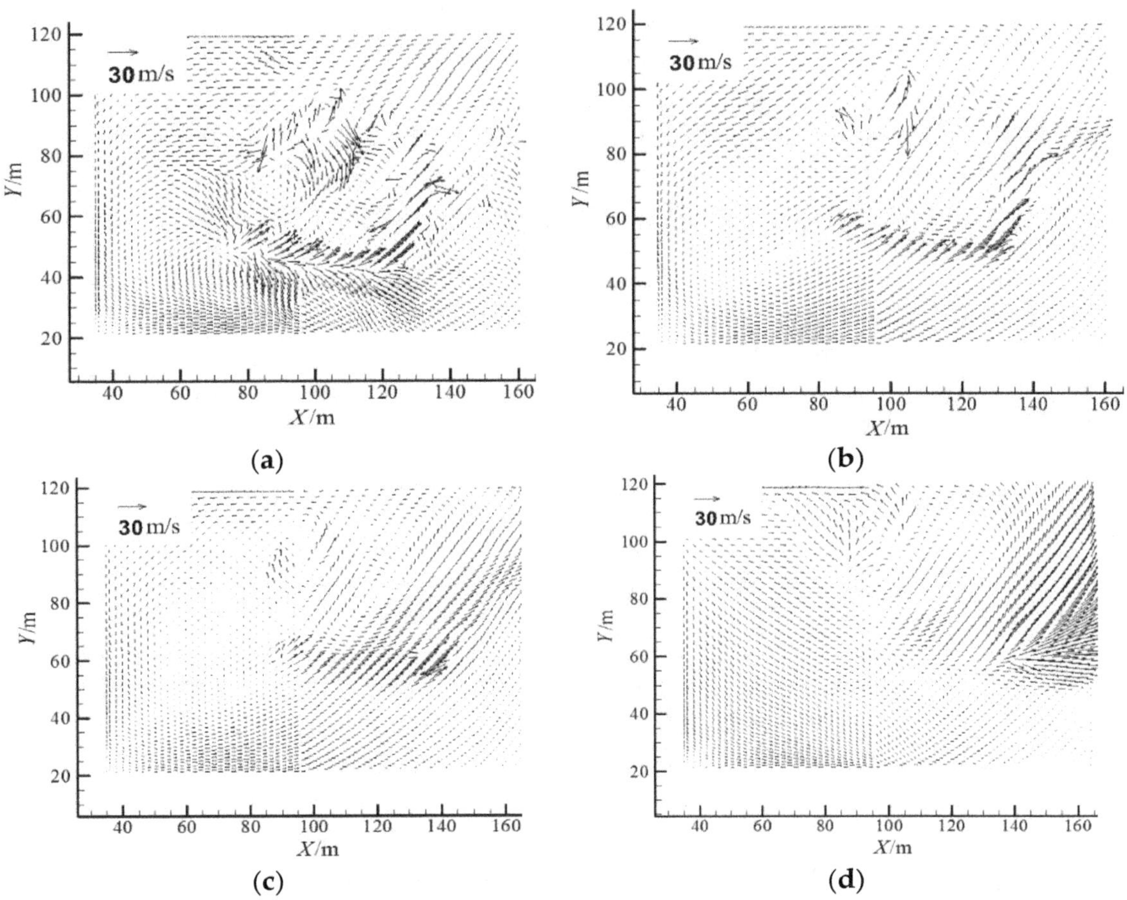

Figure 11. Velocity vector distributions in the plunge pool: (**a**) Z = −38 m; (**b**) Z = −44 m; (**c**) Z = −50 m; (**d**) Z = −58 m.

Figure 12 shows the velocity vector distributions at the cross-section (X = 100 m, the furthest lateral jet length for $\alpha = 45°$). The incident angles of nearer and further jet flow are almost the same, which are moderate to take full advantage of the water depth to dissipate the jet flow kinetic energy. The uniform distributions of velocity near the left side and the bottom wall of the plunge pool confirms that no obvious impact is caused by the ski-jump flow. The slight air–water surge is only at the dropping area and hardly influences the symmetric water flow pattern in the total plunge pool without a difference of water depth between the left and right side. This is because the thickness of the water jet dropping into the pool is thin and the aeration becomes fully developed as it diffuses into the air. Both of them will make efficient contributions to improving the energy dissipation and jet impact effect.

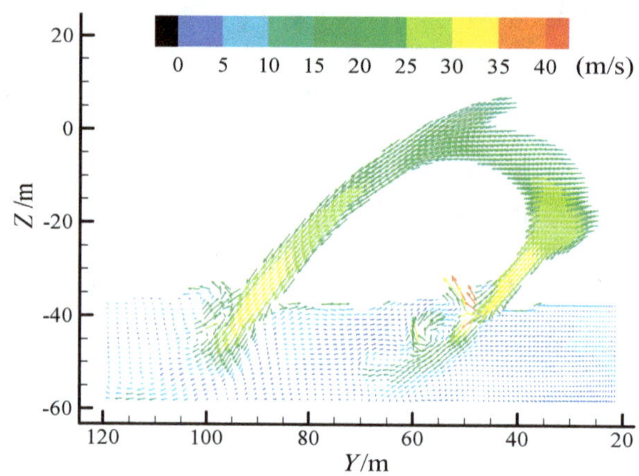

Figure 12. Lateral velocity vector profile of streamwise-lateral ski-jump flow (X = 100 m).

Based on the experimental tests and numerical simulation results, the flow characteristics of the streamwise-lateral spillway can be described as follows: (1) When the water flows on the spillway, some of the water discharge moves in a lateral direction without one side wall constraint, dropping into the plunge pool; (2) when the other water reaches the end of the spillway, a stretched ski-jump flow jet generates in the limited lateral space above the plunge pool, due to the coupling effect of the inclined floor and the anti-arch side wall; (3) when the water flow impinges into the plunge pool, submerged jet forms cause energy to dissipate through the drastic flow shearing and turbulent disturbance. The present study shows that for α = 30°~45°, the structural design of the streamwise-lateral discharge spillway performs well in controlling the ski-jump flow in terms of the high-speed and large flood discharge conditions, resulting in well jet diffusion in the air and the water discharge rate having uniform distribution.

5. Conclusions

A new type of design for streamwise-lateral ski-jump flow discharge in the spillway is introduced in this paper. In contrast to the traditional spillway, in this design, the side wall next to the plunge pool is removed, and an inclined floor is set between the horizontal bottom floor and the right-side wall with an anti-arch deflection design at the end. Lateral discharge generates along the streamwise flow discharge due to the natural pressure difference between the two sides. The free jet flow fully diffuses into the limited area, and is affected by the coupling constraint of the inclined floor and the anti-arch side wall.

The optimum free jet flow diffusion results occurred in the range of inclination angles of 30°~45°. The experimental and simulated results show the interior characteristics of the streamwise-lateral jet flow. This fully stretched free jet flow will make a beneficial contribution to the reduction in discharge per unit width dropping into the plunge pool, and an improvement in energy dissipation performance without an obvious impact on the plunge pool. Therefore, the application of the streamwise-lateral flow discharge spillway has a promising future in hydraulic engineering.

Author Contributions: This paper is a product of joint efforts of the authors who worked together on the experimental model tests. J.D. has a scientific background in applied hydraulics while W.W. and F.Z. conducted the experimental and numerical simulation investigations. J.D. and Z.T. generated the proposed effects of outlet structure on the hydraulic characteristics while J.D. and W.W. tested the methods and wrote this paper.

References

1. Xu, W.L.; Luo, S.J.; Zheng, Q.W.; Luo, J. Experimental study on pressure and aeration characteristics in stepped chute flows. *Sci. China Technol. Sci.* **2015**, *58*, 720–726. [CrossRef]
2. Chanson, H. *Energy Dissipation in Hydraulic Structures*; IAHR Monograph; CRC Press: Boca Raton, FL, USA, 2015.
3. Liu, P.Q. *Energy Dissipater Theory of Modern Dam Construction*; Science Press: Beijing, China, 2010. (In Chinese)
4. Heller, V.; Hager, W.H.; Minor, H.E. Ski jump hydraulics. *J. Hydraul. Eng.* **2005**, *131*, 347–355. [CrossRef]
5. Hager, W.H.; Boes, R.M. Hydraulic structures: A positive outlook into the future. *J. Hydraul. Res.* **2014**, *52*, 299–310. [CrossRef]
6. Chanson, H. Aeration of a free jet above a spillway. *J. Hydraul. Res.* **1991**, *29*, 655–667. [CrossRef]
7. Lenau, C.W.; Cassidy, J.J. Flow through spillway flip bucket. *J. Hydraul. Div.* **1969**, *95*, 633–648.
8. Mason, P.J.; Arumugam, K. Free jet scour below dams and flip buckets. *J. Hydraul. Eng.* **1985**, *111*, 220–235. [CrossRef]
9. Juon, R.; Hager, W.H. Flip bucket without and with deflectors. *J. Hydraul. Eng.* **2000**, *126*, 837–845. [CrossRef]
10. Erpicum, S.; Archambeau, P.; Dewals, B.; Pirotton, M. Experimental investigation of the effect of flip bucket splitters on plunge pool geometry. *Wasserwirtschaft* **2010**, *4*, 73–80.
11. Mason, P.J. Practical guidelines for the design of flip buckets and plunge pools. *Water Power Dam Constr.* **1993**, *45*, 40–45.
12. Li, S.; Liang, Z. Gravity-affected potential flows past spillway flip buckets. *J. Hydraul. Eng.* **1988**, *114*, 409–427.
13. Rajan, B.H.; Shivashankara, R.K.N. Design of trajectory buckets. *J. Irrig. Power* **1980**, *37*, 63–76.
14. Steiner, R.; Heller, V.; Hager, W.H.; Minor, H.E. Deflector Ski Jump Hydraulics. *J. Hydraul. Eng.* **2008**, *134*, 562–571. [CrossRef]
15. Alias, N.A.; Mohamed, T.A.; Ghazali, A.H.; Noor, M.J.M.M. Impact of takeoff angle of bucket type energy dissipater on scour hole. *Am. J. Appl. Sci.* **2008**, *5*, 117–121. [CrossRef]
16. Wu, J.H.; Ma, F.; Yao, L. Hydraulic characteristics of slit-type energy dissipaters. *J. Hydrodyn.* **2012**, *24*, 883–887. [CrossRef]
17. Pfister, M.; Hager, W.H. Deflector-generated jets. *J. Hydraul. Res.* **2010**, *47*, 466–475. [CrossRef]
18. Li, N.W.; Liu, C.; Deng, J.; Zhang, X.Z. Theoretical and experimental studies of the flaring gate pier on the surface spillway in a high-arch dam. *J. Hydrodyn.* **2012**, *24*, 496–505. [CrossRef]
19. Zhang, T.; Chen, H.; Xu, W.L. Allotypic hybrid type flip bucket. II: Effect of contraction ratio on hydraulic characteristics and local scour. *J. Hydroelec. Eng.* **2013**, *32*, 140–146. (In Chinese)
20. Felder, S.; Chanson, H. Scale effects in microscopic air-water flow properties in high-velocity free-surface flows. *Exp. Therm. Fluid Sci.* **2017**, *83*, 19–36. [CrossRef]
21. Xu, W.L.; Liao, H.S.; Yang, Y.Q.; Wu, C.G. Computational and Experimental Investigation on the 3-D Flow Feature and Energy Dissipation Characteristics of Plunge Pools. *Chin. J. Theor. Appl. Mech.* **1998**, *30*, 35–42.
22. Deng, J.; Xu, W.L.; Zhang, J.M.; Qu, J.X.; Yang, Y.Q. A new type of plunge pool—Multi-horizontal submerged jets. *Sci. China Technol. Sci.* **2008**, *51*, 2128–2141. [CrossRef]
23. Deng, J.; Yang, Z.L.; Tian, Z.; Zhang, F.X.; Wei, W.R.; You, X.; Xu, W.L. A new type of leak-floor flip bucket. *Sci. China Technol. Sci.* **2016**, *59*, 565–572. [CrossRef]
24. Wei, W.R.; Deng, J.; Liu, B. Influence of aeration and initial water thickness on axial velocity attenuation of jet flows. *J. Zhejiang Univ. Sci. A* **2013**, *14*, 362–370. [CrossRef]
25. Yu, T.; Deng, J.; Xia, Y.; He, C.L.; Xu, W.L. Numerical simulation of turbulent trajectory flow with turnover nappe. *Water Resour. Hydropower Eng.* **2005**, *36*, 37–39. (In Chinese)

14

A Comprehensive Method of Calculating Maximum Bridge Scour Depth

Rupayan Saha [1], Seung Oh Lee [2] and Seung Ho Hong [1,*]

[1] Department of Civil and Environmental Engineering, West Virginia University, 1306 Evansdale Drive, Morgantown, WV 26506, USA; rs0002@mix.wvu.edu

[2] Department of Civil Engineering, Hongik University, 94 Wausan-ro, Mapo-gu, Seoul 04066, Korea; seungoh.lee@hongik.ac.kr

* Correspondence: sehong@mail.wvu.edu.

Abstract: Recently, the issues of scour around a bridge have become prominent because of the recurrent occurrence of extreme weather events. Thus, a bridge must be designed with the appropriate protection measures to prevent failure due to scour for the high flows to which it may be subjected during such extreme weather events. However, the current scour depth estimation by several recommended equations shows inaccurate results in high flow. One possible reason is that the current scour equations are based on experiments using free-surface flow even though extreme flood events can cause bridge overtopping flow in combination with submerged orifice flow. Another possible reason is that the current practice for the maximum scour depth ignores the interaction between different types of scour, local and contraction scour, when in fact these processes occur simultaneously. In this paper, laboratory experiments were carried out in a compound shape channel using a scaled down bridge model under different flow conditions (free, submerged orifice, and overtopping flow). Based on the findings from laboratory experiments coupled with widely used empirical scour estimation methods, a comprehensive way of predicting maximum scour depth is suggested which overcomes the problem regarding separate estimation of different scour depths and the interaction of different scour components. Furthermore, the effect of the existence of a pier bent (located close to the abutment) on the maximum scour depth was also investigated during the analysis. The results show that the location of maximum scour depth is independent of the presence of the pier bent but the amount of the maximum scour depth is relatively higher due to the discharge redistribution when the pier bent is absent rather than present.

Keywords: bridge scour; sediment transport; submerged flow; physical hydraulic modeling

---

1. Introduction

When a bridge is constructed in a river, the flow pattern around the bridge changes because a unique flow field develops locally around the bridge pier and abutment. Furthermore, reduced flow area through the bridge opening by the existence of embankment/abutments at both and/or one side of the river results in higher flow velocity due to the acceleration. This unique flow field with the higher velocity can seriously damage bridge foundations. Thus, if the depth of the foundation is not deep enough, the chance of bridge failure becomes higher.

A bridge can fail because of several causes, such as earthquake, wind, and flooding. Among them, bridge scour is the biggest reason of bridge failure [1,2]. For example, about 60% of bridge failures of total bridge collapse in the United States since 1950 have been related to the scour of bridge foundations [3]. The Colorado Department of Transportation (CDOT) estimated a minimum of 30 state highway bridges were destroyed and twenty were seriously damaged by floods in the year of 2013 [4]. In Nepal, due to the degradation of bed materials during the 2014 flooding, the foundation of the highway bridge over the Tinau River was seriously exposed [5]. As explained in the above examples, it is justified to say that bridge scour is one of the main bridge safety problems all over the world. Thus, accurate prediction of scour at the bridge foundation becomes the primary aim of engineers for the safety of a bridge.

Since the late 1950s, numerous studies on scour around bridges have been conducted and have generated formulas for equilibrium scour depth estimation at bridge foundations [6–9]. Although using equilibrium scour depth around a bridge foundation is a reasonable practice to design a bridge, sometimes it shows an overly conservative design compared to the field measurements. Contrary to the conservative design, a study in South Carolina found that the observed scour depth was greater than the equilibrium scour depth based on using 100-year flooding even if there had not been a 100-year flooding since the bridge was built [10,11]. Repeated occurrences of smaller flooding events might cause scour that was greater than the scour estimation in South Carolina. Moreover, some researchers have used computational fluid dynamics (CFD) simulations and found that the scour depth was under-predicted by the steady-state calculations while it was over-predicted by the unsteady-state [12,13].

One of the possible reasons of lacking an accurate scour prediction method is that many of the scour predictor equations were derived from simplified laboratory studies under free flow cases. The extreme amount of water associated with huge flood events can result in a complex flow field around the bridge under bridge overtopping flow in combination with submerged orifice flow. In addition to the complex flow field, the irregular shape of river geometry as well as non-uniform sediment distribution cannot be reproduced in a simplified-idealized laboratory setting such as in rectangular flume as in the previous research. Also, most of the scour predicting equations are 2nd or 3rd order which may not accurately predict scour depth as the scouring process is a complicated scholastic phenomenon.

Another possible reason for the inaccurate prediction of scour depth is that the current practice of total scour depth assumes that contraction and local scour are independent processes. Contraction scour is the consequence of flow acceleration due to contraction in the flow area, while local scour is caused by local vortex structures around the base of the obstruction. However, when the bridges are constructed in the river, these two-flow patterns tend to occur concurrently, which make local scour and contraction scour time dependent [14]. Thus, to predict maximum scour depth around a bridge foundation, a single equation should be developed rather than two separate equations for different types of scour.

Hence, the main objective of this study is to develop a single equation that can be used to predict maximum scour depth where different types of scour occur simultaneously. A scale down laboratory experiment was conducted to find the interaction between different scour components and the result was applied to and compared with the most widely used scour equations in the US (Colorado State University (CSU) and Melville-Sheppard (M/S) equations) to suggest an improved way of calculating the maximum scour depth. These equations predict maximum pier scour depths. Basic applications include simple pier substructure configurations and riverine flow situations in alluvial sand-bed channels. The CSU equation has been used for bridge scour evaluations and bridge design for countless bridges in the U.S. and worldwide. The M/S equation combines pier geometry, shape, and angle of attack to compute an effective pier width, a^*, and also distinguishes between clear-water and live-bed flow conditions.

2. Materials and Methods

2.1. Physical River Modeling

As shown in Figure 1, a 1:60 scale down physical model was constructed in the hydraulics laboratory including full river bathymetry of the Towaliga River bridge at Macon, Georgia by using Froude number similarities between the model and the prototype [15–17]. A relatively small bridge was selected in this experiment because a large number of smaller bridges can fail during extreme hydrologic events. The drainage area of the selected site at the bridge is 816 km^2. Discharge was estimated as 1700 m^3/s by the U.S. Geological Survey for Tropical Storm Alberto in 1994 which is larger than 500 year flooding. During this historic event, severe overtopping of the bridge and significant scour around the pier bents in the left floodplain occurred, but the embankment and bridge remained sufficiently intact until repairs could be made to the scoured area. The actual scour depth measured during the storm events can be found in Hong and Sturm [16,17]. As shown in Figure 1, each of the pier bents consists of two in-line rectangular columns having a width of 1 m, and they were modeled inclusive of pile caps and piles. A bridge deck width of 13 m, in accordance with standard two-lane roads was also constructed in the laboratory.

Figure 1. Towaliga River bridge in the field and model in the laboratory.

The approach section of the bridge was 7.3 m long followed by a 6.09 m long moveable working bed section. The approach section was filled with uniform size of small gravel (d_{50} = 3.3 mm) to make a fully rough turbulent approach flow. In the moveable working bed section, the full depth was filled with sand with median diameter size (d_{50}) of 0.53 mm. The bridge model was constructed within the moveable working bed section for the scour experiment.

For selecting sediment size in the laboratory study, recently developed scour modeling methodology was applied [2,18–20]. Selecting sediment size in the laboratory was mostly related to the ratio of pier size to sediment size (a/d_{50}). In the field, the effect of sediment size, d_{50}, has not been considered to be important to pier scour because of very large values of a/d_{50} [21]. However, in the laboratory, pier scour depth to pier size ratio (d_{ps}/a) tends to increase with a/d_{50} up to a maximum at $a/d_{50} \approx 25$ and seemingly becomes independent of the ratio when a/d_{50} is greater than 50 [19]. However, another researcher has suggested that d_{ps}/a may decrease at very large values of a/d_{50} based on experiments in a large flume [22]. Therefore, the sediment size, d_{50}, was chosen such that the ratio of pier size to sediment size, a/d_{50} was in the range of 25–50 where it has negligible influence on pier scour. Furthermore, the ratio of the approach velocity to the critical velocity which concludes the condition of clear water scour regime was also an important factor to choose sediment size in the experiment. Considering all the above conditions, the sediment size for this experiment was chosen as d_{50} = 0.53 mm with σ_g = 1.2, where σ_g is geometric standard deviation of the particle size distribution.

Within the working moveable bed section, the erodible embankment/abutment was constructed at both sides of the channels by using erodible fill with rock-riprap protection in order to reproduce the same conditions as in the field [16,23]. Furthermore, the existing roadway and bridge deck were modeled based on the scale-ratio and put on top of the embankments. With this approach, submerged orifice flow with and without overtopping during the extreme amount of flooding as observed in the field could be simulated in the laboratory. As shown in Figure 2, eight pier bents were also constructed through the bridge to support the bridge deck and for the local scour experiment. Each of the pier bents consisted of two in-line rectangular columns. Abutment structures were constructed and buried at both sides of the erodible embankments.

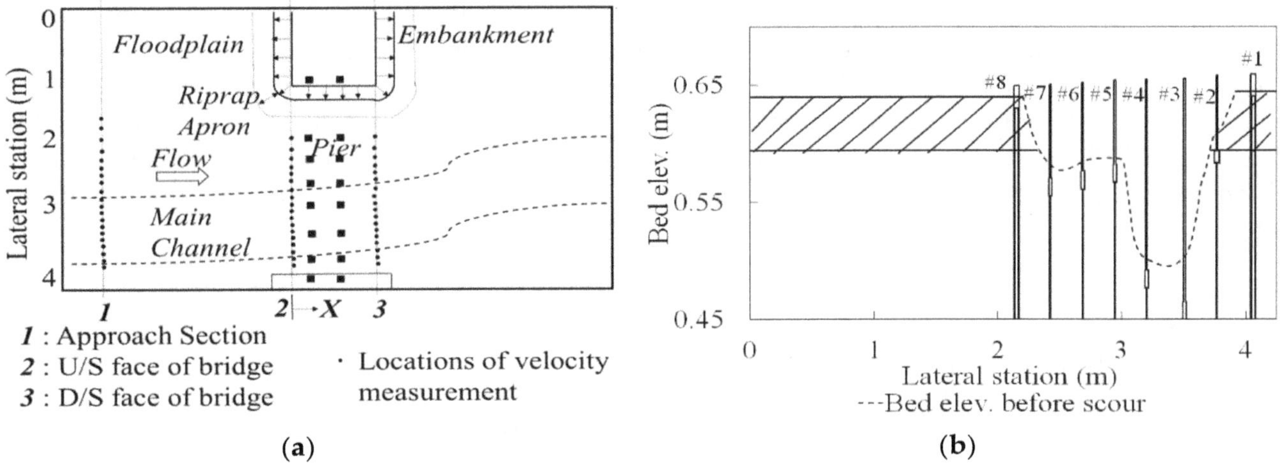

Figure 2. Geometry of compound channel for (**a**) plan view with velocity measurement locations; (**b**) cross section view at bridge.

2.2. Experimental Procedure

After completion of the model structure, the flume was slowly filled with water from a downstream supply hose to saturate the sand without disturbing the initial bottom contours. Then bottom elevations were measured in detail throughout the entire working section using an Acoustic Doppler Velocimeter (ADV). After that, a larger flow depth than the required value was set by the tailgate, then discharge was increased slowly to prevent initial scour while setting up the test discharge. Then the tailgate was lowered to achieve the desired depth of flow. In the meantime, a point gauge was used to measure the flow depth to measure the targeted water depth. Once the desired flow rate and water depth were achieved, scouring was continued for 5 to 6 days until equilibrium (change in scour depth less than 2% within 24 h) was reached. After reaching equilibrium condition, the entire bed elevation (bed elev.) was measured by a point gauge and the ADV in detail to obtain accurate contours after scour. After finishing the moveable bed experiment, the surface of the movable bed was fixed by spraying polyurethane. In the fixed bed conditions, the velocities were measured by ADV in the approach section and bridge upstream (U/S) and downstream (D/S) section. During the velocity measurements, typical correlation values in the experiments were greater than 90% and the Signal Noise Ratio (SNR) was greater than 15. The sampling frequency of the ADV was chosen to be 25 Hz with a sampling duration of 2 min at each measuring location. More detailed measurements techniques using ADV can be found in several other articles [24–29].

2.3. Assessment of Reference Scour Depth

Two well established theoretical pier scour equations were used to decide the reference scour depth at the pier for this research. One of the most commonly used pier scour equation in the United State is the CSU equation (also known as the HEC-18 equation). The CSU equation includes a correction factor for pier shape, angle of attack of flow, and the bed conditions. The CSU equation was initially developed from a laboratory data set measured by several researchers [6,30,31]. After the initial development, the CSU equation has been progressively modified over the years and is currently recommended by the Federal Highway Administration (FHWA) for estimating equilibrium scour depths at simple piers as follow [6]:

$$\frac{d_{csu}}{Y_2} = 2K_1K_2K_3\left(\frac{a}{Y_2}\right)^{0.65} Fr_2^{0.43} \tag{1}$$

where, K_1, K_2, K_3 are the correction factors for pier nose shapers, for angle of attack of flow, and for bed conditions, respectively; Fr_2 = Froude number at the pier face = $V_2/(gY_2)^{0.5}$; a = Pier width; d_{csu} = Equilibrium pier scour depth; Y_2 = Flow depth directly upstream of the pier; V_2 = Mean velocity of flow directly upstream of the pier; g = Acceleration of gravity. Several evaluations of the CSU equation through laboratory and field datasets have shown this equation may perform better than other bridge predictive equations [32–34]

Another commonly used pier scour prediction method is the Melville–Sheppard or M/S equation [6,34,35]. A new variable is introduced in the M/S equation, effective pier width, a^*, which shows the combined effect of pier geometry, shape, and angle of attack. The equation is as follows:

$$\frac{d_{ms}}{a^*} = 2.5 f_1 f_2 f_3 \tag{2}$$

where, f_1, f_2, f_3 are the factors for flow-structure interactions, for flow-sediment interactions, and for sediment-structure interactions, respectively. Here,

$$f_1 = \tanh\left[\left(\frac{Y_2}{a^*}\right)^{0.4}\right] \tag{3}$$

$$f_2 = \left[1 - 1.2\left[\ln\left(\frac{V_2}{V_c}\right)\right]^2\right] \tag{4}$$

$$f_3 = \left[\frac{\left(\frac{a^*}{d_{50}}\right)^{1.13}}{10.6 + 0.4\left(\frac{a^*}{d_{50}}\right)^{1.33}}\right] \tag{5}$$

where, V_c is the critical velocity for movement of d_{50}, calculated by Keulegan's equation.

Using the provided theoretical pier scour equation, pier scour depth was predicted with the variables measured in the experiments. Because theoretical equations were derived from a simple laboratory set-up in a rectangular flume, the result can be used as a reference pier scour depth without any effect of flow contraction. Then, the reference pier scour depth was compared with the measured maximum scour depth around the pier where the flow contraction and local vortex structure interact and lead to the maximum scour depth. The results showed the effect of flow contraction on the pier scour and were used to suggest precise estimation of pier scour depth [8].

3. Results and Discussion

A total of eight experiments were conducted for this research and the experimental conditions are summarized in Table 1 where Q is the total discharge and q_2/q_1 is the discharge contraction ratio which can be used as a key-independent variable that accounts for flow redistribution and resulting flow acceleration through a bridge section. V_1, Y_1, and V_2, Y_2 are cross-sectional mean velocity, and depth of floodplain in the approach and at the upstream face of the bridge section, respectively. Also, maximum scour depth expressed as flow depth measured at the location of maximum scour (Y_m) and longitudinal (flow direction) distance measured from the upstream face of the bridge to the location of maximum scour depth (X) (see in Figure 2a) are presented in Table 1. A better idea of the scour distribution at the bridge cross section in these experiments can be seen in Figure 6 as a reference. Run 9 and 10 were selected from the previous study [25] and used to validate the results in this research so that the suggested method can be used for different hydraulic and geometry conditions. As shown in Table 1, runs 1, 4, 5, and 6 were performed in free (F) flow condition; whereas, the remaining runs were conducted in the submerged flow conditions with overtopping (OT) and without overtopping (SO).

Close to the toe of the abutment where the abutment side slope ends and the first pier is introduced, these locations are the most vulnerable to scour because abutment scour, pier scour, and contraction scour occur simultaneously. Thus, to find the effect of pier on the maximum scour depth, pier bent #7 was removed from the river model and experiments were conducted in runs 7 and 8 with the exact same flow conditions as in runs 2 and 3. Even if the velocities and scour depths were measured in the entire working moveable bed section, only the floodplain flow variables and scour depths are presented in this paper because the maximum scour depth occurred on the floodplain in all our cases.

For the velocity measurements at each cross-section, point velocities were measured along multiple vertical transects through the entire cross-section, and at each vertical transect. Point velocities were measured at minimum of four points and maximum ten points vertically depending on the depth of water. Then, the depth-averaged velocity was determined by the best fit of the logarithmic velocity profile in each vertical transect. Based on the depth-averaged velocity, the cross-sectional mean velocity of the floodplain was calculated for the approach and bridge section as presented in Table 1.

Table 1. Experimental conditions and location of maximum scour depth measurement.

Run	Flow Type	Q (m^3/s)	q_2/q_1	V_1 (m/s)	Y_1 (m)	V_2 (m/s)	Y_2 (m)	X (m)	Y_m (m)
1	F	0.029	1.45	0.082	0.125	0.134	0.110	0.122	0.174
2	SO	0.038	1.53	0.082	0.122	0.177	0.088	0.094	0.201
3	OT	0.053	1.37	0.076	0.149	0.183	0.088	0.076	0.219
4	F	0.046	1.46	0.085	0.146	0.140	0.131	0.049	0.207
5	F	0.053	1.71	0.076	0.149	0.146	0.137	0.049	0.210
6	F	0.038	1.58	0.082	0.122	0.146	0.110	0.049	0.186
7	SO	0.038	1.53	0.082	0.122	0.177	0.088	0.186	0.202
8	OT	0.053	1.37	0.076	0.149	0.183	0.088	0.094	0.229
9	F	0.105	1.71	0.165	0.073	0.302	0.070	-	0.180
10	OT	0.198	1.05	0.219	0.149	0.390	0.088	-	0.274

Notes: F = free flow, SO = Submerged orifice flow, OT = Overtopping.

3.1. *Measurement of the Maximum Scour Depths*

Initial bottom elevations were measured throughout the test (moveable bed) section before the experiment and then the final bottom elevations were measured at the same locations in equilibrium conditions. Then, to find the location and magnitude of the maximum scour depth, bed elevations after and before the scouring were compared. The maximum scour depth in all of the laboratory experimental runs was found within the bridge and close to the bridge pier (Table 1).

3.2. *Prediction of Maximum Scour Depth*

The main assumption in this research is that maximum scour depth consists of theoretical pier scour depth and additional scour due to flow contraction. Based on this assumption, the maximum scour depth can be calculated by the equation,

$$\text{Max. Scour depth} = \text{Theoretical pier scour} + \text{Additional scour by flow contraction} \tag{6}$$

The schematic diagram is given in Figure 3 to define the variables used for the analysis. The theoretical pier scour depth can be decided using the CSU (d_{csu}) or M/S equation (d_{ms}) with measured flow variables and the results are shown in Table 2. Then, the "Additional scour by flow contraction" can be determined by measured the water depth at the deepest point and theoretical pier scour depth as follows;

$$Y_{m\text{-}csu} = Y_m - d_{csu} \tag{7}$$

$$Y_{m\text{-}ms} = Y_m - d_{ms} \tag{8}$$

where, $Y_{m\text{-}csu}$ and $Y_{m\text{-}ms}$ are the calculated water depths at the deepest location subtracted from the total water depth at the point by the theoretical pier scour depth of CSU (d_{csu}) or M/S equation (d_{ms}) respectively, which stands for the resulting water depth due to the additional scour. The variables required for calculation of the reference pier scour depth using the CSU and M/S equations are listed in Tables 1 and 2. Here, θ is the flow angle of attack at the pier face which is required to estimate correction factor, K_2 and effective pier width, a^* for the CSU and M/S equations, respectively. The additional scour components are also shown in Table 2 in terms of non-dimensional variables.

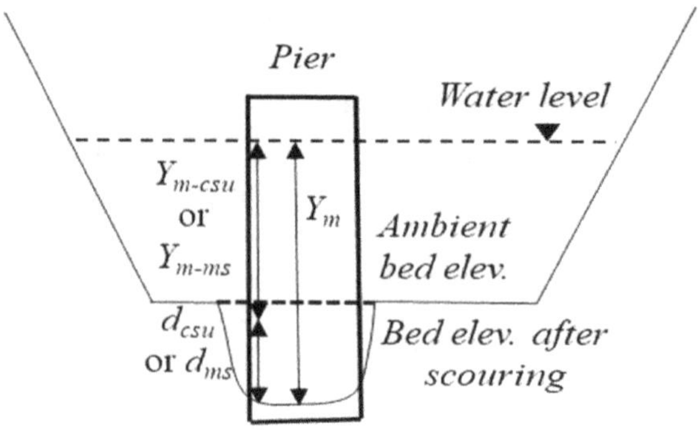

Figure 3. Schematic diagram for calculation of maximum scour depth.

Table 2. Summary of experimental results to calculate maximum scour depth.

Run	Fr_2	θ	a^* (m)	V_2/V_c	d_{csu} (m)	d_{ms} (m)	$\frac{Y_{m-ms}}{Y_1}$	$\frac{Y_{m-csu}}{Y_1}$
1	0.13	20	0.009	0.46	0.029	0.006	1.34	1.15
2	0.19	12	0.009	0.63	0.027	0.014	1.51	1.41
3	0.20	8	0.008	0.64	0.024	0.014	1.36	1.29
4	0.12	15	0.009	0.46	0.027	0.006	1.38	1.23
5	0.13	15	0.009	0.49	0.028	0.007	1.35	1.21
6	0.14	22	0.009	0.50	0.031	0.009	1.44	1.27
7	0.19	12	0.009	0.63	0.027	0.014	1.51	1.41
8	0.20	8	0.008	0.64	0.024	0.014	1.42	1.35
9	0.36	15	0.045	0.66	0.083	0.075	1.40	1.30
10	0.42	15	0.045	0.68	0.095	0.081	1.29	1.19

If the assumption is correct, the effect of additional scour due to flow contraction can be represented by the values of $Y_{m\text{-}csu}$ and $Y_{m\text{-}ms}$. Thus, the non-dimensional value of additional scour components calculated using the CSU and M/S equation are compared with the measured discharge contraction ratio in Figure 4a,b respectively. As the value of flow contraction ratio increases, the normalized value of additional scour depth gradually increases. The results clearly reveal that the effect of flow contraction on the additional scour term becomes higher as the value of q_2/q_1 increases. Runs 9 and 10 were also plotted in Figure 4 to validate the findings in different hydraulic and geometry conditions and the result favors the validation as it shows a similar trend of runs 9 and 10 with pressure and free flow data respectively. For both cases using the CSU equation and M/S equation, the pressure flow line has a steeper slope than for the free flow cases and lies above. Because of the vertical flow contraction in addition to the existing lateral flow contraction in pressure flow, the resulting additional scour depth due to the flow contraction is higher than in the free flow cases. Currently, additional experiments are being conducted. With more experimental conditions as well as the data set in Figure 4, the best fit equation will be provided for calculation of scour due to flow contraction under free and pressure flow by à least square regression analysis

It is interesting to note that, the regression analysis results using only with the data set in Figure 4 show higher values of exponents of q_2/q_1 in the case with the M/S pier scour equation for both of the pressure and free flow cases. This finding illustrates that the M/S equation shows a smaller value of pier scour depth in the same flow conditions compared to the CSU equation. The ratios between the theoretical pier scour depths calculated using CSU (d_{csu}) and M/S (d_{ms}) are plotted with the value of flow intensity factor (V_2/V_c) in Figure 5. As shown in Figure 5, the M/S equation shows almost 50% to

75% underestimated values compared to the CSU results. The CSU pier scour equation was developed for live-bed scour and thus ignores the effect of different values of flow intensity and assumes the value is equal to 1; but in the M/S equation, V_2/V_c is considered as an individual factor to account for the effect of clear water conditions and varies from 0.3 to 1. So, it is obvious that the CSU pier scour equation shows a larger value when applied to the clear water condition. However, there is another important difference between the CSU and M/S equations, which is the consideration of bed material size. However, in our case, the effect of sediment size cannot be considered because all of the experiments are conducted with same sediment size.

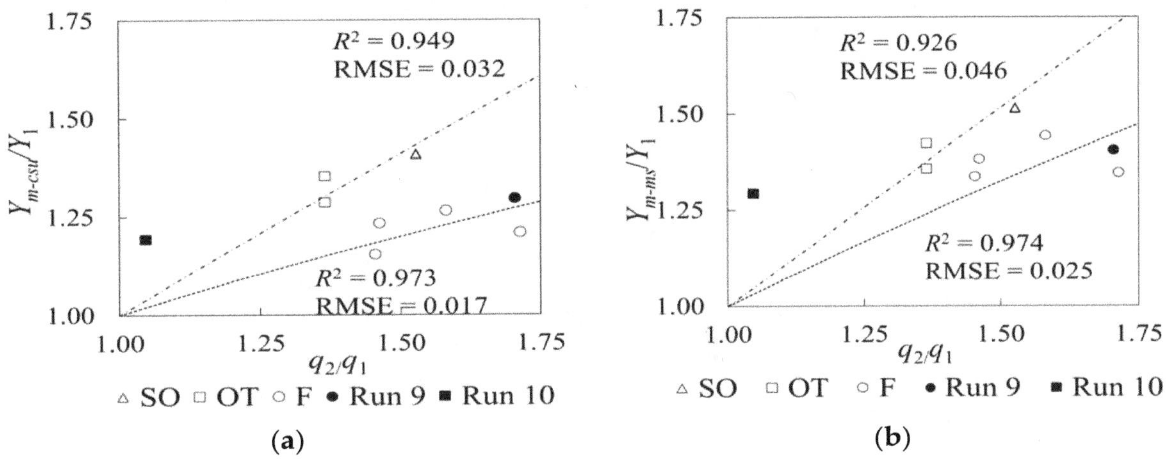

Figure 4. Effect of flow contraction on additional scour components using (**a**) Colorado State University (CSU) and (**b**) Melville-Sheppard (M/S) equations.

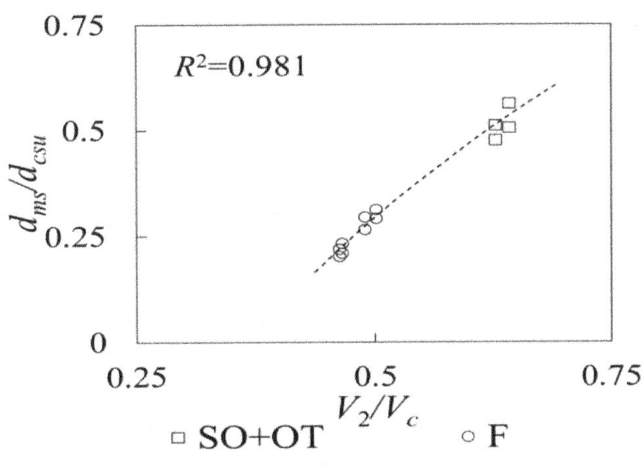

Figure 5. Comparison of CSU and M/S pier scour depth in terms of flow intensity.

Finally, a comprehensive procedure for predicting the maximum scour depth with respect to bridge design is introduced:

1. Collecting field geometry data including sediment size;
2. Calculate the flow variables using software or hydraulic laboratory modeling;
3. Compute the theoretical pier scour depth from established pier scour equations (CSU or M/S equations);
4. The additional scour depth due to the effect of flow contraction can be estimated with the equations developed by regression analysis with the data in Figure 4 as well as more laboratory/field data sets;

5. Adding the results from steps 3 and 4 to predict the maximum scour depth for bridge design.

In addition to suggest a comprehensive procedure, the effect of pier bent (located near to the abutment) on maximum scour depth was investigated qualitatively. As shown in Table 1, runs 7 and 8 were conducted with the exact same flow condition as in runs 2 and 3 respectively, but removing pier bent #7 in the bridge section. Figure 6 shows the cross-section comparisons between runs 3 and 8 after scouring. The maximum scour depth occurred at pier bent #6 for both runs 3 and 8. So, the location was independent of the presence of the closest pier bent (#7). However, the amount of maximum scour depth was slightly higher in the case of run 8 where pier bent #7 was absent. A similar trend was observed for runs 2 and 8. Discharge redistribution over the time and interaction between each pier can be the result for this difference of maximum scour depth and needs further study.

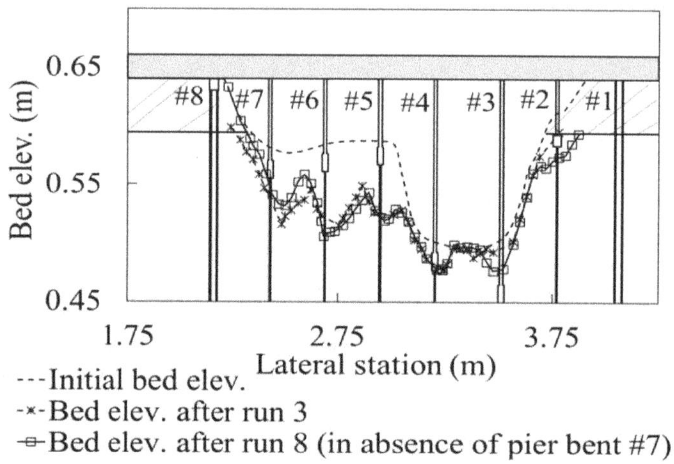

Figure 6. Comparison of cross-sections for runs 3 and 8.

4. Conclusions

Many investigations have been made attempting to estimate the maximum scour depth and to understand the mechanism of scour around bridge piers. Most of the previous investigations were based on lab experiments using a rectangular channel under free flow. However, due to the recent extreme rainfall events, submerged orifice flow and overtopping flow occur at the bridge frequently, where the flow field around the bridge substructure is more complex than in free flow because of vertical flow contraction in addition to existing lateral flow contraction. Furthermore, most of the natural channel shapes are not rectangular. Also, the current guidelines recommended by HEC-18 assumed that contraction and local scour processes are independent and so they can be determined separately and summed to estimate total scour depth. However, during large flooding events, local scour and contraction scour occur simultaneously and a separate calculation of local scour and contraction scour results in inaccurate scour depth. To overcome the weak points that the current methodology has, laboratory experiments were carried out in a scaled down physical model and a single equation was developed to predict maximum scour depth which can be used without separate calculation of different types of scour components in pressure flow as well as in free flow cases.

Based on the basic assumption of maximum scour depth as summation of theoretical pier scour depth and additional scour depth due to flow contraction, a comprehensive way of predicting maximum scour depth in clear water conditions was suggested. Furthermore, the results demonstrate that the contraction effect on maximum scour depth increases as the flow contraction ratio increases. Also, due to additional vertical contraction in the pressure flow case, the effect of flow contraction on the maximum scour depth shows up larger than in free flow. Another outcome from our investigation concludes that the location of the maximum scour depth is independent of the existence of the closest pier bent but the absence of the closest pier bent increases the scour depth.

Even though this study suggested an improved method for the scour depth prediction in clear water conditions, a well-designed physical model is recommended to investigate the scour characteristics under live bed conditions. In addition, different sediment sizes and non-uniform size sediment should be incorporated in the future research, as natural rivers generally consist of non-uniform sediment. Finally, scour modelling using computational fluid dynamics (CFD) should also be explored to refine the proposed equation for design purposes.

Author Contributions: S.H.H. and S.O.L. provided background data and motivation. S.H.H. and S.O.L. designed the experiments and determined the results; R.S. analyzed the data; R.S. and S.H.H. contributed to motivation, data analysis input, and interpretation of results; R.S. and S.H.H. interpreted the results and wrote the paper; and all authors participated in final review and editing of the paper.

Acknowledgments: The laboratory data used in this paper were measured in Georgia Tech and the permission was granted by Seung Ho Hong and Seung Oh Lee.

References

1. Kattell, J.; Eriksson, M. *Bridge Scour Evaluation: Screening, Analysis, and Countermeasures*; General Technical Reports 9877 1207-SDTDC; U.S. Department of Agriculture: San Dimas, CA, USA, 1998.
2. Melville, B.W.; Coleman, S.E. *Bridge Scour*; Water Resources Publications, LLC: Highlands Ranch, CO, USA, 2000.
3. Shirhole, A.M.; Holt, R.C. *Planning for A Comprehensive Bridge Safety Program*; Transportation Research Record No. 1290; Transportation Research Board, National Research Council: Washington, DC, USA, 1991.
4. Novey, M. Cdot Assessing 'Millions and Millions' in Road Bridge Damage. Available online: www.coloradoan.com (accessed on 15 September 2013).
5. Shrestha, C.K. Bridge Pier Flow Interaction and Its Effect on the Process of Scouring. Ph.D. Thesis, University of Technology Sydney (UTS), Ultimo, Australia, 2015.
6. Arneson, L.A.; Zevenbergen, L.W.; Lagasse, P.F.; Clopper, P.E. *Evaluating Scour at Bridges*, 15th ed.; FHWA-HIF-12-003, HEC-18; Department of Transportation, Federal Highway Administration: Washington, DC, USA, 2012.
7. Lee, S.O. Physical Modeling of Local Scour Around Complex Bridge Piers. Ph.D. Thesis, School of Civil and Environmental Engineering, Georgia Institute of Technology, School of Civil and Environmental Engineering, Georgia Institute of Technology, Atlanta, GA, USA, 2006.
8. Sheppard, D.; Melville, B.; Demir, H. Evaluation of Existing Equations for Local Scour at Bridge Piers. *J. Hydraul. Eng.* **2014**, *140*, 14–23. [CrossRef]
9. Sturm, T.W.; Ettema, R.; Melville, B.M. *Evaluation of Bridge-Scour Research: Abutment and Contraction Scour Processes and Prediction*; NCHRP 24-27; National Co-operative Highway Research Program: Washington, DC, USA, 2011.
10. Melville, B.; Chiew, Y. Time Scale for Local Scour at Bridge Piers. *J. Hydraul. Eng.* **1999**, *125*, 59–65. [CrossRef]
11. Shatanawi, K.M.; Aziz, N.M.; Khan, A.A. Frequency of discharge causing abutment scour in South Carolina. *J. Hydraul. Eng.* **2008**, *134*, 1507–1512. [CrossRef]
12. Alemi, M.; Maia, R. Numerical Simulation of the Flow and Local Scour Process around Single and Complex Bridge Piers. *IJCE* **2018**, *16*, 475. [CrossRef]
13. Sajjadi, S.A.H.; Sajjadi, S.H.; Sarkardeh, H. Accuracy of numerical simulation in asymmetric compound channels. *IJCE* **2018**, *16*, 155. [CrossRef]
14. Hong, S. Interaction of Bridge Contraction Scour and Pier Scour in a Laboratory River Model. Master's Thesis, School of Civil and Environmental Engineering, Georgia Institute of Technology, Atlanta, GA, USA, 2005.
15. Hong, S.; Lee, S.O. Insight of Bridge Scour during Extreme Hydrologic Events by Laboratory Model Studies. *KSCE J. Civ. Eng.* **2017**, *22*, 1–9. [CrossRef]
16. Hong, S.; Sturm, T.W. Physical modeling of abutment scour for overtopping, submerged orifice, and free surface flows. In Proceedings of the 5th Conference on Scour and Erosion, San Francisco, CA, USA, 7–10 November 2010.

17. Hong, S.; Sturm, T.W. Physical model study of bridge abutment and contraction scour under submerged orifice flow conditions. In Proceedings of the 33rd IAHR Congress: Water Engineering for a Sustainable Environment, Vancouver, BC, Canada, 9–14 August 2009.
18. Fael, C.M.S.; Simarro-Grande, G.; Martin-Vide, J.P.; Cardoso, A.H. Local scour at vertical wall abutments under clear-water flow conditions. *Water Res. Res.* **2006**, *10*, 1–12. [CrossRef]
19. Hong, S.; Abid, I. Physical Model Study of Bridge Contraction Scour. *KSCE J. Civ. Eng.* **2016**, *20*, 2578–2585. [CrossRef]
20. Lee, S.O.; Sturm, T.W.; Gotvald, A.; Landers, M. Comparison of laboratory and field measurements of bridge pier scour. In Proceedings of the Second International Conference on SCOUR and EROSION-ICSE, Meritus Mandarin, Singapore, 14–17 November 2004; pp. 231–239.
21. Lee, S.O.; Sturm, T.W. Effect of sediment size scaling on physical modeling of bridge pier scour. *J. Hydraul. Eng.* **2009**, *135*, 793–802. [CrossRef]
22. Sheppard, D.; Odeh, M.; Glasser, T. Large Scale Clear-Water Local Pier Scour Experiments. *J. Hydraul. Eng.* **2004**, *130*, 957–963. [CrossRef]
23. Ettema, R.; Kirkil, G.; Muste, M. Similitude of large-scale turbulence in experiments on local scour at cylinders. *J. Hydraul. Eng.* **2006**, *132*, 33–40. [CrossRef]
24. Hong, S.; Sturm, T.W.; Stoesser, T. Clear Water Abutment Scour in a Compound Channel for Extreme Hydrologic Events. *J. Hydraul. Eng.* **2015**, *141*, 1–12. [CrossRef]
25. Hong, S. Prediction of Clear Water Abutment Scour Depth in Compound Channel for Extreme Hydrologic Events. Ph.D. Thesis, School of Civil and Environmental Engineering, Georgia Institute of Technology, Atlanta, GA, USA, 2011.
26. Lane, S.N.; Biron, P.M.; Bradbrook, K.F.; Butler, J.B.; Chandler, J.H.; Crowell, M.D.; McLelland, S.J.; Richards, K.S.; Roy, A.G. Three-dimensional measurement of river channel flow processes using acoustic Doppler velocimetry. *Earth Surf. Process. Landf.* **1998**, *23*, 1247–1267. [CrossRef]
27. SonTek. *Acoustic Doppler Velocimeter (ADV) Principles of Operation*; SonTek Technical Notes; SonTek: San Diego, CA, USA, 2001.
28. Wu, P.; Hirshifield, F.; Sui, J. ADV measurements of flow field around bridge abutment under ice cover. CGU HS Committee on River Ice Processes and the Environment. In Proceedings of the 17th Workshop on River Ice, Edmonton, AB, Canada, 21 July 2013.
29. Ben Meftah, M.; Mossa, M. Scour holes downstream of bed sills in low-gradient channels. *J. Hydraul. Res.* **2006**, *44*, 497–509. [CrossRef]
30. Chabert, J.; Engeldinger, P. *Etude Des Affonillements Author Des Piles Des Ponts*; Laboratoire National d'Hydraulique de Chatou: Chatou, France, 1956.
31. Shen, H.W.; Schneider, V.R.; Karaki, S. Local scour around bridge piers. *J. Hydraul. Div.* **1969**, *95*, 1919–1940.
32. Gaudio, R.; Grimaldi, C.; Tafarojnoruz, A.; Calomino, F. Comparison of formulae for the prediction of scour depth at piers. In Proceedings of the 1st IAHR European Division Congress, Edinburgh, UK, 4–6 May 2010.
33. Gaudio, R.; Tafarojnoruz, A.; Bartolo, S.D. Sensitivity analysis of bridge pier scour depth predictive formulae. *J. Hydroinform.* **2013**, *15*, 939–951. [CrossRef]
34. Ferraro, D.; Tafarojnoruz, A.; Gaudio, R.; Cardoso, A.H. Effects of pile cap thickness on the maximum scour depth at a complex pier. *J. Hydraul. Eng.* **2013**, *139*, 482–491. [CrossRef]
35. Melville, B.W. Pier and abutment scour: Integrated approach. *J. Hydraul. Eng.* **1997**, *123*, 125–136. [CrossRef]

15

Study of Cavitation Bubble Collapse near a Wall by the Modified Lattice Boltzmann Method

Yunfei Mao, Yong Peng * and Jianmin Zhang

State Key Laboratory of Hydraulics and Mountain River Engineering, Sichuan University, Chengdu 610065, China; maoyunfeigy@163.com (Y.M.); zhangjianmin@scu.edu.cn (J.Z.)

* Correspondence: pengyongscu@foxmail.com.

Abstract: In this paper, an improved lattice Boltzmann Shan-Chen model coupled with Carnahan-Starling equation of state (C-S EOS) and the exact differential method (EDM) force scheme is used to simulate the cavitation bubble collapse in the near-wall region. First, the collapse of a single cavitation bubble in the near-wall region was simulated; the results were in good agreement with the physical experiment and the stability of the model was verified. Then the simulated model was used to simulate the collapse of two cavitation bubbles in the near-wall region. The main connection between the two cavitation bubble centre lines and the wall surface had a 45° angle and parallel and the evolution law of cavitation bubbles in the near-wall region is obtained. Finally, the effects of a single cavitation bubble and double cavitation bubble on the wall surface in the near-wall region are compared, which can be used to study the method to reduce the influence of cavitation on solid materials in practical engineering. The cavitation bubble collapse process under a two-dimensional pressure field is visualized, and the flow field is used to describe the morphological changes of cavitation bubble collapse in the near-wall region. The improved lattice Boltzmann Method (LBM) Shan-Chen model has many advantages in simulating cavitation problems, and will provide a reference for further simulations.

Keywords: collapse near a wall; double cavitation bubble; tilt distribution cavitation; parallel cavitation; pseudopotential lattice Boltzmann model

---

1. Introduction

When a liquid is heated at a constant temperature or depressurized by static or dynamic methods at a constant temperature, steam bubbles or vapour-filled cavitation bubbles appear and develop over time. Cavitating water in a low-pressure zone involves a large amount of vapour forming a two-phase flow, and when the water flows through a region with a higher pressure downstream, the cavitation bubbles collapse under the effect of the pressure or temperature. In the collapse event, the cavitation phenomenon includes the nascent development and collapse of cavitation bubbles, which is an unsteady compressible multiphase turbulent flow phenomenon involving mass transfer between the gas and liquid phases. During the collapse of the cavitation bubble, high pressure of up to thousands of atmospheres is generated. When the collapse of the cavitation bubble occurs within a certain distance from a solid sidewall, the sidewall is subjected to continuous impact, causing fracture or fatigue damage of the material, which leads to erosion and cavitation. Corrosion is associated with a series of problems such as mechanical efficiency reduction and equipment damage in devices. Therefore, the formation and collapse of cavitation bubbles near a wall has been a focus of cavitation research.

Cavitation phenomena were first observed experimentally using a flow field. Kling and Hammitt [1] and Lauterborn [2,3] used high-speed photography techniques to study spark- and laser-induced cavitation bubble collapse processes. Lauterborn [3] studied the collapse of cavitation bubbles near a wall. Some scholars have also conducted in-depth physical research on cavitation [4–13]. However, although the bubble collapse process can be observed through this test method, data such as pressure field and velocity field are not available. With the development of numerical simulation technology, more methods have emerged to simulate the bubble collapse process, such as the interface tracking method, volume-of-fluid (VOF) method, and level set method. LBM is another new and efficient numerical simulation technology that has been developed as a digital model. Relative to traditional computational fluid dynamics (CFD), LBM has substantial advantages. Sukop and Or [14] first simulated the expansion and collapse of cavitation bubbles using the lattice Boltzmann Shan-Chen model; Chen et al. [15] used the lattice Boltzmann method to simulate two-dimensional cavitation within static and shear flows. The Rayleigh-Plesset equation results were compared with their simulation results. Shan et al. [16] and Zhou et al. [17] used LBM to simulate the growth and collapse of a single bubble under a two-dimensional pressure field and the sound field at a plane rigid wall. Kucera and Blake [18] used the mirror image method to simulate cavitation erosion of a plane rigid interface upon single bubble collapse near the interface and simulated bubbles near a rigid interface at different angles but did not consider the influence of cavitation on the flow field. Li et al. [19] used shadow photography and light deflection to investigate the motion of a single cavitation bubble core and the collapse time under different cone angles. Zhang et al. [20] investigated the three-dimensional (3D) cavitation bubble phenomenon at a low liquid pressure and successfully reproduced bubble growth in low-pressure water. Mishra et al. [21] investigated the coupling between the hydrodynamics of a collapsing cavity and the resulting solute chemical species introduced by cavitation based on the Shan-Chen multiphase model. Other scholars have carried out some numerical simulations on cavitation [22–33].

Previous simulations of cavitation have focused on the relationship between bubble collapse and the position of a single cavitation bubble relative to the boundary or the collapse of multiple cavitation bubbles, but no simulation of the density field or pressure field has been performed. In this paper, based on the Shan-Chen lattice Boltzmann method, a C-S state equation and an exact difference method that can accurately derive the external force term are coupled. The non-equilibrium extrapolation format and pressure boundary are used to simulate the collapse of a bubble in the near-wall region. First, the collapse evolution law of a single cavitation bubble is obtained, which is consistent with the physical experiment. On this basis, the collapse evolution law of two cavitation bubbles in the near-wall region is studied, and the collapse evolution law and flow field changes of two cavitation bubbles at different pressures are obtained.

2. Basic Principle of the LBM

LBM is considered to be one of the most effective methods to solve multiphase flow problems. Commonly used two-dimensional models are D2Q7, D2Q9, etc., where D is the dimension of space and Q is the number of discrete velocities. In this paper, the two-dimensional case is simulated, and in order to ensure the calculation accuracy, nine discrete velocities are used, so the D2Q9 model is used for simulation. The specific arrangement is shown in Figure 1.

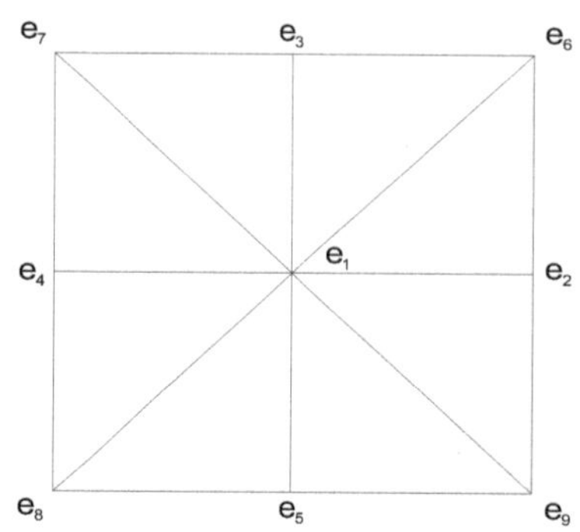

Figure 1. D2Q9 model.

In the Boltzmann model, discrete particle distribution function f_i is used to replace the fluid particle distribution function:

$$f_i(x + ce_i\Delta t, t + \Delta t) = f(x,t) - \frac{1}{\tau}[f_i(x,t) - f_i^{eq}(x,y)] + F_i(x,t) \quad (1)$$

where $f_i(x,t)$ is a single particle density distribution function, $f_i^{eq}(x,t)$ is the equilibrium particle distribution function, τ is the relaxation time, and the kinematic viscosity is $\nu = c_s^2(\tau - 0.5)\Delta t$. The discrete speed can be defined as

$$[e_1, e_2, e_3, e_4, e_5, e_6, e_7, e_8, e_9] = c\begin{bmatrix} 0 & 1 & 0 & -1 & 0 & -1 & -1 & 1 \\ 0 & 0 & 1 & 0 & -1 & 1 & -1 & -1 \end{bmatrix}, \quad (2)$$

where $c = \Delta x/\Delta t$ is the grid velocity and Δx and Δt are the grid step and time step, respectively. The lattice sound velocity is $c_s = c/\sqrt{3}$. In the D2Q9 model, the equilibrium distribution function can be expressed as

$$f_i^{eq}(x,t) = \omega_i\rho[1 + \frac{e_iu}{c_s{}^2} + \frac{(e_iu)^2}{2c_s{}^4} - \frac{u^2}{2c_s{}^2}], \quad (3)$$

where u is the fluid velocity and ω_i is the weighting factor of the equilibrium distribution function, where $\omega_1 = 4/9$, $\omega_i = 1/9(i = 2,3,4,5)$, $\omega_i = 1/9(i = 6,7,8,9)$. The discrete velocity weight coefficient is related to the discrete strategy. In the discretized velocity space, the macroscopic velocity and density of the fluid can be expressed as

$$\rho = \sum_i f_i(x,t) \quad (4)$$

$$\rho u = \sum_i f_i(x,t)e_i. \quad (5)$$

The single-component multiphase flow interaction force $F_i(x,t)$ is modelled as follows:

$$F(X,t) = -G\psi(X,t)\sum_{i=2}^{9} \xi\psi(X + e_i\Delta t, t)e_i, \quad (6)$$

where G is the interaction strength and $\xi_i = 1/9(i = 2,3,4,5)$, $\xi_i = 1/36(i = 6,7,8,9)$. According to Yuan and Schaefer (2006) [34], the interaction potential can be expressed as $\psi = \sqrt{2(p - \rho c_s)/Gc_s^2}$, where $c_s = \sqrt{RT}$ is the lattice sound velocity. Through this interaction potential, different equations of state can be applied.

The C-S state equation can be expressed as follows:

$$p = \rho RT\frac{1 + b\rho/4 + (b\rho/4)^2 - (b\rho/4)^3}{(1 - b\rho/4)^3} - a\rho^2, \quad (7)$$

where $a = 0.4963R^2T_c^2/p_c$ and $b = 0.18727RT_c/p_c$. The critical parameters can be expressed as follows:

$$T_c = \frac{0.3773a}{bR} \quad (8)$$

$$p_c = \frac{0.0706}{b^2} \quad (9)$$

The forces of the fluid act all over the objects placed in it, and how to integrate these forces into the LBM is very important. An EDM with second-order accuracy is suitable. The Boltzmann equation is discretized in the velocity space, and the term in accordance with the LB equation is derived by the EDM. The external force term in Equation (1) can be expressed as

$$F_i = f_i^{eq}(\rho, u + F\Delta t/\rho) - f_i^{eq}(\rho, u), \quad (10)$$

where F is the total interaction force between the fluid and the solid.

3. Physical Model

In order to obtain the evolution of cavitation bubbles in the near-wall region, the physical models are shown in Figure 2.

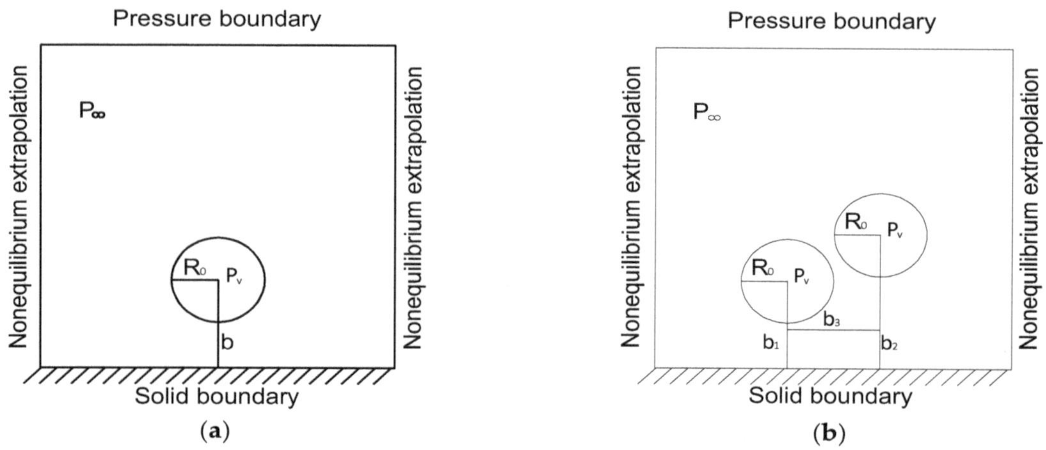

Figure 2. Physical model. (**a**) (single bubble model); (**b**) (double bubbles model) (R_0—bubble initial radius; b_i—distance; P_v—vapour pressure in bubble; P_∞—ambient pressure).

The figure above shows the initial layout of the simulation. Here, (a) is the physical model of the study of the evolution of a single cavitation bubble in the near-wall region. R_0 is the radius of the cavitation bubble, b is the distance from the centre of the cavitation bubble, P_v is the pressure in the cavitation bubble, and P_∞ is the pressure outside the cavitation bubble. The left and right boundaries are infinite areas, the upper boundary is the pressure inlet, and the lower boundary is the rigid wall. (b) is the physical model for the study of the evolution of the double cavitation bubble in the near-wall region. b_1 is the distance from the centre of the left cavitation bubble to the rigid wall, b_2 is the distance from the centre of the right cavitation bubble to the rigid wall, and b_3 is the horizontal distance of the centre of the two cavitation bubbles. The other settings are the same as those of the physical model of a single cavitation bubble in the study of the collapse of the near-wall region.

4. Simulation Content and Parameter Initialization Settings

In this paper, the numerical simulation of cavitation collapse in the near-wall region mainly includes three parts: First, the evolution law of the collapse of a single cavitation bubble in the near-wall region is obtained by numerical simulation, and the bubble deformation is compared with the results of classical experiments to verify the model. Second, the pressure field of the bubble is simulated and theoretically study the mechanism of the erosion of the bubble on the solid sidewall. Third, the previously validated model is used to study the evolution of the double cavitation bubble in the near-wall region. In this paper, if there is no special explanation, the unit in the text always uses the grid unit, the length unit is lu, the time unit is ts, the mass unit is mu, and the temperature unit is tu; thus, the density unit is mu·lu^{-3}, and the pressure unit is mu·lu^{-1}ts^{-2}. The simulated initial layout is the same as the physical model. In the simulation, $T/T_c = 0.689$ is used to simulate and the equilibrium pressure $p = 0.0028$ mu·lu^{-1}ts^{-2} is gotten by the equal area rule is. In the C-S state of equation, $a = 1$,

b = 4 and R = 1 [34] are adopted. The initial temperature is set to a specific temperature, the velocity is zero, and the density field is initialized as follows:

$$\rho(x,y) = \left(\rho_{liquid} + \rho_{gas}\right)/2 + \left(\rho_{liquid} - \rho_{gas}\right)/2 \cdot [\tanh((2(\sqrt{(x - x_1)^2 + (y - y_1)^2} - R_0)/W)], \quad (11)$$

where x_1 and y_1 is the location of the middle of the bubble at the initial moment, the hyperbolic tangent function $tanh = (e^x - e^{-x})/(e^x + e^{-x})$ and the phase interface width is W = 4. Since the cavitation bubble radius is approximately ten times the width of the phase interface, it can obtain better numerical stability of the model, but it is essential to ensure that the bubble collapse is not affected by other boundaries and that the calculation cost is reduced. Therefore, our simulation calculation area is 401 $\times$ 401.

Through the C-S state equation, the P-V curve of the gas-liquid isotherm curve can be obtained, and then the Maxwell construction should be used, which can be stated as

$$\int_{V_{m,l}}^{V_{m,g}} P dV_m = p_0(V_{m,g} - V_{m,l}), \quad (12)$$

where P is the pressure in the EOS and p_0 is a constant pressure. When the equation is established, p_0 is the equilibrium pressure, and Vm,g and $Vm,_l$ are the physical quantities that characterize the equilibrium gas pressure and the equilibrium liquid pressure, respectively. In addition, the coexistence densities ρ_v and ρ_l of gas and liquid, respectively, can also be determined by phase separation simulation with a slight random disturbance of the initial density. In the calculation process, ρ_l need to be slightly adjusted to ensure ρ_l has the same density as the pressure boundary so that an additional pressure difference between the inside and the outside of the bubble is obtained after the fluid balance in the entire calculation domain.

The collapse of cavitation bubbles under the influence of a single wall surface is studied, so infinite areas are needed on both the left and right sides to ensure the computing domain is unaffected by the boundary fluctuations. Under the premise of the minimum calculation area, the unbalanced extrapolation format is well suited for our needs. Therefore, the non-equilibrium extrapolation format is used for the left and right borders. The bottom boundary uses a standard bounce-back format. In addition, a pressure boundary condition is applied at the inlet, and the liquid pressure in the calculation zone is equal to the pressure boundary pressure.

5. Study of the Evolution of a Single Cavitation Bubble

Detailed experimental data have been obtained for the evolution of cavitation bubble collapse at different distances from the wall [35]. However, traditional experimental data and numerical simulation have great limitations. For example: Plesset and Chapman [36] made six assumptions, including negligible surface tension, constant vapour pressure and ambient pressure, an incompressible liquid, non-viscous flow, and no permanent gas, which are challenging to satisfy in experiments and LBM simulations, resulting in inapplicability to specific practical problems. In this paper, using the improved Shan-Chen model, numerical simulation data of cavitation bubble collapse at different locations are obtained at a specific temperature. This part of the study mainly consists of two parts: the first part verifies the simulation results by comparison with physical experiments, and the second part obtains the evolution law of the collapse of a single cavitation bubble in detail. The following is a comparative analysis of density field images of LBM numerical simulations and experimental images of cavitation bubbles collapsing at two different positions with the same radius and pressure. Through comparison with the experimental data of Philipp [35], the LBM calculation results which are shown in Figures 3 and 4 are found to be consistent with the qualitative analysis of the experimental data which are shown in Figures 5 and 6.

Figure 3 shows the change of density field during the collapse of cavitation bubbles. In this simulation, a dimensionless quantity $\lambda = b/R_0$ was introduced, which is the amount that characterizes the distance from the centre of the bubble to the wall. The figures below show cases with λ = 1.6,

$R_0 = 80$ lu and $\lambda = 2.5$, $R_0 = 80$ lu. The cavitation bubble is initially a circle. The bubble size and the thickness of the gas-liquid boundary layer are controlled by Equation (11). Since the pressure difference exists inside and outside the bubble, the bubble is deformed by extrusion. Due to the influence of the bottom rigid sidewall, the longitudinal flow is blocked, and a negative pressure forms under the bubble to induce longitudinal expansion of the bubble. Due to the shrinkage and deformation of the bubbles, the volume is decreasing, and the surrounding liquid fills the space created by the bubbles, resulting in a decrease in the density and pressure around them. Then, the pressure of the upper pressure boundary is first transmitted to the upper surface of the bubble, and a high-pressure zone is formed in the upper part of the bubble that acts together with the low-pressure zone of the cavitation bubble to form a depression at the upper portion ($t = 470$). Due to the rebound effect of the liquid and the relatively high speed of movement of the upper portion of the bubble, a relatively large conical high-pressure region is formed in the upper portion, which is crucial in the subsequent deformation. Over time, the depression continues to expand ($t = 530$), and with the influence of the surrounding high pressure, the bubble gradually shrinks, assuming a crescent shape ($t = 570$). When the sag causes the upper surface of the bubble to touch the lower surface, a large pressure difference directly breaks down the empty bubble, forming a micro shock wave, which has a destructive effect on the wall surface ($t = 588$). At the same time, a complex sound field is generated, which causes additional damage to the rigid wall. By comparison, the morphology of cavitation bubble collapse differs for different dimensionless parameters λ and ΔP is confirmed. In our simulation, a crescent-shaped bubble is formed at $\lambda = 1.6$ and is then broken down to form two bubbles, which generates a micro-shock; however, at $\lambda = 2.5$, the bubble is squashed directly. A crescent-shaped bubble is formed, but the bubble does not break in the middle and finally collapses in the form of a small bubble. The calculation results are consistent with the results of Philipp [35]. When the bubble is too far from the wall surface, the sidewall has a small retarding effect on the bubble, and a strong negative pressure is unlikely to form under the bubble, so when the bubble collapses, the bubbles are gradually crushed and collapsed, and the impact of the formed pressure on the wall surface is also alleviated.

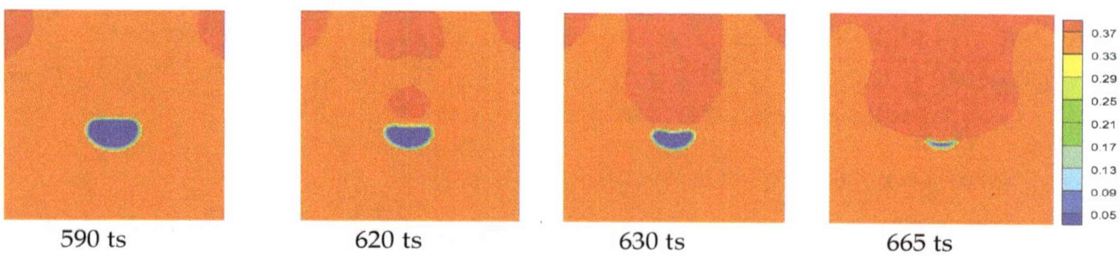

Figure 3. Density field of the LBM simulation: $\lambda = 2.5$, $R_0 = 80$ lu.

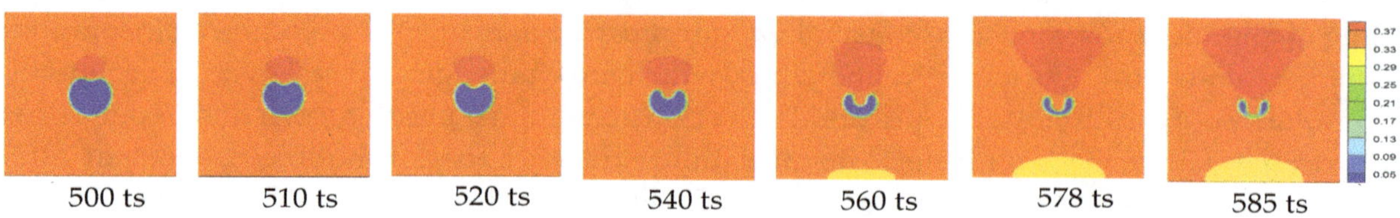

Figure 4. Density field of the LBM simulation: $\lambda = 1.6$, $R_0 = 80$ lu.

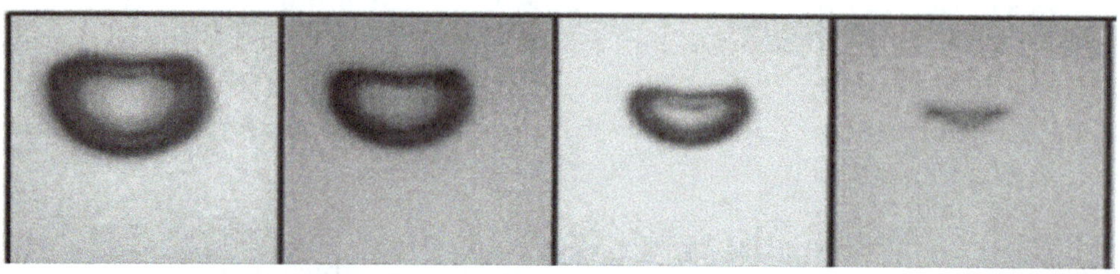

Figure 5. Experimental images: $\lambda = 2.5$, $R_0 = 1.45 \times 10^{-3}$ m.

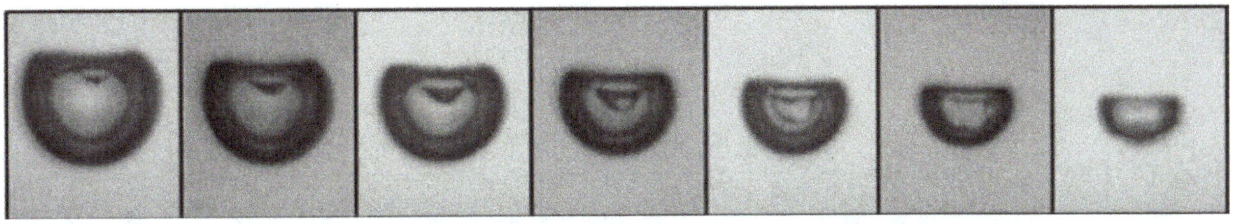

Figure 6. Experimental images: $\lambda = 1.6$, $R_0 = 1.45 \times 10^{-3}\ m$.

6. Study of the Evolution of a Double Cavitation Bubble

In the previous section, the collapse evolution of a single cavitation bubble in the near-wall region was discussed, which was found to be in good agreement with the physical experiment, thus verifying the stability of the numerical simulation model. However, in actual engineering, cavitation bubbles do not appear alone. Cavitation clouds typically form in cavitation-concentrated areas, and hundreds of cavitation bubbles interact and collapse under the extra pressure; thus, the erosion of the wall is more complex. This research area merits further study and is instructive for the possibility of reducing cavitation damage. In our study, using the previously validated mathematical model, the collapse evolution law by simulating the collapse of two cavitation bubbles in the near-wall region under pressure induction was obtained. Taking two of these cases as examples, the evolution of the two cavitation bubbles in the near-wall region under pressure induction was analysed. For convenience of description, the left and right bubbles are designated left bubble (LB) and right bubble (RB), respectively. The spatial direction is set as shown in Figure 1. For example, the upper right is the e_6 direction. Similar to the previous simulation, the dimensionless quantities λ_1, λ_2, and λ_3 is introduced, where $\lambda_1 = b_1/R_0$, which is the amount that characterizes the distance of the LB centre from the wall; $\lambda_2 = b_2/R_0$, which is the amount that characterizes the distance of the LB centre from the wall surface; and $\lambda_3 = b_3/R_0$, which is the amount that represents the horizontal distance between the LB and RB centres.

6.1. Case 1: Numerical Simulation of Tilt Distribution Cavitation with Two Cavitation Bubbles

When $\lambda_1 = 1.2$, $\lambda_2 = 2.7$, and $\lambda_3 = 1.5$, the line connecting the centres of the two cavitation bubbles is at an angle of 45° to the wall surface. The density field and the pressure field coupling with the velocity field of the cavitation bubble collapse process are shown in Figures 7 and 8, respectively.

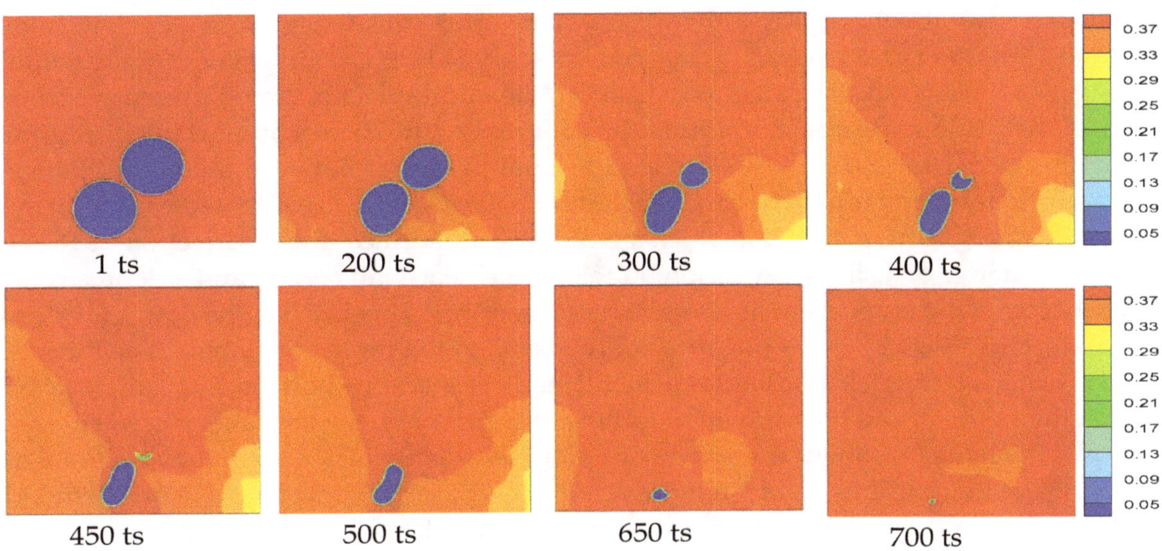

Figure 7. Density field of cavitation bubble collapse.

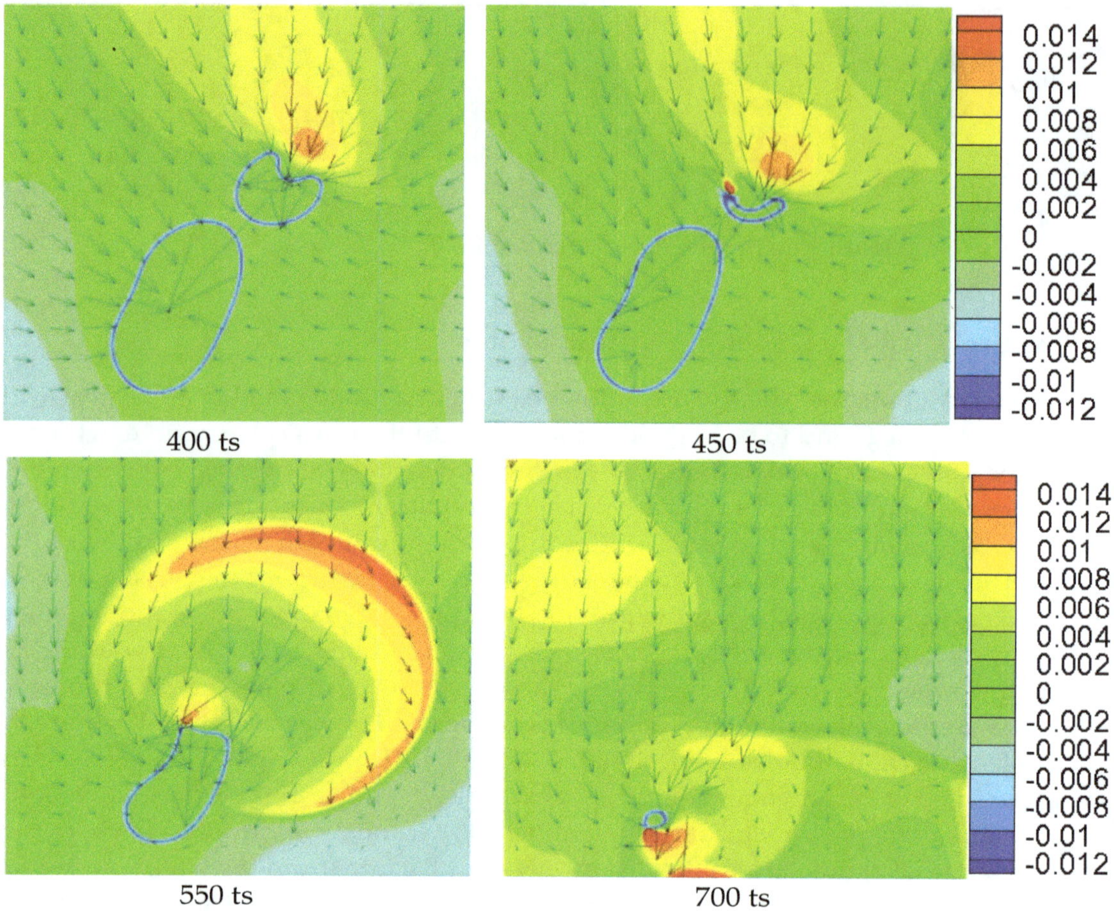

Figure 8. Pressure field and velocity field of cavitation bubble collapse.

The cavitation bubble distribution at the initial moment is as shown by $t = 1$ ts, and the specific parameter settings are the same as the previous settings. The collapse process of each cavitation bubble was analysed. When the simulation proceeds to $t = 200$ ts, the change of LB is similar to the collapse of a single cavitation bubble. Due to the blockage of the wall surface, a low-pressure zone is formed in the bottom region, and LB exhibits an elongated deformation state in the e_5 direction. The e_6 direction of the LB and the e_8 direction of the RB are attracted to each other, and the opposite direction is deformed. If the gas density and the liquid density are in equilibrium at this time, the two cavitation bubbles would attract each other and eventually merge into a single bubble. However, since the liquid has been pressurized during the simulated initialization, the additional generated pressure prevents the emergence of the two bubbles. As the simulation progresses, the deformation is further aggravated, and the bubbles decrease under the action of additional pressure. Between them, the change of LB is the most easily detected. When the time reaches $t = 350$ ts, the LB is elongated and deformed due to the low pressure resulting from the blockage of the wall surface; the e_6 direction of RB appears the same as the change of a single cavitation bubble in the near-wall region, and a depression occurs. The change becomes more apparent at 400 ts. Around the shrinking bubble, the liquid fills the newly available space, and a slightly lower pressure appears around the cavitation bubble. The transition can be observed in the velocity field. Around the bubble, the velocity direction tends to support cavitation. Note that in the figure of the pressure field and the velocity field, for the sake of convenience, only the velocity of the liquid is shown and we do not plot the velocity inside the bubble. The LB exhibits an elliptical shape that exhibits an inclination in the e_6–e_8 direction under the action of the sidewall, the attraction of RB and the external extreme pressure. The RB is also elliptical due to the attraction of the two bubbles, but the deformation is not as strong as with the bubble on the left. At this time, relative with the collapse rule of a single cavitation bubble in the near-wall region, LB is equivalent to rigidity avoidance for RB, so a low-pressure zone is generated between the two bubbles, and high pressure is generated by the upper pressure boundary. A pressure difference is generated in the e_6–e_8 direction of the RB caused the deformation to continue to produce. Inspection of the velocity field reveals that the maximum velocity occurs in the same direction of the RB. At t = 450 ts, the cavitation bubble is crescent-shaped. The bubble is strongly compressed from the upper part of the depression until the entire bubble is completely collapsed from the upper part of the bubble, which occurs because the flow field is not completely symmetrical on both sides of the e_6–e_8 direction. The upper part is first crushed by the pressure transferred from the pressure boundary. The other reason is that the most attractive part of LB is located at the nearest position of the two bubbles. The velocity field indicates that the maximum velocity direction has been deflected, and RB begins to collapse from top to bottom under the pressure difference, producing a huge jet and complex sound field. The huge pressure generated by the collapse acts in conjunction with other factors to promote the continued collapse of LB. At t = 550 ts, the pressure field and velocity field images indicate that the flow field changes due to the collapse pressure of RB, and the micro-jet generated by RB in the e_6 direction collides with the downward flowing liquid to generate a high voltage and noise, and energy begins to be consumed; furthermore, the micro-jet in the e_7 direction overlaps with the original flow field, and the flow state becomes complex. As the simulation proceeds, the flow field exhibits vortices, and LB begins to collapse. First, the cavitation bubble e_6 direction begins to shrink under the influence of the pressure of RB, which causes a protrusion in the upper portion of the cavitation bubble. At this time, the energy generated by the RB collapse is insufficient to continue to compress LB. Under the joint action of the upper pressure boundary pressure and the high pressure generated by the RB collapse, a high-pressure region (t = 550 ts) is formed at the upper convex portion, and collapse from the upper portion of the cavitation bubble is induced. As RB is destroyed, the e_9 direction finally collapses, and a jet from the e_7 to the e_9 direction is generated in the flow field, so that the entire flow field exhibits vortices. Finally, LB collapses under the combined action of various factors, and the generated pressure impacts the wall surface to avoid impact. The pressure field diagram of t = 700 ts indicates that a high pressure is generated on the wall surface. Note that after RB collapse, LB collapse occurs under the influence of various factors, and the mechanism of action is relatively complex. The LB high-pressure zone rotates around the cavitation bubble. This visualization is a powerful way to illustrate the complex vortices and other phenomena in the flow field, involving other physical quantities. The parameters of this study cannot describe the more detailed characteristics. The specific mechanism needs further study.

6.2. Case 2: Numerical Simulation of Two Parallel Cavitation Bubbles

When λ_1 = 1.2, λ_2 = 1.2, and λ_3 = 2.4, that is, the line connecting the centres of the two cavitation bubbles is parallel to the wall surface, and the density field of the cavitation bubble collapse process is shown in Figure 9 as well as the pressure field and the velocity field shown in Figure 10.

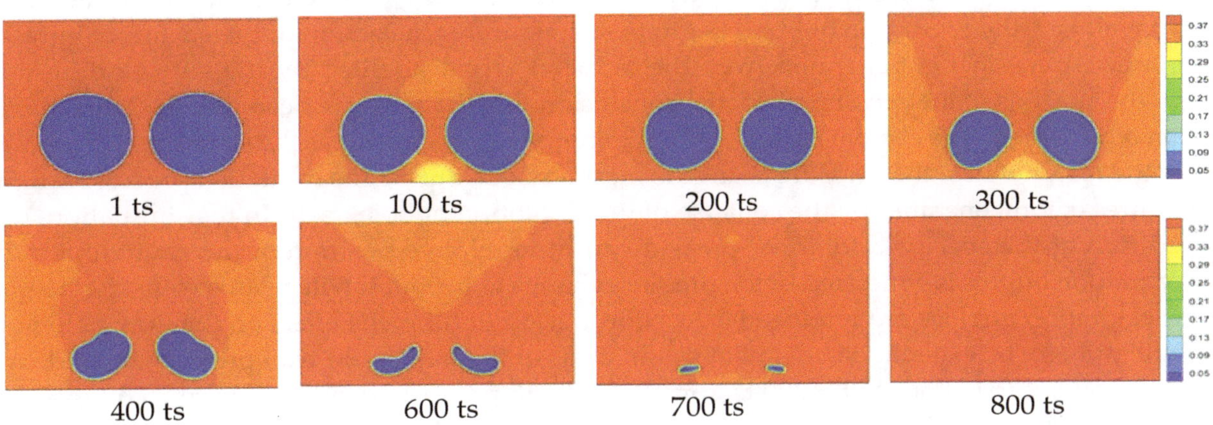

Figure 9. Density field of cavitation bubble collapse.

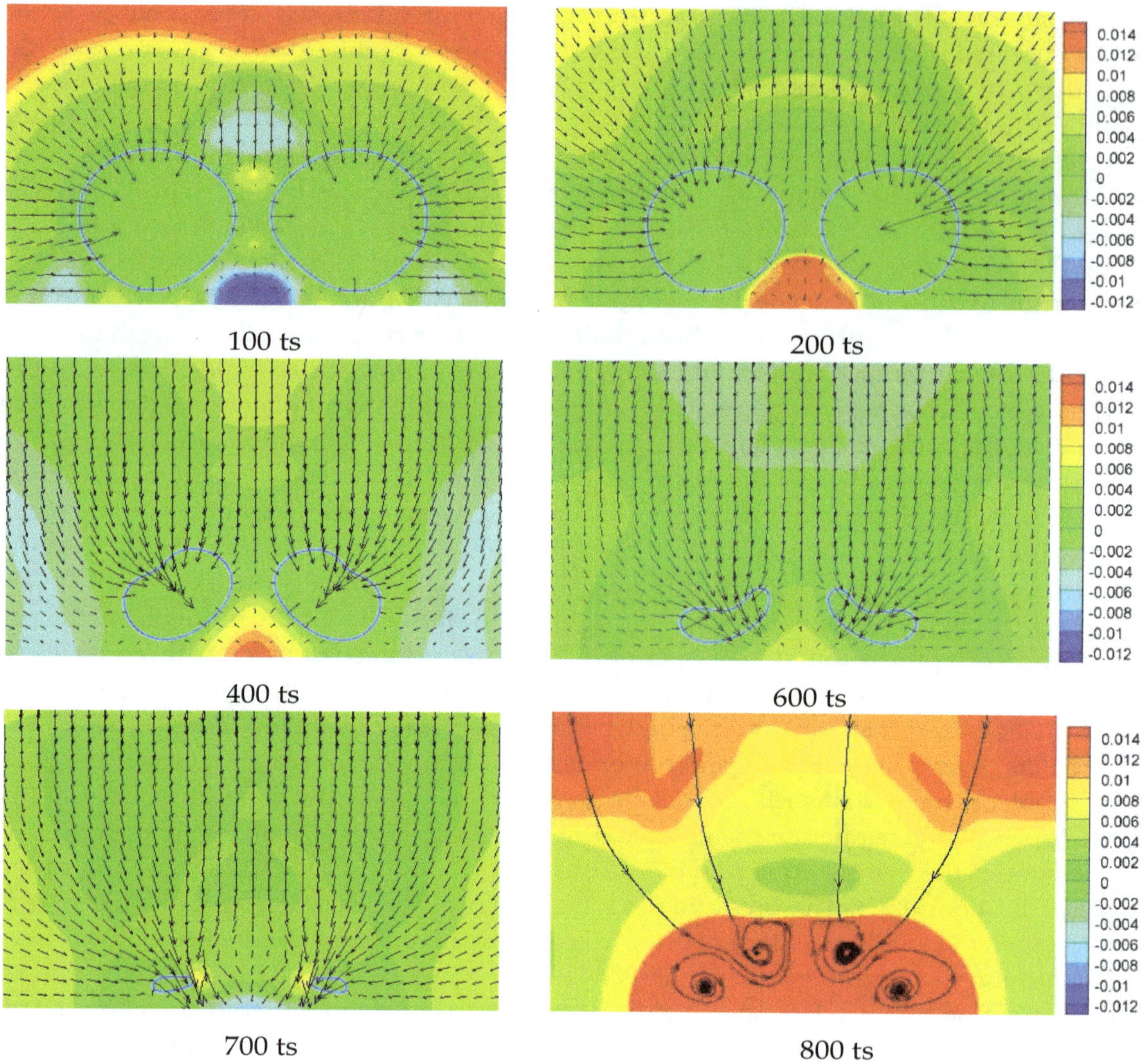

Figure 10. Pressure field and velocity field of cavitation bubble collapse (800 ts is the pressure field and streamline diagram).

The figure shows the change in the two cavitation bubbles in the near-wall region and the initial boundary and density settings are the same as described above. Since the changes of the two bubbles are basically the same, the LB is used as an example for analysis. At t = 100 ts, low pressure is created at the bottom of the cavitation bubbles due to the retardation of the rigid wall. At the same time, the two cavitation bubbles are attracted to each other. The cavitation bubbles that are not attracted to the rigid wall and the cavitation bubbles begin to shrink under the action of the additional pressure, and the new space that is generated from the reduced area was filled by the surrounding liquid. Through the analysis of the pressure field, the conclusion that the pressure decreases in the e_9 direction of LB can be obtained. The bottom region of the symmetry axis is defined as the pressure change region, which is the region where the pressure change is most obvious during the cavitation process, except for the pressure region generated by the collapse of the cavitation bubble itself. In the simulation, under the condition of the specific additional pressure and the relative position of the cavitation bubble, a new cavitation bubble is generated in the pressure change zone, but it will collapse quickly. Although the new cavitation bubbles are produced for a short period of time, the effect on the flow field is very important and will be explained in detail in the analysis of the maximum wall pressure behind. When the simulation is carried out to t = 200 ts, for LB, the high pressure generated by the collapse of the new cavitation bubble in the pressure change zone continues to act on the e_9 direction of the cavitation bubble, showing a slightly lifted shape. As the simulation progresses to t = 400 ts, the deformation continues; the deformation of the cavitation bubbles is already evident, and the cavitation bubbles are inclined at an inclination angle of 45°, and the cavitation bubbles become crescent-shaped. This is mainly due to the interaction between the cavitation bubbles, the interaction between the cavitation bubbles and the wall surface, and the collapse of the new cavitation bubbles in the pressure change zone. In addition, the velocity field indicates that the velocity of the LB in the e_7 direction is the largest. At t = 600 ts, the LB appears asymmetric on the axis in the e_7–e_9 direction. The upper part is strongly influenced by the pressure boundary and the lower part is strongly influenced by the rigid wall. Therefore, the position where the pressure difference between the upper side and the lower side of the LB is the largest is located at the centre, so that the cavitation bubbles collapse from the upper portion until they completely collapse. For each bubble, a complex eddy current phenomenon occurs in the flow field because the collapse does not occur instantaneously but spreads from the top to the adjacent region. The 800 ts streamline diagram clearly shows the eddy currents in the flow.

7. Maximum Wall Pressure

The maximum wall pressure of a single cavitation bubble and two parallel distributed cavitation bubbles collapsed in the near-wall region was identified in this study. The initial conditions are set as follows: For the case of a single cavitation bubble, the value of λ (from 1.05–2.0) with a series of gradients is used to simulate the maximum wall pressure resulting from single cavitation bubble collapse under different initial conditions. Since the maximum pressure of a single cavitation bubble on the wall is only likely to occur below the cavitation bubble (e_5 direction), the pressure of the wall below the cavitation bubble is recorded to find the maximum value. For the case of a single cavitation bubble, a lambda value (from 1.05–2.0) with a series of gradients is used to simulate the maximum wall pressure resulting from a single cavitation bubble collapse under different initial conditions. Since the maximum pressure of a single cavitation bubble on the wall is only likely to occur below the cavitation bubble (e_5 direction), the pressure of the wall below the cavitation bubble is recorded to find the maximum value. For the case of two cavitation bubbles, $\lambda_3 = 2.4$ is used and, likewise, the value of λ (from 1.05 to 2.0) with a series of gradients was used to simulate the maximum wall pressure resulting from the collapse of two cavitation bubbles under different initial conditions, where $\lambda = \lambda_1 = \lambda_2$. Since the cavitation bubble collapse process is complicated, the maximum wall pressure is not generated at the same position for the collapse process under different initial conditions. The wall pressure below the cavitation bubble (e_5 direction) and the pressure change zone is recorded to find the maximum value The comparison of the maximum wall pressure under different initial conditions is shown in Figure 11.

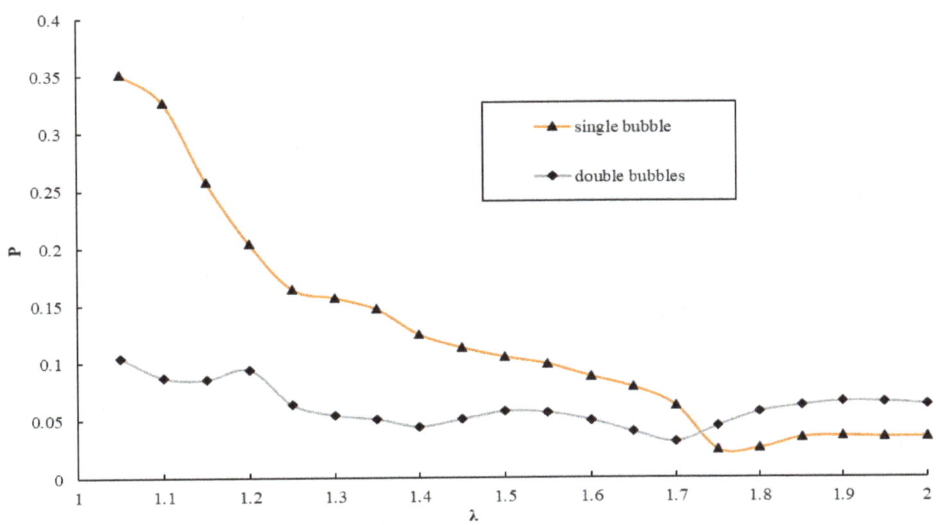

Figure 11. Comparison of the maximum wall pressure under different initial conditions.

Some conclusions are given by analysis:

- The maximum wall pressure generated by the collapse of a single cavitation bubble in the near-wall region decreases with increasing distance from the wall surface, and it first drops sharply, then slowly decreases and finally becomes relatively stable. This is because the closer to the wall, the smaller the thickness of the fluid that the micro shock wave generated by cavitation passes to the wall and the smaller the blockage effect of the flow field, the greater the wall pressure generated.
- It is complex for the case where a double parallel cavitation bubble collapses in the near-wall region because the maximum wall pressure produces a different position under different initial conditions. When $\lambda = 1.05$–1.10, the wall below the cavitation bubble is subjected to the maximum wall pressure, because the cavitation bubble is closer to the wall surface, and the generated micro-shock is transmitted to the wall surface almost unimpeded and the cavitation bubble of the pressure change region is not fully developed, which is not enough to generate create a huge pressure with the lifting force to bubble. When $\lambda = 1.15$–1.25, the maximum wall pressure occurs in the pressure change zone where new cavitation bubbles are generated and collapse rapidly, resulting in a large wall pressure exceeding the maximum wall pressure at other locations. When $\lambda = 1.40$–2.0, the area where the maximum wall pressure is generated is the pressure change zone, but the pressure generation at this time is not when the new cavitation bubble collapses but the pressure after the collapse of the two cavitation bubbles overlaps each other on the wall. The change in wall pressure depends mainly on the size of the new cavitation bubble induced and the pressure generated by the collapse and the lifting force of the bubbles.
- When $\lambda < 1.7$, the wall pressure generated by the collapse of a single cavitation bubble is relatively large. This is because, for the case where the double cavitation bubble collapses in the near-wall region, the pressure generated by the collapse of the new cavitation bubble induced in the pressure change region has an effect of lifting force on the cavitation bubble. Thereby, the bottom pressure of the cavitation bubble collapse is increased, and the blockage effect of the wall surface is weakened, which is equivalent to an increase of λ, and the resulting wall pressure is relatively small. When $\lambda > 1.7$, the wall pressure of the pressure change zone is formed by the superposition of pressure generated by the collapse of two cavitation bubbles, so the generated wall pressure is greater than the pressure generated by the collapse of a single cavitation bubble.

In summary, the maximum wall pressure generated by the collapse of the two cavitation bubbles in the near-wall region is smaller than the single cavitation bubble at $\lambda < 1.7$; the closer the bubble is to the wall, the greater the influence on the wall surface. When $\lambda > 1.7$, the distance between the cavitation bubble pair and the wall surface has little effect on the maximum wall pressure. This is a point worthy of further study with different initial conditions. It can be used as a reference for reducing cavitation damage in practical engineering.

8. Conclusions

Based on the improved lattice Boltzmann Shan-Chen model coupled with C-S EOS, the collapse phenomenon of cavitation bubbles in the near-wall region is studied. In this paper, the collapse of a single bubble in the near-wall region is simulated and compared with the physical experiment; the results are in good agreement. Then, using the verified mathematical model, the simulation of the collapse of two cavitation bubbles in the near-wall region under different pressure-induced conditions is carried out, and the collapse evolution law is obtained. And analysed the effect of wall pressure on different initial conditions. By analysing in detail the density field changes and pressure field changes during bubble collapse, the following conclusions can be drawn:

1. For the case where a single cavitation bubble collapses in the near-wall region, when $\lambda = 1.6$, a crescent-shaped bubble is formed that is broken down to form two bubbles, and a micro-shock is generated; when $\lambda = 2.5$, the bubbles are crushed to form crescent-shaped bubbles, but the

bubbles do not break in the middle but rather ultimately collapse in the form of a small bubble. The shape of the cavitation bubble is related to the distance of the cavitation bubble from the rigid wall.

2. For the numerical simulation of tilted distribution cavitation with two cavitation bubbles, in the early stage of simulation, the collapse behaviour is similar to that of the single-bubble case. Subsequently, the e_6 direction of RB has a concave deformation, which is very important in the collapse of the two bubbles. The velocity field indicates that the maximum velocity appears in the depression of RB. The tremendous pressure generated by this velocity directly penetrates the bubble; thereafter, the bubble is crescent-shaped until collapsing completely. Due to the asymmetry of the collapse, the flow field becomes complex after RB collapses and particularly so after LB collapses, and vortices appear in the flow field.

3. For the numerical simulation of parallel distribution cavitation of two bubbles, the bubble collapses under the blocking effect of the rigid wall and the attraction of the two bubbles, and the two bubbles collapse simultaneously. Therefore, the mutual influence during the collapse process is smaller than that of Case 1; in addition, the erosion effect of the bubble collapse on the wall surface is the result of superimposing the pressure fields formed by the collapse of the two bubbles. However, for each bubble, since the collapse does not occur instantaneously but is collapsed from the upper part of the closest position of the two cavitation bubbles, a complex vortex phenomenon occurs in the flow field.

4. By comparing the maximum wall pressure generated by cavitation under different initial conditions, the factors affecting the maximum wall pressure are obtained. For a single cavitation bubble, the distance from the wall is the most important factor. For two cavitation bubbles, the lifting effect of the new induced cavitation bubble collapse is the most important factor.

The results indicate that the improved LBM Shan-Chen model has many advantages in simulating cavitation problems, providing a reference for further simulation.

All the abbreviations are explained in the Appendix A (Table A1).

Author Contributions: Conceptualization, Y.P. and J.Z.; Methodology, Y.P. and J.Z.; Software, Y.M.; Investigation, Y.M.; Data Curation, Y.M.; Writing-Original Draft Preparation, Y.P. and J.Z.; Writing-Review & Editing, Y.M.; Visualization, Y.M.; Funding Acquisition, Y.P.

Appendix A

Table A1. All variables with definitions.

f_i	**Single Particle Density Distribution Function**
f_i^{eq}	equilibrium particle distribution function
τ	relaxation time
ν	kinematic viscosity
c	grid velocity
Δx	grid step
Δt	time step
c_s	lattice sound velocity
ω_i	weighting factor
u	fluid velocity
F_i	interaction force
ψ	interaction potential
T_c	critical temperature
P_c	critical pressure
F	total interaction force

Table A1. *Cont.*

f_i	Single Particle Density Distribution Function
P_v	pressure of the cavitation bubble
P_∞	the pressure outside the cavitation bubble
R_0	radius of the cavitation bubble
b (b_1, b_2, b_3)	distance between corresponding points
T	lattice temperature
a (in Equation (11))	parameter of the C-S state of equation
b (in Equation (11))	parameter of the C-S state of equation
R (in Equation (11))	parameter of the C-S state of equation
$V_{m,g}$	the equilibrium gas pressure
$V_{m,l}$	the equilibrium liquid pressure
$\lambda(\lambda_1, \lambda_2, \lambda_3)$	dimensionless value that characterizes the distance

References

1. Kling, C.L.; Hammitt, F.G. A photographic study of spark-induced cavitation bubble collapse. *J. Basic Eng.* **1972**, *94*, 825–832. [CrossRef]
2. Lauterborn, W. High-speed photography of laser-induced breakdown in liquids. *Appl. Phys. Lett.* **1977**, *31*, 663–664. [CrossRef]
3. Lauterborn, W.; Bolle, H. Experimental investigations of cavitation-bubble collapse in the neighbourhood of a solid boundary. *J. Fluid Mech.* **1975**, *72*, 391–399. [CrossRef]
4. Reuter, F. Electrochemical wall shear rate microscopy of collapsing bubbles. *Phys. Rev. Fluids* **2018**, *3*, 063601. [CrossRef]
5. Watanabe, R.; Yanagisawa, K.; Yamagata, T.; Fujisawa, N. Simultaneous shadowgraph imaging and acceleration pulse measurement of cavitating jet. *Wear* **2016**, *358–359*, 72–79. [CrossRef]
6. Reuter, F.; Cairos, C.; Mettin, R. Vortex dynamics of collapsing bubbles: Impact on the boundary layer measured by chronoamperometry. *Ultrason. Sonochem.* **2016**, *33*, 170–181. [CrossRef] [PubMed]
7. Cui, P.; Zhang, A.M.; Wang, S.P.; Khoo, B.C. Ice breaking by a collapsing bubble. *J. Fluid Mech.* **2018**, *841*, 287–309. [CrossRef]
8. Rossello, J.M.; Urteaga, R.; Bonetto, F.J. A novel water hammer device designed to produce controlled bubble collapses. *Exp. Therm. Fluid Sci.* **2018**, *92*, 46–55. [CrossRef]
9. Pal, A.; Joseph, E.; Vadakkumbatt, V.; Yadav, N.; Srinivasan, V.; Maris, H.J.; Ghosh, A. Collapse of Vapor-Filled Bubbles in Liquid Helium. *J. Low Temp. Phys.* **2017**, *188*, 101–111. [CrossRef]
10. Daou, M.M.; Igualada, E.; Dutilleul, H.; Citerne, J.-M.; Rodriguez-Rodriguez, J.; Zaleski, S.; Fuster, D. Investigation of the collapse of bubbles after the impact of a piston on a liquid free surface. *AIChE J.* **2017**, *63*, 2483–2495. [CrossRef]
11. Ma, X.J.; Huang, B.A.; Zhao, X.; Wang, Y.; Chang, Q.; Qui, S.; Fu, X.; Wang, G. Comparisons of spark-charge bubble dynamics near the elastic and rigid boundaries. *Ultrason. Sonochem.* **2018**, *43*, 80–90. [CrossRef] [PubMed]
12. Gong, S.W.; Ohl, S.W.; Klaseboer, E.; Khoo, B.C. Interaction of a spark-generated bubble with a two-layered composite beam. *J. Fluids Struct.* **2018**, *76*, 336–348. [CrossRef]
13. Oh, J.; Yoo, Y.; Seung, S.; Kwak, H.-Y. Laser-induced bubble formation on a micro gold particle levitated in water under ultrasonic field. *Exp. Therm. Fluid Sci.* **2018**, *93*, 285–291. [CrossRef]
14. Sukop, M.C.; Or, D. Lattice Boltzmann method for homogeneous and heterogeneous cavitation. *Phys. Rev. E* **2005**, *71*. [CrossRef] [PubMed]
15. Chen, X.P.; Zhong, C.W.; Yuan, X.L. Lattice Boltzmann simulation of cavitating bubble growth with large density ratio. *Comput. Math. Appli.* **2012**, *61*, 3577–3584. [CrossRef]
16. Shan, M.L.; Zhu, C.P.; Zhou, X.; Yin, C.; Han, Q.B. Investigation of cavitation bubble collapse near rigid boundary by lattice Boltzmann method. *J. Hydrodyn.* **2016**, *28*, 442–450. [CrossRef]
17. Zhou, X.; Shan, M.L.; Zhu, C.P.; Chen, B.Y.; Yin, C.; Ren, Q.G.; Han, Q.B.; Tang, Y.B. Simulation of Acoustic Cavitation Bubble Motion by Lattice Boltzmann Method. *Appl. Mech. Mater.* **2014**, *580–583*, 3098–3105. [CrossRef]

18. Kucera, A.; Blake, J.R. Approximate methods for modeling cavitation bubbles near boundaries. *Bull. Austral. Math. Soc.* **1990**, *41*, 1–44. [CrossRef]
19. Li, B.B.; Zhang, H.C.; Han, B.; Chen, J.; Ni, X.W.; Lu, J. Investigation of the collapse of laser-induced bubble near a cone boundary. *Acta Phys. Sin.* **2012**, *61*, 21.
20. Zhang, X.M.; Zhou, C.Y.; Islam, S.; Liu, J.Q. Three-dimensional cavitation simulation using lattice Boltzmann method. *Acta Phys. Sin.* **2009**, *58*, 8406–8414.
21. Mishra, S.K.; Deymier, P.A.; Muralidharan, K.; Frantziskonis, G.; Pannala, S.; Simunovic, S. Modeling the coupling of reaction kinetics and hydrodynamics in a collapsing cavity. *Ultrason. Sonochem.* **2010**, *17*, 258–265. [CrossRef] [PubMed]
22. Schanz, D.; Metten, B.; Kurz, T.; Lauterborn, W. Molecular dynamics simulations of cavitation bubble collapse and sonoluminescence. *New J. Phys.* **2012**, *14*. [CrossRef]
23. Du, T.; Wang, Y.; Huang, C.; Liao, L. A numerical model for cloud cavitation based on bubble cluster. *Theor. Appl. Mech. Lett.* **2017**, *7*, 231–234. [CrossRef]
24. Wang, Q.; Yao, W.; Quan, X.; Cheng, P. Validation of a dynamic model for vapor bubble growth and collapse under microgravity conditions. *Int. Commun. Heat Mass Transf.* **2018**, *95*, 63–73. [CrossRef]
25. Ogloblina, D.; Schmidt, S.J.; Adams, N.A. Simulation and analysis of collapsing vapor-bubble clusters with special emphasis on potentially erosive impact loads at walls. *EPJ Web Conf.* **2018**, *180*, 9. [CrossRef]
26. Chen, Y.; Lu, C.J.; Chen, X.; Li, J.; Gong, Z.X. Numerical investigation of the time-resolved bubble cluster dynamics by using the interface capturing method of multiphase flow approach. *J. Hydrodynam.* **2017**, *29*, 485–494. [CrossRef]
27. Ming, L.; Zhi, N.; Chunhua, S. Numerical Simulation of Cavitation Bubble Collapse within a Droplet. *Comput. Fluids* **2017**, *152*, 157–163. [CrossRef]
28. Peng, K.; Tian, S.; Li, G.; Huang, Z.; Yang, R.; Guo, Z. Bubble dynamics characteristics and influencing factors on the cavitation collapse intensity for self-resonating cavitating jets. *Petrol. Explor. Dev.* **2018**, *45*, 343–350. [CrossRef]
29. Xue, H.H.; Shan, F.; Guo, X.S.; Tu, J.; Zhang, D. Cavitation Bubble Collapse near a Curved Wall by the Multiple-Relaxation-Time Shan-Chen Lattice Boltzmann Model. *Chin. Phys. Lett.* **2017**, *34*, 83–87. [CrossRef]
30. Tagawa, Y.; Peters, I.R. Bubble Collapse and Jet Formation in Corner Geometries. *Phys. Rev. Fluids* **2018**, 2. [CrossRef]
31. Goncalves, E.; Zeidan, D. Numerical study of bubble collapse near a wall. *AIP Conf. Proc.* **2018**, *1978*. [CrossRef]
32. Peng, Y.; Mao, Y.F.; Wang, B. Study on C–S and P–R EOS in pseudo-potential lattice Boltzmann model for two-phase flows. *Int. J. Mod. Phys. C* **2017**, *28*. [CrossRef]
33. Peng, Y.; Wang, B.; Mao, Y. Study on Force Schemes in Pseudopotential Lattice Boltzmann Model for Two-Phase Flows. *Math. Probl. Eng.* **2018**. [CrossRef]
34. Yuan, P.; Schaefer, L. Equations of state in a lattice Boltzmann model. *Phys. Fluids* **2006**, *18*. [CrossRef]
35. Philipp, A.; Lauterborn, W. Cavitation erosion by single laser-produced bubbles. *J. Fluid Mech.* **2000**, *361*, 75–116. [CrossRef]
36. Plesset, M.S.; Chapman, R.B. Collapse of an initially spherical vapour cavity in the neighbourhood of a solid boundary. *J. Fluid Mech.* **1970**, *47*, 283–290. [CrossRef]

16

Influence of Flushing Velocity and Flushing Frequency on the Service Life of Labyrinth-Channel Emitters

Zhangyan Li [1], Liming Yu [1,*], Na Li [1], Liuhong Chang [2] and Ningbo Cui [3]

1 Faculty of Modern Agriculture Engineering, Kunming University of Science and Technology, Kunming 650500, China; zhangyanli4851@sina.com (Z.L.); kjclina@163.com (N.L.)
2 School of Hydraulic Engineering, Changsha University of Science and Technology, Changsha 410114, China; claire886@163.com
3 State Key Laboratory of Hydraulics and Mountain River Engineering and College of Water Resource and Hydropower, Sichuan University, Chengdu 610065, China; cuiningbo@scu.edu.cn
* Correspondence: liming16900@sina.com

Abstract: Dripline flushing is an effective way to relieve emitter clogging and extend the longevity of drip irrigation systems. This laboratory study was conducted at Kunming University of Science and Technology to evaluate the effect of three targeted flushing velocities (0.3, 0.6, and 0.9 m/s) and four flushing frequencies (no flushing, flushing daily, and flushing every three or five days) on the emitter's service life and the particle size distribution of the sediment discharged from emitters and trapped in an emitter channel. The gradation of particle size was analyzed by a laser particle size analyzer. The experiment results suggested that flushing velocity and flushing frequency had a significant effect on the service life of emitters, and the emitter's service life was extended by 30.40% on average under nine different flushing treatments. Flushing can effectively reduce the accumulation of sediments in the dripline and decrease the probability of coarse particles flowing into emitters and fine particles aggregating and cementing in the labyrinth channel, thus relieving the emitter clogging. Therefore, dripline flushing can effectively slow down clogging in muddy water drip irrigation system. The recommended flushing velocity should be set at 0.6 m/s, and the flushing intervals should be shortened.

Keywords: clogging; drip irrigation; flushing; particle size distribution; sediments

---

1. Introduction

Dripline flushing is a necessary maintenance practice for micro-irrigation systems. It removes particles that are not strained by the micro-irrigation system filters and that accumulate in the driplines [1,2]. Physical clogging caused by solid particles is considered the most common emitter clogging category of emitters [3–5]. For drip irrigation systems in the Yellow River irrigation areas of Ningxia and Inner Mongolia, the average sediment concentration in the water abstracted from the Yellow River reaches 35 kg/m^3. A large quantity of sand enters into drip irrigation systems and results in emitter clogging even after deposition and prefiltration treatment measures are taken [6]. Therefore, flushing is required to ensure a long economic service life of drip irrigation systems [7]. On the other hand, numerical simulations on shallow water flows and river flushing have been carried out by Peng et al. [8–14].

In order to achieve an optimal flushing effect, drip irrigation systems should be designed properly. Flushing must be done often enough and at an appropriate velocity to dislodge and remove the accumulated sediments [15]. A minimum flushing velocity of 0.3 m/s was recommended by the

American Society of Agricultural and Biological Engineers Engineering Practice, EP-405 [16], but some researchers have advised that a flushing velocity of 0.5–0.6 m/s is necessary when larger particles need to be discharged; for example, it is useful when coarser filters are used in drip irrigation systems [17,18]. In a 30-day field study in which the target flushing velocities were ranged from 0.23 to 0.61 m/s, Puig-Bargués et al. [18] did not find remarkable effects of flushing velocity on the emitter discharge. However, the higher the flushing velocities were set, the more solids were removed from the laterals. Increasing the velocity of flushing may need more costly system designs (e.g., larger supply pipe main, higher pumping requirements, reduced zone sizes) and labor requirements for flushing in irrigation systems should be increased [14].

Different studies explored different flushing frequencies: daily [19], weekly [20], every two weeks [17,21], and monthly [18]. Li et al. [21] claimed that emitter clogging was minimized in cases of biweekly flushing. However, Puig-Bargués et al. [22] reported no obvious differences between flushing frequencies at a velocity of 0.6 m/s, and emitter clogging was primarily affected by the interactions between emitter type, emitter location, and frequency of flushing. Thus, a general agreement on the optimum flushing frequency is lacking at this time, and related studies on flushing frequency and flushing velocity are scarce.

Therefore, the objective of this study was to analyze the influence of three flushing velocities and four flushing frequencies on emitter clogging in drip irrigation systems when using muddy water with full tests. Additionally, particle size distributions of the discharged sediments and residual sediments in the emitter were investigated to determine appropriate flushing schemes to extend the service life of drip irrigation systems.

2. Materials and Methods

2.1. *Emitter Characteristics*

The drip tape with non-pressure-compensating emitter manufactured by Dayu Water Conservation Ltd. in Jiuquan City, Gansu Province, China, which is widely used in agricultural irrigation, was applied in the experiments. Its external diameter was 16 mm, and the thickness of the wall was 0.40 mm. Figure 1 presents the structure of the labyrinth-channel emitter. The working pressure was 0.1 MPa. The rated discharge (q) was 1.37 L/h. The flow path had a width (W) of 0.94 mm, length (S) of 37.8 mm, and depth (D) of 0.60 mm. The sectional area (A) was 0.56 mm^2. The angle between the bevel edge and baseline was 38.2°. Space width (L) and height (H) of the tooth were 1.50 and 0.88 mm, respectively. The manufacturing variation of the emitter was 0.028. The flow exponent (x) was 0.49, and the discharge coefficient (k) was 1.12.

The test driplines were cut randomly from the same roll of irrigation drip tapes. The emitter's discharge was measured using pressures of 0.01, 0.02, 0.03, 0.04, 0.05, 0.06, 0.08, and 0.10 MPa to calculate the discharge coefficient and flow exponent.

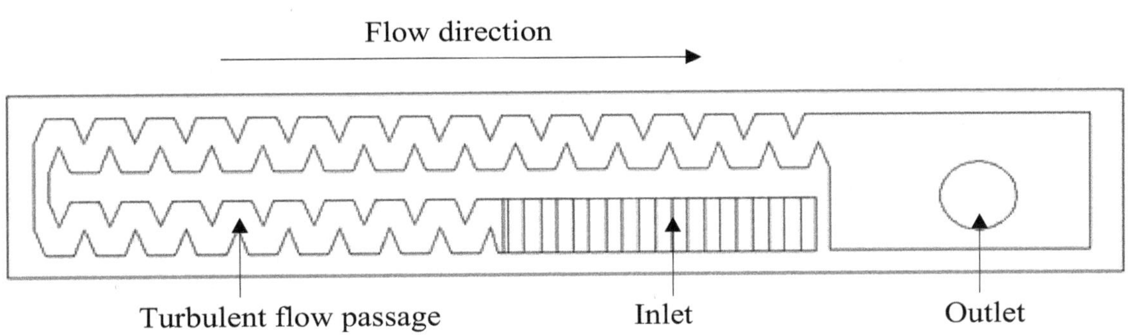

(a)

Figure 1. *Cont.*

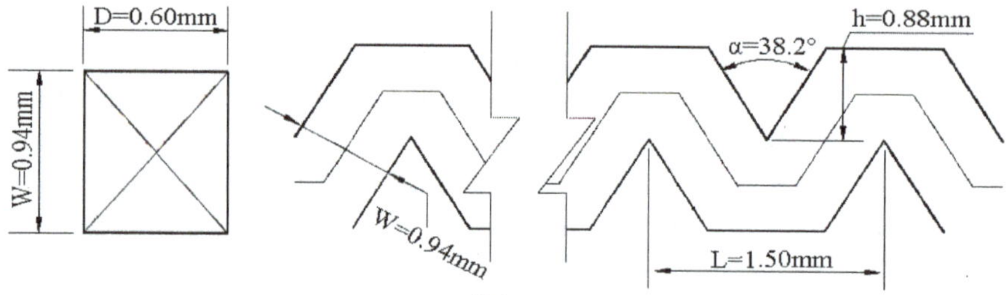

(**b**)

Figure 1. Emitter used in drip tape. (**a**) Labyrinth-channel emitter; (**b**) parameters of the flow path in the emitters. W, tooth space width; D, depth; α, angle of the bevel edge and baseline; L, tooth space width; h, tooth height.

2.2. Experimental Setup

Figure 2 shows that clean water was stored in a 100-L tank and muddy water was stored in a 150-L tank. Each was equipped with a 1.8-kW submersible pump with a 42-m rated head and 1.8-m^3/h rated discharge to provide the working pressure. The pressure adjustment valve and manometer I were installed on the three parallel tubes, which were connected to the submerged pump in the clean water tank, to adjust the different targeted flushing velocities. Manometer II was used to monitor the irrigation pressure. The pressure in manometer I and manometer II was 0.06 and 0.25 MPa, respectively, and their precision was 0.25%. A test platform (1.5 m in width, 4.8 m in length, and 1.0 m in height) was applied to support the laterals and emitters. Ten driplines were arranged on the test platform and all of them were equipped with control valves at the front and back ends. The spacing between each lateral was 0.16 m. Each lateral pipe had 15 emitters, with a spacing of 0.3 m.

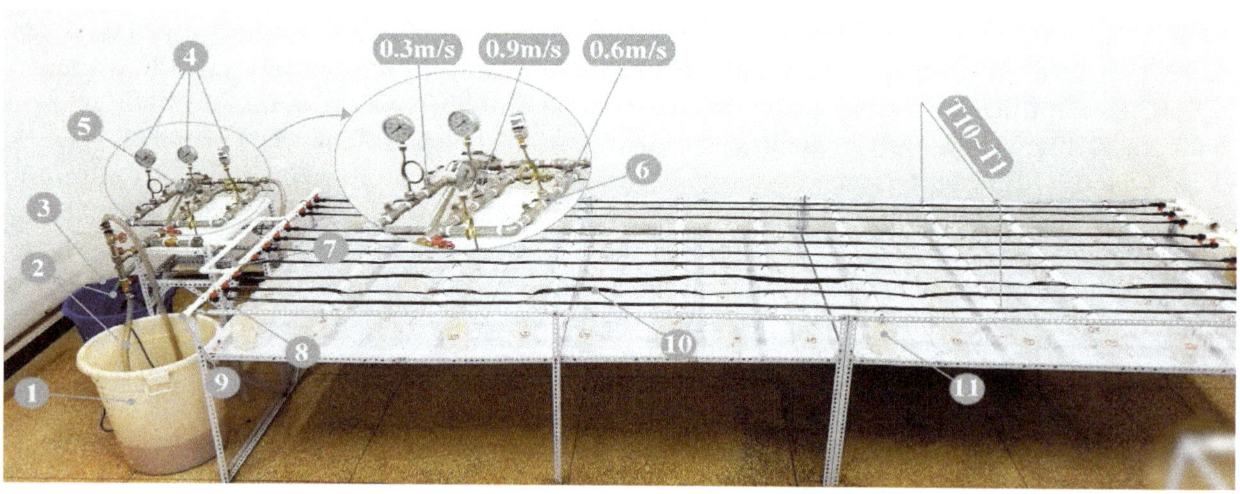

Figure 2. Experimental setup. 1, muddy water tank; 2, submersible pump; 3, clean water tank; 4, manometer I; 5, manometer II; 6, pressure adjustment valve; 7, polyvinyl chloride (PVC) pipe; 8, control valve; 9, PVC choke plug; 10, lateral; 11, measuring cup.

2.3. Measurement of Particle Size Distribution and Water Source

The distribution of particles size was analyzed by the Malvern laser particle size analyzer 2000 (Malvern Instruments Ltd., Malvern, UK). The measuring range of the particle analyzer was 0.02–2000 μm, and the particles were arranged in order of increasing size. When the accumulation volumes of sediment reached 10%, 50%, and 90%, the largest values of particle size were noted as D10, D50, and D90, respectively.

Tap water was used for lateral flushing in this study. Based on the maximum sediment concentration of irrigation water (0.8 g/L), the sand content in the muddy water was set at 2.5 g/L to accelerate the clogging process and reduce the experimental period. The test sediment was sandy loam soil obtained from Kunming, Yunnan Province of China, which was filtered using screens with 160 meshes after natural air drying. (In this study, 120 mesh screens were not selected for filtration to avoid insufficient test numbers caused by untimely clogging and the deposition of coarse particles.) Five sediment samples were randomly taken from the filtered sand after mixing well and the test sediment particle size distribution was presented based on their average values (see Table 1). The cumulative distribution and differential distribution of the tested sediments are shown in Figure 3.

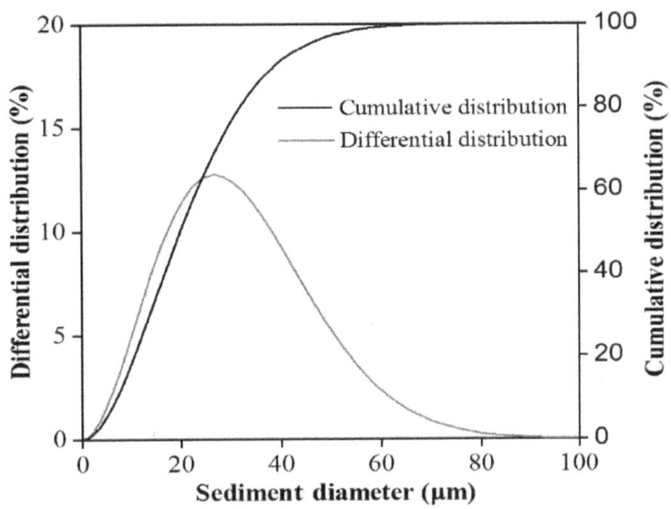

Figure 3. Differential distribution and cumulative distribution of sediments.

Table 1. Size distribution of sediment particles.

	D10	D20	D30	D40	D50	D60	D70	D80	D90	D100
Particle size	6.92	10.54	13.72	16.79	19.93	23.35	27.29	32.18	39.51	97.45
Standard deviation	0.34	0.45	0.55	0.63	0.72	0.82	1.00	1.27	1.65	1.33

2.4. Test Procedures

An experimental setup was conducted in a laboratory at Kunming University of Science and Technology, Kunming, Yunnan Province of China. In order to prevent the emitter flow rate and anti-clogging performance from being influenced by temperature variation, the experiments in this study were conducted between 13:00 and 16:00 CST, from 19 April to 7 September in 2017. Repeat experiments were carried out in three periods due to the limited width of the platform. The first repeated test was conducted between 19 April and 2 June, the second between 5 June and 19 July, and the third between 25 July and 7 September. The average temperature of the environment for the three test periods was 21.39 °C, 22.27 °C, and 22.08 °C, respectively.

2.4.1. Muddy Water Irrigation

The test conditions were as follows: normal irrigating pressure, 0.1 MPa; duration of irrigation event of each test, 20 min; initial volume of muddy water, 100 L; and total volume for each irrigation event, approximately 68.5 L. The muddy water was well stirred manually to avoid sedimentation. The emitter's discharge was measured every day using a marked 1000-mL measuring cup which was placed exactly beneath all the emitters. After each irrigation event, the weight of each measuring cup containing emitted muddy water was measured by a digital balance. The resolution of the digital balance was 0.01-g.

The emitter discharge was recorded as volume per unit time (L/h). For the emitter discharge less than 75% of the nominal rated flow, it was deemed to be clogged. The tested muddy water was replaced with new muddy water every day.

2.4.2. Sampling and Testing

The irrigation test was conducted with 45 events and 10 driplines, and the average discharge of all emitters in each lateral was taken into consideration. Therefore, 10 samples of each irrigation were measured three times. In all the tests, a total of 1350 (i.e., $45 \times 10 \times 3 = 1350$) muddy water samples were evaluated. The distribution of particle size was measured with all emitters in each dripline. Ten samples were assigned to each irrigation event, and the process was repeated three times. A total of 1350 samples were measured.

2.4.3. Clean Water Flushing

The present study evaluated the effect of two factors: (1) the flushing frequency, either daily (F_1), once per three days ($F_{1/3}$), or once per five days ($F_{1/5}$); and (2) the flushing velocity, at either 0.3 m/s ($V_{0.3}$), 0.6 m/s ($V_{0.6}$), or 0.9 m/s ($V_{0.9}$). Table 2 shows 10 treatments consisting of a control non-flushed treatment (T10) and a combination of targeted flushing velocities and flushing frequencies (T1–T9). There was only one dripline being flushed in every experiment, and each flushing event lasted for 5 min.

Table 2. Flushing velocity and flushing frequency treatment.

Treatment No.	Flushing Velocity (V)	Flushing Frequency (F)	Average Irrigation Events Before Emitter Clogging (d)
T1	$V_{0.3}$	F_1	32.67
T2	$V_{0.3}$	$F_{1/3}$	30.67
T3	$V_{0.3}$	$F_{1/5}$	27.33
T4	$V_{0.6}$	F_1	39.67
T5	$V_{0.6}$	$F_{1/3}$	33.67
T6	$V_{0.6}$	$F_{1/5}$	29.00
T7	$V_{0.9}$	F_1	41.33
T8	$V_{0.9}$	$F_{1/3}$	36.67
T9	$V_{0.9}$	$F_{1/5}$	30.67
T10	No flushing		23.33

2.5. *Statistics*

The data were analyzed using SPSS software for Windows, version 22.0 (IBM Corp., Chicago, IL, USA). Based on main effects analysis of variance (ANOVA), the significance of the differences between treatments for different response variables was evaluated. Furthermore, the results were categorized as "descriptive statistics" (such as mean) and Levene's test of equality of error variances. In addition, multiple comparisons were carried out based on Fisher's least significant difference tests because significant differences ($p < 0.05$) were indicated by ANOVA.

3. Results

3.1. *Variations in Emitter Discharge*

Figure 4 illustrates the value changes of the emitter discharge versus irrigating events. The level of straight line was 75% of the initial discharge, which was used as the criterion for emitter inefficiency or severe emitter clogging. Ten discharges of the emitters treated under diversified conditions showed a descendant trend with increasing irrigation events, suggesting that clogging with different degrees was seen in emitters until their service life ended. In the figure, the bold black line shows changes of the emitter discharge in T10, the control experiment. The emitter discharge in treatment T10 reached 75% of the rated discharge first, and its average discharge at the end of the test was the minimum without any flushing treatments made after irrigation. According to the irrigating events (Table 2), T10 provided only 23.33 irrigation events before emitters were severely plugged. However, for the remaining nine flushing treatments, the average normal irrigating event was 33.52, which indicated that the emitter's average service life in the flushed driplines after irrigation increased by 30.40% in comparison with those in the non-flushed T10. In the nine flushing treatments, treatment T7 had the longest irrigation event 41.33 and the universally largest average discharge; treatment T3 had the shortest irrigation event 27.33 and the universally lowest average discharge after 28 irrigation events. The service life of treatments T7 and T3 increased by 43.55% and 14.64%, respectively, compared with the non-flushing group T10. Additionally, treatments T7 and T3 were flushed at the frequency of F_1 and $F_{1/5}$ and at the velocity of $V_{0.9}$ and $V_{0.3}$, respectively, demonstrating that the emitters had a longer service life under high flushing velocities and high flushing frequencies.

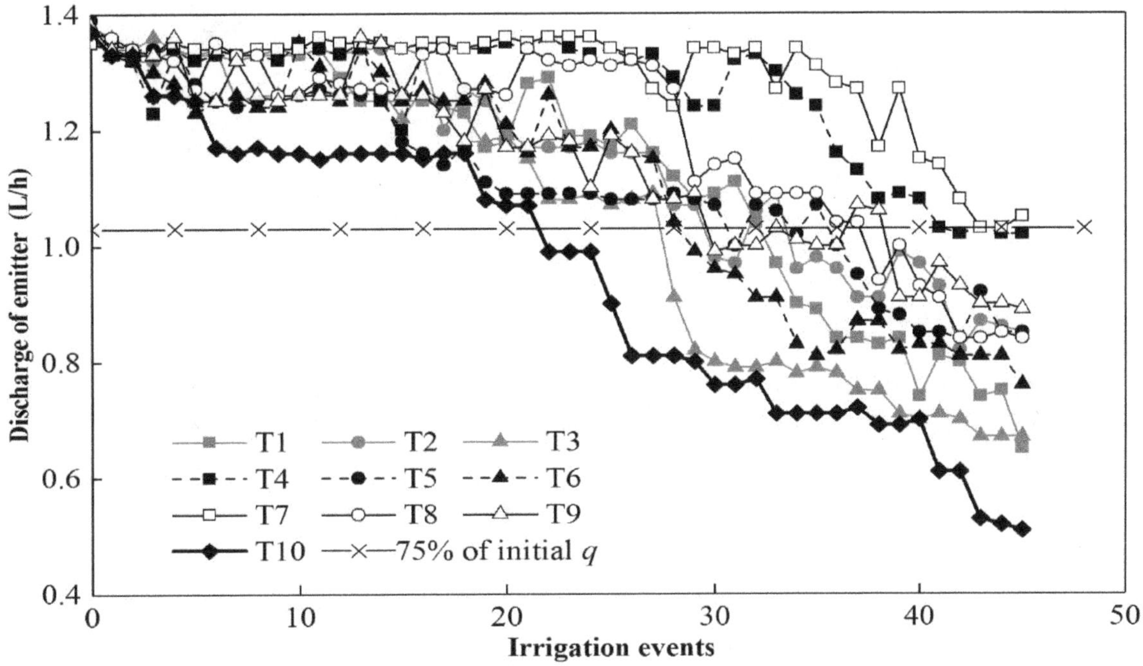

Figure 4. Average change in the emitter discharge versus irrigating events.

3.2. *Variance Analysis of the Two Flushing Factors on the Service Life of Emitters*

Flushing velocity and flushing frequency have significant effects on the service life of emitters, and the interactions were negligible (see Table 3). Figure 5a shows that the service life of emitters increased with the increase in the flushing velocity. The average service life of emitters was 22.80%, 31.60%, and 35.59% higher under $V_{0.3}$, $V_{0.6}$, and $V_{0.9}$ conditions, respectively, compared with the non-flushing conditions. Moreover, the service life of emitters under $V_{0.6}$ and $V_{0.9}$ conditions was similar, which meant that the flushing velocities in excess of 0.6 m/s would only help improve the service life to an insubstantial degree. Therefore, 0.6 m/s proved to be the optimal flushing velocity.

Table 3. Variance analysis of the effects of flushing velocity and flushing frequency on the service life of emitters.

Factors	*F* Value
Flushing velocity	21.94 **
Flushing frequency	46.82 **
Flushing velocity × Flushing frequency	2.03

* $p < 0.05$; ** $p < 0.01$.

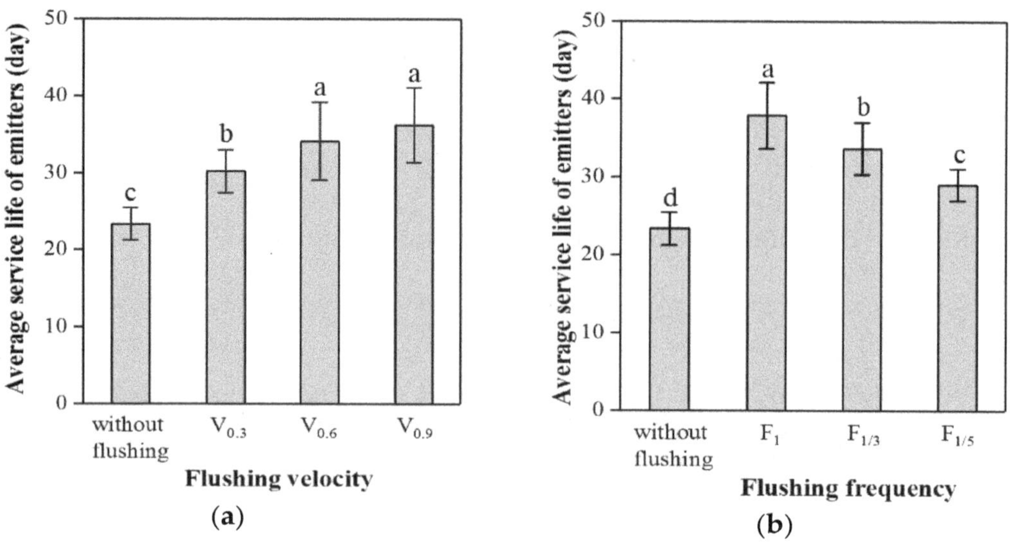

Figure 5. Average service life and standard errors of emitters at different flushing (**a**) velocities and (**b**) frequencies.

On the contrary, the service life of emitters was significantly reduced as the flushing frequency decreased (Figure 3b). The average service life of emitters was 38.43%, 30.71%, and 19.55% higher under F_1, $F_{1/3}$, and $F_{1/5}$ conditions, respectively, compared with the non-flushing treatment, indicating that a decrease in the flushing frequency effectively extended the service life of emitters.

3.3. Particle Size of Discharged Sediments

Table 4 shows that the effect of flushing velocity on D90 was significant, indicating that coarse particles passing through the labyrinth channel were affected by the flushing velocity, whereas fine particles were not. The flushing frequency had no effect on the particle size of discharged sediments. Figure 6 shows that D10, D50, and D90 mean values under non-flushing measures were 3.00, 7.16, and 12.65 μm, respectively, which were significantly higher than the different processing levels of flushing velocity and flushing frequency. This can be attributed to the fact that residual sediments in the lateral were discharged by flushing, resulting in a reduced concentration of sediments at the inlet of the labyrinth channel. Figures 7 and 8 show large amounts of residual sediments in the non-flushing lateral. The sediments mixed together, and more coarse particles flew into the labyrinth channel with the turbulence of the flow. Therefore, flushing effectively reduced the concentration of residual sediments in the lateral and decreased the probability of large particles entering into the labyrinth channel. Meanwhile, Figure 6b shows that the degree of D90 increased slightly with the decrease in the flushing frequency for the same reason.

Table 4. Variance analysis of the effects of flushing velocity and flushing frequency on the particle size of discharged sediments.

Factors	***F* Value**		
	D10	**D50**	**D90**
Flushing velocity	2.15	3.38	4.18 *
Flushing frequency	0.66	1.20	3.41
Flushing velocity × Flushing frequency	0.14	0.39	0.91

* $p < 0.05$; ** $p < 0.01$.

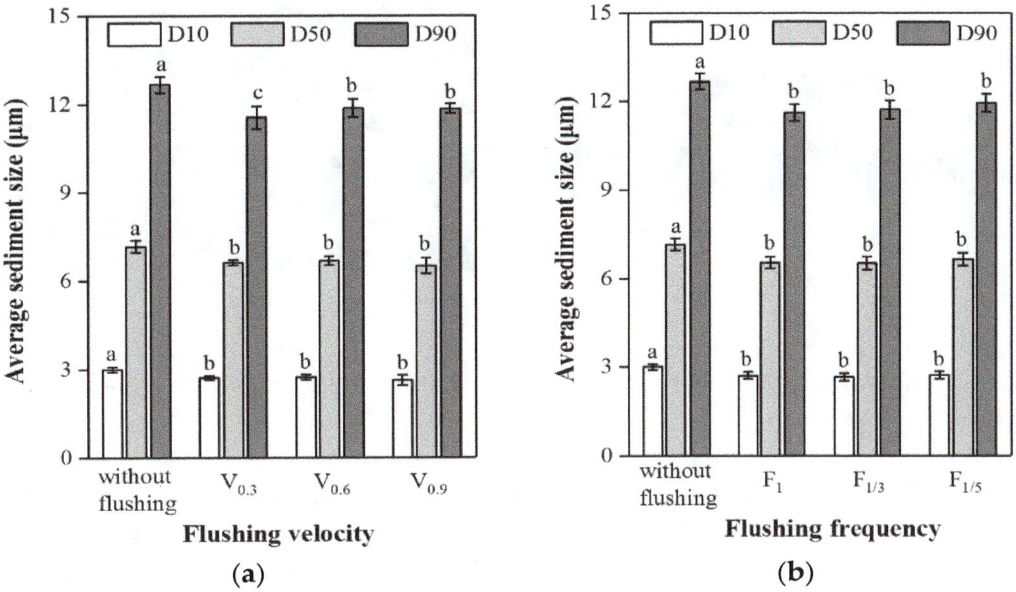

Figure 6. Average and standard errors of D10, D50, and D90 at different flushing (**a**) velocities and (**b**) frequencies.

3.4. Residual Sediments in Drip Tape and Emitters

Emitters and pipe sampled from the driplines were cut open to provide visual proof of clogging when experiments were finished (Figures 7 and 8). A few sediments were observed on the inner walls of the lateral with flushing treatment, whereas substantial sediment accumulations were observed in the lateral without flushing (Figure 5). Hence, flushing can effectively prevent particles from accumulating in the drip tape and emitters. Figure 8 shows that the sediment particles were seldom deposited in the labyrinth channel and the outlet of emitters due to flow turbulence with flushing, whereas clogging particles were found along the labyrinth channel of emitters without flushing, especially at the outlet. Additionally, other impurities such as plant root debris, which were not or seldom observed in the labyrinth channel of emitters with flushing treatment, were widely distributed in the labyrinth channel of emitters without flushing. In summary, flushing could effectively remove residual sediments and other impurities in the labyrinth channel, thus enhancing the anti-clogging performance of emitters.

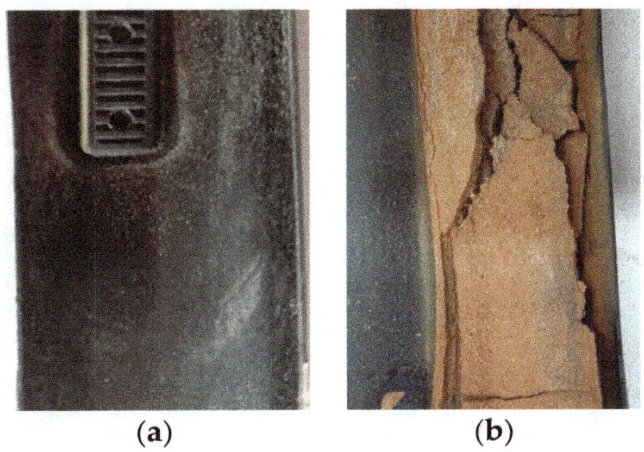

Figure 7. Residual sediments in the drip tape after tests. (**a**) Drip tape with flushing treatment; (**b**) drip tape without flushing treatment.

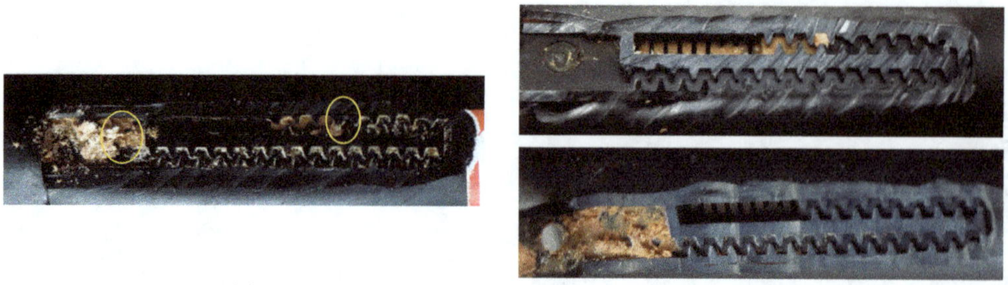

(a) (b)

Figure 8. Residual sediments in the labyrinth channel of emitters (yellow parts indicate that the clogging substances are other impurities, not sediments). (**a**) Emitters with flushing treatment; (**b**) emitters without flushing treatment.

Table 5 shows the variance analysis of clogging substances in emitters. The flushing frequency and flushing velocity had significant effects on D10, D50, and D90. Figure 9 shows that D50 and D90 were the highest under non-flushing conditions, being 30% and 52% higher than the averages under the corresponding flushing conditions, respectively, indicating that flushing prevented large particles from entering into the labyrinth channel. The figure also shows that D50 and D90 decreased as the flushing velocity increased. For instance, D10, D50, and D90 under $V_{0.9}$ conditions were the minimum, being, respectively, 25.76%, 38.55%, and 47.14% lower than those under non-flushing conditions. In summary, a higher flushing velocity contributed to the discharge of coarse particles in emitters, thereby lowering risks of emitter clogging.

Table 5. Variance analysis of the effects of flushing velocity and frequency on the particle size of residual sediments in emitters.

Factors	*F* Value		
	D10	**D50**	**D90**
Flushing velocity	18.07 **	7.48 **	10.85 **
Flushing frequency	10.06 **	10.36 **	19.58 **
Flushing velocity × Flushing frequency	2.81	2.47	2.67

* $p < 0.05$; ** $p < 0.01$.

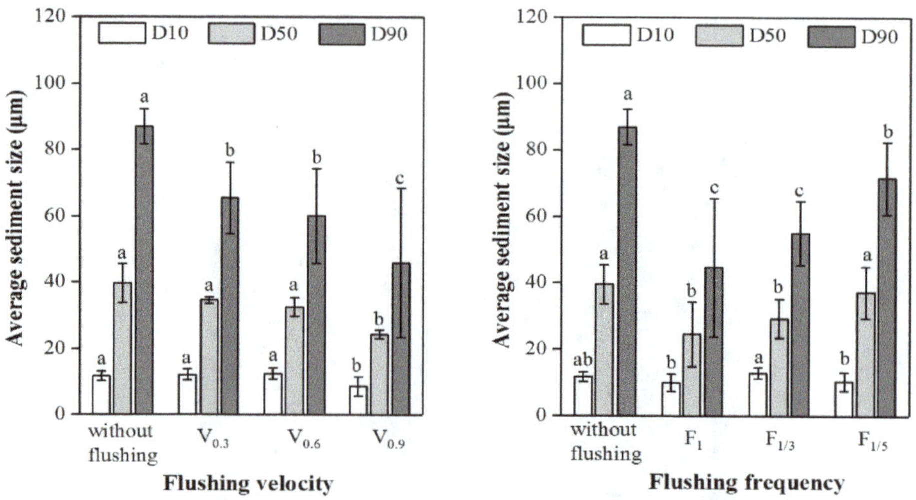

(a) (b)

Figure 9. Average and standard error of D10, D50, and D90 at different flushing (**a**) velocities and (**b**) frequencies.

Figure 9 further shows that D50 and D90 increased as the flushing frequency decreased and was maximized under non-flushing conditions. D90 under F_1, $F_{1/3}$, and $F_{1/5}$ conditions was 48.50%, 36.63%, and 17.66%, respectively, significantly lower than that under non-flushing conditions. D50 under F_1 and $F_{1/3}$ was 37.60% and 25.85%, respectively, significantly lower than that under non-flushing conditions. However, D10 differed slightly between the F_1, $F_{1/3}$, and $F_{1/5}$ conditions. A low flushing frequency caused the accumulation of sediments in the lateral and increased the probability of large particles being trapped in the labyrinth channel, resulting in increased particle sizes of residual sediments in the channel. It also favored the agglomeration of sediments, resulting in a higher degree of D50 and D90. Combining these two effects, the flushing frequency had a significant effect on the size distributions of residual sediments in the labyrinth channel.

4. Discussion

4.1. *Influence of Flushing Treatment on Emitter Clogging*

For drip irrigation systems, appropriate flushing can effectively prevent emitter clogging by hindering the agglomeration of sediment particles or their adhesion to organic residuals to generate large particles [23,24]. This study demonstrated that lateral flushing using clean water after the irrigation of muddy water could enhance the service life of emitters by 14.64–43.55% compared with non-flushing cases, and both the flushing velocity and flushing frequency significantly affected the service life of emitters. However, Puig-Bargués et al. [18] reported that the flushing frequency and flushing velocity had little effect on the resulting emitter discharges in subsurface drip irrigation systems measured at the end of the study. The results were different from the findings of the present study mainly due to different irrigation systems and flushing operations.

Recent studies found that the service life of emitters increased with the increase in the flushing velocity. For instance, the service life of emitters flushed at the velocities of 0.6 and 0.9 m/s was extended by more than 30% compared with that without flushing. This paper held that the flushing velocity of 0.6 m/s would appear to be adequate. In a laboratory study simulating a dripline with a transparent PVC pipe, Puig-Bargués and Lamm [25] suggested that the minimum flushing velocity of 0.3 m/s appeared sufficient for most micro-irrigation systems working under typical conditions. This was not consistent with the findings in this paper, which could be attributed to the fact that the researchers did not fully consider the complexities of the flow regime that might occur in the emitter. Zhang et al. [26] compared the influence of constant pressure and pulse pressure on the anti-clogging performance of the labyrinth emitter. According to their findings, the pulse pressure could alleviate clogging and reduce discharge of the emitter. Some studies [27,28] demonstrated that the emitter's anti-clogging performance could be enhanced by increasing the flushing pressure. This was consistent with the conclusion of this study because the pressure of flushing was proportional to the velocity of flushing.

In this study, the service life of emitters degraded significantly in the order of $F_1 > F_{1/3} > F_{1/5} >$ without flushing. Li et al. [21] conducted an in situ surface drip irrigation experiment under three conditions of dripline flushing frequency, in which the recycled water from a sewage disposal plant was used. They reported that lateral flushing could significantly relieve the emitter clogging of the irrigation systems using reclaimed water. Elberry et al. [23] reported that clogging at a high flushing frequency was significantly lower than that at a low flushing frequency. The aforementioned study results are consistent with the findings of this study. Nevertheless, Feng et al. [29] found that emitter clogging was not prevented with five different frequencies of flushing in the drip irrigation experiment using salty groundwater, which is primarily affected by flushing operations and water quality.

4.2. *Influence of Flushing Treatment on Particle Size Distributions*

Flushing can effectively enhance the anti-clogging performance of emitters by removing residual sediments from the dripline and maintaining the cleanliness and the passability of the labyrinth channel [29,30]. The results of this study indicated that the passability of large sediment particles in the labyrinth channel was the dominant factor for emitter clogging. In this study, the corresponding pressure in emitters was 0.004, 0.008, and 0.016 MPa, which was much less than 0.1 MPa, the normal irrigation pressure under the flushing velocity of 0.3, 0.6, and 0.9 m/s, respectively. Nevertheless, this study demonstrated that these relatively low pressures still had a significant effect on the particle size of residual sediments in the labyrinth channel. Kou et al. [31] reported that a disordered system of granular materials, such as powders, sand, and foams, formed stable structures when unperturbed, but that they "relax" in the presence of external influences such as shear or tapping, becoming fluid in nature. Similarly, the slightest perturbation of a particle system due to flushing destabilized some of the particles trapped in the labyrinth channel of emitters, increasing the flowability of particles in the channel. Hence, the passability of the labyrinth channel was enhanced, thus extending the service life of emitters. In this study, D90 of discharged sediments was significantly affected by the flushing velocity. Specifically, D90 under $V_{0.9}$ was significantly larger than that under $V_{0.3}$ and $V_{0.6}$ and without flushing, indicating that the perturbation of the particles increased with the increase in the flushing velocity and resulted in the increased trafficability of coarse particles.

A certain concentration of particles is indispensable for the follow-up collision and flocculation of particles. As particle concentration in the influent water increased, the local concentration distribution of particles in the labyrinth channel was improved [32–34]. In this study, the particle size of discharged sediments increased with the increase in the flushing frequency and reached its maximum under non-flushing conditions. As the flushing frequency decreased, fine particles were continuously discharged from the emitters, resulting in an increase in the relative contents of coarse particles and concentration of sediments, thereby improving the probability of collision, sedimentation, and accumulation of large sediment particles. This was why D50 and D90 of residual sediments in emitters increased with the increase in the frequency of flushing.

4.3. *Influence of Particle Size on Emitter Clogging*

The clogging mechanism in the emitter is dependent on the particle size of sediments. Coarse particles readily underwent collision and deposition in the vortices of the water flow due to bigger sizes and the larger drag force of the particles. Therefore, the particles could not escape from the vortices. On the other hand, the major forces causing the fine particles to clog came from their surface tension and adhesion to surrounding substances. In this study, D10 and D50 of discharged sediments were not affected by flushing velocity or flushing frequency, indicating that fine particles had better flow characteristics due to their lower drag force in the flow. In other words, this study demonstrated that coarse particles tended to be trapped in the labyrinth channel, which was consistent with the findings of previous studies [35–38].

Wu et al. [39] pointed out that it was difficult to cause clogging when the particle size was less than 20 mm. Particles <20 mm took up 50% of the muddy water (see Table 1), whereas D90 of discharged sediments was significantly lower than 20 μm (Table 6). Moreover, D100 of discharged sediments was only 27.18 μm. This indicated that particles larger than 20 μm tended to stay in the labyrinth channel and that particles smaller than 20 μm could easily flow out with the water. Table 6 shows that D50 of residual sediments in emitters was 24.38–39.67 μm, which was 22–99% higher than that in the initial water. In addition, D90 of residual sediments in emitters was 44.74–86.89 μm, which was 13–120% higher than that in the initial irrigation water resource.

Generally, the agglomeration and flocculation of fine particles occurred during the flow, and the upper-limit particle size was 10 μm [40]. Table 6 shows that the percentage of particle sizes below 10 μm was 10% in the residual sediments in the labyrinth channel but 90% in the discharged sediments. Hence, the emitter plugging in this study was caused mainly by large residual sediment particles in the emitter's labyrinth channel, instead of the agglomeration and flocculation of small sediment particles. This could be attributed to high flushing velocity and high flushing frequency.

Table 6. Average particle size of residual sediments in emitters and discharged sediments (μm).

		D10	D50	D90	D100
Discharged sediments	With flushing	2.68 ± 0.07	6.56 ± 0.14	11.73 ± 0.25	25.30 ± 0.50
	Without flushing	2.98 ± 0.29	7.17 ± 0.42	12.70 ± 0.74	27.18 ± 3.05
Residual sediments in emitters	With flushing	11.14 ± 2.30	30.49 ± 8.14	57.12 ± 14.68	69.63 ± 9.67
	Without flushing	11.74 ± 1.41	39.67 ± 5.92	86.88 ± 5.32	89.68 ± 6.71

5. Conclusions

The nine flushing measures led to an enhancement of 30.40% in the service life of emitters on average, and both flushing velocity and flushing frequency had significant effects on the service life of emitters. Although the life of emitters flushed at 0.9 m/s was longer than that of emitters flushed at 0.6 m/s, the numerical differences were small (<4%). Therefore, the flushing velocities setting at around 0.6 m/s appeared to be sufficient. Increasing the flushing frequency would be an economical way to achieve the optimal flushing effect without providing a higher flushing velocity, which might increase the system cost. Moreover, the inner walls of the lateral were clean in the presence of flushing and only local clogging was observed in the labyrinth channel. Severe sedimentation was observed in the lateral in the absence of flushing, and the clogging by sediments was observed along the labyrinth channel. Furthermore, flushing decreased the accumulation of sediments in the lateral, which prevented large sediment particles from entering into the labyrinth channel and lowered the probability for small sediment particles to agglomerate and flocculate into large particles, thus reducing the risks of emitter clogging.

Author Contributions: Software, Z.L.; Methodology, N.L.; Writing—original draft preparation, Z.L.; Writing—review and editing, L.Y.; Funding acquisition, L.Y., L.C., and N.C.

Acknowledgments: The authors would like to express their appreciation for the financial support received from the National Natural Science Foundation of China (NSFC) (Nos. 51769009, 51379024, 51809022) and the Funding Program for University Engineering Research Center of the Yunnan Province in China.

References

1. Adin, A.; Sacks, M. Dripper-clogging factors in wastewater irrigation. *J. Irrig. Drain. Eng.* **1991**, *117*, 813–826. [CrossRef]
2. Ravina, I.; Paz, E.; Sofer, Z.; Marcu, A.; Shisha, A.; Sagi, G. Control of emitter clogging in drip irrigation with reclaimed wastewater. *Irrig. Sci.* **1992**, *13*, 129–139. [CrossRef]
3. Nakayama, F.S.; Bucks, D.A. Water quality in drip/trickle irrigation: A review. *Irrig. Sci.* **1991**, *12*, 187–192. [CrossRef]
4. Pitts, D.J.; Haman, D.Z.; Smajstrla, A.G. Causes and prevention of emitter plugging in micro-irrigation systems. *Bull. Fla. Cooperative Ext. Serv.* **2011**, *40*, 201–218.
5. Yan, D.Z.; Bai, Z.H.; Mike, R.; Gu, L.K.; Ren, S.M.; Yang, P.L. Biofilm structure and its influence on clogging in drip irrigation emitters distributing reclaimed wastewater. *J. Environ. Sci.* **2009**, *21*, 834–841. [CrossRef]
6. Li, Y.K.; Liu, Y.Z.; Li, G.B.; Xu, T.W.; Liu, H.S.; Ren, S.M.; Yan, D.Z.; Yang, P.L. Surface topographic characteristics of suspended particulates in reclaimed wastewater and effects on clogging in labyrinth drip irrigation emitters. *Irrig. Sci.* **2012**, *30*, 43–56. [CrossRef]
7. Lamm, F.R.; Camp, C.R. Chapter 13: Subsurface drip irrigation. In *Micro-Irrigation for Crop Production: Design, Operation, and Management*, 1st ed.; Lamm, F.R., Ayars, J.E., Nakayama, F.S., Eds.; Elsevier: Amsterdam, The Netherlands, 2007; pp. 473–551.
8. Peng, Y.; Zhou, J.G.; Burrows, R. Modelling the free surface flow in rectangular shallow basins by lattice Boltzmann method. *J. Hydraul. Eng.* **2011**, *137*, 1680–1685. [CrossRef]
9. Peng, Y.; Zhou, J.G.; Burrows, R. Modelling solute transport in shallow water with the lattice Boltzmann method. *Comput. Fluids* **2011**, *50*, 181–188. [CrossRef]
10. Peng, Y.; Zhang, J.M.; Zhou, J.G. Lattice Boltzmann model using two-relaxation-time for shallow water equations. *J. Hydraul. Eng.* **2016**, *142–149*. [CrossRef]

11. Peng, Y.; Zhang, J.M.; Meng, J.P. Second order force scheme for lattice Boltzmann model of shallow water flows. *J. Hydraul. Res.* **2017**, *55*, 592–597. [CrossRef]
12. Peng, Y.; Zhou, J.G.; Zhang, J.M.; Burrows, R. Modeling moving boundary in shallow water by LBM. *Int. J. Mod. Phys. C* **2013**, *24*, 1–17. [CrossRef]
13. Peng, Y.; Zhou, J.G.; Zhang, J.M.; Liu, H.F. Lattice Boltzmann modelling of shallow water flows over discontinuous beds. *Int. J. Numer. Meth. Fluids* **2014**, *75*, 608–619. [CrossRef]
14. Peng, Y.; Zhou, J.G.; Zhang, J.M. Mixed numerical method for bed evolution. *Proc. Inst. Civ. Eng. Water Manag.* **2015**, *168*, 3–15. [CrossRef]
15. Nakayama, F.S.; Boman, B.J.; Pitts, D.J. Chapter 11: Maintenance. In *Micro-Irrigation for Crop Production: Design, Operation, and Management*, 1st ed.; Lamm, F.R., Ayars, J.E., Nakayama, F.S., Eds.; Elsevier: Amsterdam, The Netherlands, 2007; pp. 389–430.
16. ASAE Standards. *EP405.1: Design and Installation of Micro-Irrigation Systems*; ASAE: St. Joseph, MI, USA, 2003.
17. Brenes, M.J.; Hills, D.J. Micro-irrigation of wastewater effluent using drip tape. *Appl. Eng. Agric.* **2001**, *17*, 303–308.
18. Puig-Bargués, J.; Lamm, F.R.; Trooien, T.P.; Clark, G.A. Effect of dripline flushing on subsurface drip irrigation systems. *Trans. ASABE* **2010**, *53*, 147–155. [CrossRef]
19. Ravina, I.; Paz, E.; Sofer, Z.; Marm, A.; Schischa, A.; Sagi, G.; Yechialy, Z.; Leve, Y. Control of clogging in drip irrigation with stored treated municipal sewage effluent. *Agric. Water Manag.* **1997**, *33*, 127–137. [CrossRef]
20. Hills, D.J.; Tajrishy, M.A.; Tchobanoglous, G. The influence of filtration on ultraviolet disinfection of secondary effluent for micro-irrigation. *Trans. ASAE* **2000**, *43*, 1499–1505. [CrossRef]
21. Li, Y.; Song, P.; Pei, Y.; Feng, J. Effects of lateral flushing on emitter clogging and biofilm components in drip irrigation systems with reclaimed water. *Irrig. Sci.* **2015**, *33*, 235–245. [CrossRef]
22. Puig-Bargués, J.; Arbat, G.; Elbana, M.; Duran-Ros, M.; Barragán, J.; Ramírezde Cartagena, F.; Lamm, F.R. Effect of flushing frequency on emitter clogging in micro-irrigation with effluents. *Agric. Water Manag.* **2010**, *97*, 883–891. [CrossRef]
23. Elberry, A.M.; Abuarab, M.E.; Elebaby, F.G. Effect of flushing frequency with sulfuric and phosphoric acids on emitter clogging. In Proceedings of the ASABE Annual International Meeting, Louisville, KY, USA, 7–10 August 2011.
24. Li, Y.F.; Li, J.S.; Zhao, W.X.; Wang, Z. Review on irrigation technology applying sewage effluent-advances and prospects. *Trans. Chin. Soc. Agric. Mach.* **2015**, *46*, 102–110.
25. Puig-Bargués, J.; Lamm, F.R. Effect of flushing velocity and flushing duration on sediment transport in micro-irrigation driplines. *Trans. ASABE* **2013**, *56*, 1821–1828.
26. Zhang, L.; Wu, P.T.; Zhu, D.L. Effect of pulsating pressure on labyrinth emitter clogging. *Irrig. Sci.* **2017**, *35*, 267–274. [CrossRef]
27. Yu, L.; Li, N.; Liu, X.; Yang, Q.; Long, J. Influence of flushing pressure, flushing frequency and flushing time on the service life of a labyrinth-channel emitter. *Biosyst. Eng.* **2018**, *172*, 154–164. [CrossRef]
28. Yu, L.; Li, N.; Yang, Q.; Liu, X. Influence of flushing pressure before irrigation on the anti-clogging performance of labyrinth channel emitters. *Irrig. Drain.* **2017**, *67*, 191–198. [CrossRef]
29. Feng, D.; Kang, Y.; Wan, S.; Liu, S. Lateral flushing regime for managing emitter clogging under drip irrigation with saline groundwater. *Irrig. Sci.* **2017**, *35*, 217–225. [CrossRef]
30. Song, P.; Li, Y.; Li, J.; Pei, Y. Chlorination with lateral flushing controlling drip irrigation emitter clogging using reclaimed water. *Trans. Chin. Soc. Agric. Eng.* **2017**, *33*, 80–86.
31. Kou, B.; Cao, Y.; Li, J.; Xia, C.; Li, Z.; Dong, H.; Zhang, A.; Zhang, J.; Kob, W.; Wang, Y. Granular materials flow like complex fluids. *Nature* **2017**, *551*, 360–363. [CrossRef] [PubMed]
32. Wei, Z.; Tang, Y.; Wen, J.; Lu, B. Two-phase flow analysis and experimental investigation of micro-PIV and anti-clogging for micro-channels of emitter. *Trans. Chin. Soc. Agric. Eng.* **2008**, *24*, 1–9.
33. Wei, Q.; Lu, G.; Liu, J.; Shi, Y.; Dong, W.; Huang, S. Evaluation of emitter clogging in drip irrigation by two-phase flow simulation and laboratory experiments. *Comput. Electron. Agric.* **2008**, *63*, 294–303.
34. Li, J.; Chen, L.; Li, Y. Field evaluation of emitter clogging in subsurface drip irrigation system. *J. Hydraul. Eng.* **2008**, *39*, 1272–1278.
35. Niu, W.Q.; Liu, L.; Chen, X. Influence of fine particle size and concentration on the clogging of labyrinth emitters. *Irrig. Sci.* **2013**, *31*, 545–555. [CrossRef]

36. Zhang, J.; Zhao, W.; Wei, Z.; Tang, Y.; Lu, B. Numerical and experimental study on hydraulic performance of emitters with arc labyrinth channels. *Comput. Electron. Agric.* **2007**, *56*, 120–129. [CrossRef]
37. Wang, W.; Wang, F.; Niu, W.; Hu, X. Numerical analysis of influence of emitter channel structure on suspended granule distribution. *Trans. Chin. Soc. Agric. Eng.* **2009**, *25*, 1–6.
38. Liu, L.; Niu, W.; Bob, Z. Influence of sediment particle size on clogging performance of labyrinth path emitters. *Trans. Chin. Soc. Agric. Eng.* **2012**, *28*, 87–93.
39. Wu, Z.; Zhang, Z.; Zhang, K.; Luo, C.; Niu, W.; Yu, L. Influence of particle size and concentration of sediment on clogging of labyrinth channels emitters. *Trans. Chin. Soc. Agric. Eng.* **2014**, *30*, 99–108.
40. Qian, N.; Wan, Z. *Kinetics of Sediments*; Science Press: Beijing, China, 1983.

PERMISSIONS

All chapters in this book were first published in MDPI; hereby published with permission under the Creative Commons Attribution License or equivalent. Every chapter published in this book has been scrutinized by our experts. Their significance has been extensively debated. The topics covered herein carry significant findings which will fuel the growth of the discipline. They may even be implemented as practical applications or may be referred to as a beginning point for another development.

The contributors of this book come from diverse backgrounds, making this book a truly international effort. This book will bring forth new frontiers with its revolutionizing research information and detailed analysis of the nascent developments around the world.

We would like to thank all the contributing authors for lending their expertise to make the book truly unique. They have played a crucial role in the development of this book. Without their invaluable contributions this book wouldn't have been possible. They have made vital efforts to compile up to date information on the varied aspects of this subject to make this book a valuable addition to the collection of many professionals and students.

This book was conceptualized with the vision of imparting up-to-date information and advanced data in this field. To ensure the same, a matchless editorial board was set up. Every individual on the board went through rigorous rounds of assessment to prove their worth. After which they invested a large part of their time researching and compiling the most relevant data for our readers.

The editorial board has been involved in producing this book since its inception. They have spent rigorous hours researching and exploring the diverse topics which have resulted in the successful publishing of this book. They have passed on their knowledge of decades through this book. To expedite this challenging task, the publisher supported the team at every step. A small team of assistant editors was also appointed to further simplify the editing procedure and attain best results for the readers.

Apart from the editorial board, the designing team has also invested a significant amount of their time in understanding the subject and creating the most relevant covers. They scrutinized every image to scout for the most suitable representation of the subject and create an appropriate cover for the book.

The publishing team has been an ardent support to the editorial, designing and production team. Their endless efforts to recruit the best for this project, has resulted in the accomplishment of this book. They are a veteran in the field of academics and their pool of knowledge is as vast as their experience in printing. Their expertise and guidance has proved useful at every step. Their uncompromising quality standards have made this book an exceptional effort. Their encouragement from time to time has been an inspiration for everyone.

The publisher and the editorial board hope that this book will prove to be a valuable piece of knowledge for researchers, students, practitioners and scholars across the globe.

LIST OF CONTRIBUTORS

Jill H. Slinger
Faculty of Technology Policy and Management, Delft University of Technology, Jaffalaan 5, 2628 BX Delft, The Netherlands
Institute for Water Research, Rhodes University, Artillery Road, Grahamstown 6140, South Africa

Heleen S. I. Vreugdenhil
Faculty of Technology Policy and Management, Delft University of Technology, Jaffalaan 5, 2628 BX Delft, The Netherlands
Deltares, Boussinesqweg 1, 2629 HV Delft, The Netherlands

Eziz Atdayev and Ron C. K. Wong
Department of Civil Engineering, University of Calgary, Calgary, AB T2N 1N4, Canada

David W. Eaton
Department of Geoscience, University of Calgary, Calgary, AB T2N 1N4, Canada

Yurong Wang
State Key Laboratory of Hydraulics and Mountain River Engineering, Sichuan University, Chengdu 610065, China

Yongfei Qi
State Key Laboratory of Hydraulics and Mountain River Engineering, Sichuan University, Chengdu 610065, China
Department of Foundation Studies, Xinjiang Institute of Engineering, Urumqi 830023, China

Guosheng Duan, Anping Shu, Shu Wang and Fuyang Zhu
School of Environment, Key Laboratory of Water and Sediment Sciences of MOE, Beijing Normal University, Beijing 100875, China

Matteo Rubinato
Department of Civil and Structural Engineering, The University of Sheffield, Sir Frederick Mappin Building, Mappin Street, Sheffield S1 3JD, UK

Haifei Liu
The Key Laboratory of Water and Sediment Sciences of Ministry of Education, Beijing Normal University, Beijing 100875, China

Zhexian Zhu, Jingling Liu and Qiang Liu
School of Environment, Beijing Normal University, Beijing 100875, China

Hao Yuan, Weilin Xu, Rui Li, Yanzhang Feng and Yafeng Hao
State Key Laboratory of Hydraulics and Mountain River Engineering, Sichuan University, Chengdu 610065, China

Wenjun Liu, Bo Wang, Yunliang Chen, Chao Wu and Xin Liu
State Key Laboratory of Hydraulics and Mountain River Engineering, Sichuan University, Chengdu 610065, China

Tomasz Dysarz
Department of Hydraulic and Sanitary Engineering, Poznan University of Life Sciences, ul. Wojska Polskiego 28, 60-637 Poznan, Poland

Haijin Huang, Xingxing Zhang, Senpeng Lv and Rengmin Li
Key Laboratory of Hydraulic and Waterway Engineering of the Ministry of Education, Chongqing Jiaotong University, Chongqing 400074, China

Ming Chen
Key Laboratory of Hydraulic and Waterway Engineering of the Ministry of Education, Chongqing Jiaotong University, Chongqing 400074, China
Key Laboratory of Navigation Structure Construction Technology, Ministry of Transport, Nanjing 210029, China

Chunying Shen
State Key Laboratory of Hydraulics and Mountain River Engineering, Sichuan University, Chengdu 610065, China
Faculty of Electrical Power Engineering, Kunming University of Science and Technology, Kunming 650500, China

Shihua He and Yimin Xu
Faculty of Electrical Power Engineering, Kunming University of Science and Technology, Kunming 650500, China

Zhong Tian, Wei Wang, Ruidi Bai and Nan Li
State Key Laboratory of Hydraulics and Mountain River Engineering, Sichuan University, Chengdu 610065, China

Junping Liu, Xingye Zhu, Shouqi Yuan and Alexander Fordjour
Research Center of Fluid Machinery Engineering and Technology, Jiangsu University, Zhenjiang 212013, China

Jun Deng, Wangru Wei and Faxing Zhang
State Key Laboratory of Hydraulics and Mountain River Engineering, Sichuan University, Chengdu 610065, China

Rupayan Saha and Seung Ho Hong
Department of Civil and Environmental Engineering, West Virginia University, 1306 Evansdale Drive, Morgantown, WV 26506, USA

Seung Oh Lee
Department of Civil Engineering, Hongik University, 94 Wausan-ro, Mapo-gu, Seoul 04066, Korea

Yunfei Mao, Yong Peng and Jianmin Zhang
State Key Laboratory of Hydraulics and Mountain River Engineering, Sichuan University, Chengdu 610065, China

Zhangyan Li, Liming Yu and Na Li
Faculty of Modern Agriculture Engineering, Kunming University of Science and Technology, Kunming 650500, China

Liuhong Chang
School of Hydraulic Engineering, Changsha University of Science and Technology, Changsha 410114, China

Ningbo Cui
State Key Laboratory of Hydraulics and Mountain River Engineering and College of Water Resource and Hydropower, Sichuan University, Chengdu 610065, China

Index